**W9-CYY-888**

HANDBOOK OF
EMERGING COMMUNICATIONS TECHNOLOGIES
The Next Decade

The CRC Press
Advanced and Emerging Communications
Technologies Series

Series Editor-in-Chief: Saba Zamir

Data and Telecommunications Dictionary, Julie K. Petersen

Handbook of Sonet Technology and Applications, Steven S. Gorshe

The Telecommunications Network Management Handbook, Shervin Erfani

Handbook of Emerging Communications Technologies: The Next Decade, Rafael Osso

ADSL: Standards, Implementation, and Architecture, Charles K. Summers

Protocols for Secure Electronic Commerce, Ahmed Sehrouchni and Mostafa Hashem Sherif

After the Y2K Fireworks: Business and Technology Strategies, Bhuvan Unhelkar

Web-Based Systems and Network Management, Kornel Terplan

Intranet Management, Kornel Terplan

Multi-Domain Communication Management Systems, Alex Galis

Mobile Intelligent Agents Applied to Communication Management Systems, Alex Galis and Stefan Covaci

HANDBOOK OF

EMERGING COMMUNICATIONS TECHNOLOGIES

The Next Decade

Editor
Rafael Osso

Series Editor-in-Chief
Saba Zamir

CRC Press
Boca Raton London New York Washington, D.C.

Library of Congress Cataloging-in-Publication Data

Osso, Rafael.
    Handbook of communications technology : the next decade / Rafael Osso.
      p. cm. — (Advanced and emerging communications technologies)
    ISBN 0-8493-9594-1 (alk. paper)
    1. Telecommunications—Technological innovations. I. Title.
TK5105.062 1999
621.382—dc21
                                              99-25427
                                                CIP

Preface

The field of communications technologies is an extremely volatile one, with new technologies emerging even as I write this preface. This *Handbook* provides comprehensive coverage of the upcoming developments for those technologies that are expected to have a major impact in this field. The *Handbook* encompasses a broad spectrum of topics: RSVP, video transmission, JMAPI, electronic commerce, WDM, MBONE, MPLS, optical networks, HDTV, Digital Audio Broadcasting, to mention a few. The articles written provide up-to-date descriptions and salient features of these emerging communications technologies along with comprehensive reference lists on where to acquire additional information.

Although the majority of the articles are technically oriented, the "casual" or "non-technical" reader can also benefit from the *Handbook*, since there are several chapters (DVD, HDTV, Internet Commerce, Broadcasting in the Internet age) that present the subject matter in an informative but easy-going, non-technical style. There is no specific ordering to the articles, and it is not necessary to read the *Handbook* from start to finish; simply turn to the chapter of interest and enjoy.

I am pleased to say that the chapters have been contributed by some of the most prominent and seasoned professionals immersed in the world of voice, data, and video technology today, and they have delivered excellent discourses in their specific areas of expertise. I have included a chapter on emerging security testing and standards evaluation. Although this is not an "emerging communication technology" in itself, it is important to note that the progress and impact of emerging technologies is in many ways dependent on the existence of accurate security testing and solid, uniform, standards hence the importance of this chapter. This is the last chapter of the *Handbook*.

I trust you will find the *Handbook* to be an invaluable reference.

Rafael Osso

Editor

Rafael Osso is a senior technology executive, with specific expertise in areas of network and systems management. Mr. Osso has held senior key management positions with major financial, brokerage, and manufacturing services where he gained extensive experience in implementing cutting-edge technology systems. He has also been responsible for the successful implementation of NetWORKS®, a remote network management services program at Unisys. Mr. Osso's work has been reviewed by major industry analysts such as Meta, Forester, Dataquest, etc. and has received acclaim through articles written for *Enterprise Systems Journal* and *Unisphere*. He is currently working on another major project relating to outsourcing, which is scheduled for release in late 1999.

.

Contributors

Mark Allen
Williams Network
Tulsa, OK

Albert Azzam
Alcatel Telecom
Raleigh, NC

Krishna Bala
Tellium Inc.
Oceanport, NJ

Saleem N. Bhatti
Computer Science
University College London
London, England

Eric Bouillet
Electrical Engineering
Columbia University
New York, NY

Paul J. Brusil
Strategic Management Directions
Beverly, MA

Ted Carlin
Department of
  Communications/Journalism
Shippensburg University of
  Pennsylvania

Richard V. Ducey
Research & Information Group
National Association of Broadcasters
Washington, D.C.

Sonia Fahmy
Department of Computer and
  Information Science
Ohio State University
Columbus, OH

Douglas A. Ferguson
Perrysburg, OH

V. Hardman
Computer Science
University College London
London, England

O. Hodson
Computer Science
University College London
London, England

Raj Jain
Department of Computer and
  Information Science
Ohio State University
Columbus, OH

Mario Francois Jauvin
Stittsville, Ontario, Canada

L. Arnold Johnson
National Institute of Standards and
  Technology
Gaithersburg, MD

Ahmad Khanifar
Electronic and Electrical Engineering
University College London
London, England

Masoud Khansari
Hewlett-Packard Laboratories
Palo Alto, CA

Bruce Klopfenstein
Department of Telecommunications
Bowling Green State University
Bowling Green, OH

Chunlei Liu
Department of Computer and
  Information Science
Ohio State University
Columbus, OH

Andrew G. Malis
Ascend Communications
Westford, MA

Antonio Ortega
Department of Electrical Engineering
  Systems
University of Southern California
Los Angeles, CA

George Scheets
Department of Electrical Engineering
Oklahoma State University
Stillwater, OK

Edwin F. Steeble
National Institute of Standards and
  Technology
Gaithersburg, MD

Acknowledgments

The compilation and completion of a handbook of this nature is a unique experience. Its ultimate success is dependent on dozens of people, and thanks are owed to all. I am grateful to each of my authors for giving selflessly of their time and their expert contributions that have made this the valuable reference it was intended to be. Many thanks are owed to my special guest advisors, Saba Zamir and Jay Ranade, for making this Handbook a success. I wish to thank Dawn Mesa, for keeping track of the status of the project, contributor agreements, permission forms, and much more.

In conclusion, I must thank Saba Zamir, a very unique dear friend, very loyal and sincere, for her dedicated support and encouragement in times of need. Also, I cannot forget my friend Jay Ranade (we have faced numerous challenges together) for being a true friend, always; and most important, my spiritual second mother, Gertrude, for her unique wisdom, love, and valuable advice in helping me understand why and how. Thanks to all.

Rafael Osso

Table of Contents

Dedication

----

*To my beloved wife, my best friend, who taught me how
to love and trust, and who has given me the courage
to reach my goals in life.*

*To my children, Jacob, Jarod, Christine, and Rafael Jr.,
whose unconditional love has enriched my life, taught me
the meaning of love and hope and made me a better person
today. May your lives be filled with success and love.*

*To my dear friend Saba Zamir, whose persistence,
faith, unique kindness, and friendship gave me the
inspiration to write this book.*

*To my good friend Jay Ranade, whose loyalty
is one of a kind.*

*And to my mentor and second mom Gertrude Kleinman,
who has inspired and enriched my life, for which
I am indebted.*

1 Resource ReSerVation Protocol (RSVP)

Sonia Fahmy and Raj Jain

CONTENTS

0-8493-9594-1/00/$0.00+$.50

1.1   INTRODUCTION

In 1993, the Internet Engineering Task Force (IETF) chartered the Resource ReSerVation Protocol (RSVP) working group to specify the signaling protocol to set up resource reservations for the new (real-time) services to be provided in the Internet. The RSVP is a state-establishment protocol. RSVP will enable the Internet to support real-time and multimedia applications, such as teleconferencing and videoconferencing applications. These applications require reservations to be made in the Internet routers, and RSVP is the protocol to set up these reservations.

The Internet protocol (IP) currently supports a best effort service, where no delay or loss guarantees are provided. This service is adequate for nontime-critical applications, or time-critical applications under light load conditions. Under highly overloaded conditions, however, buffer overflows and queuing delays cause the real-time communication quality to quickly degrade.

To support real-time applications, a new service model was designed for the Internet. In this model, both real-time and nonreal-time applications share the same infrastructure, thus benefiting from statistical multiplexing gains. Applications specify their traffic characteristics and their quality of service requirements. Admission control is employed to determine whether those requirements can be met, and reservations are made. Packet scheduling services different applications with different priorities to ensure that the quality of service requirements are met.

RSVP can also transport other messages in addition to reservation messages, such as policy control and traffic control messages. The key features of RSVP include flexibility, robustness, scalability through receiver control of reservations, sharing of reservations, and use of IP multicast for data distribution.

This chapter gives an overview of RSVP and its use to support integrated services in the Internet. We first discuss RSVP goals, features, messages, and interfaces. Then we describe RSVP management and security. We also discuss how RSVP can be used with the integrated services, and how it can be used on specific link layers, such as ATM and IEEE 802 networks. Finally, we discuss the interoperability of RSVP with non RSVP routers and examine a number of issues currently under investigation, such as aggregation and differentiated services, policy control, label switching, routing, and diagnostics.

1.2   WHAT IS RSVP?

RSVP is the means by which applications communicate their requirements to the network in an efficient and robust manner. RSVP does not provide any network service; it simply communicates any end-system requirement to the network. Thus

RSVP can be viewed as a 'switch state establishment protocol,' rather than just a resource reservation protocol.

RSVP is developed to support traffic requiring a guaranteed quality of service over both IP unicast and multicast. Figure 1.1 shows the IP multicast model. The Internet Group Management Protocol (IGMP) is used to handle group membership requests. IGMP is the protocol through which hosts indicate to their local routers their interest in joining a group. The Internet multicast routing protocols are employed in Internet routers to set up the multicast trees used for forwarding the data packets to the appropriate group members.

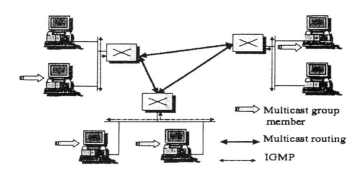

Figure 1.1 Multicasting in the Internet

Through RSVP, applications that receive real-time traffic inform networks of their needs, while applications that send real-time traffic inform the receivers about their traffic characteristics. RSVP is the signaling protocol that installs and maintains reservation state information at each router along the path of a stream. RSVP transfers reservation data as opaque data; it can also transport policy control and traffic control messages. RSVP operates on top of IP (both version 4 and version 6), and it is concerned only with the quality of service (QoS) of the packets forwarded according to routing.

The term *session* will be used throughout the chapter, since RSVP operates on a per-session basis. In the context of RSVP in the Internet, an RSVP session is a simplex data stream from a sending application to a set of receiving applications, usually defined by the triple (DestAddress, ProtocolId [, DstPort]). DestAddress is the IP destination address of the data packets and may be a unicast or multicast address. ProtocolId is the IP protocol identifier. The optional DstPort parameter could be defined by a UDP/TCP destination port field, by an equivalent field in another transport protocol, or by some application-specific information.

1.2.1 COMPONENTS OF AN RSVP-CAPABLE ROUTER

Each router in the new Internet model must contain several components, as illustrated in Figure 1.2. These components interact through explicit interfaces to improve the

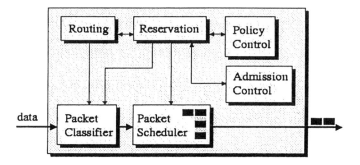

Figure 1.2 RSVP-capable routers

modularity and independence of the scheme. In addition to the routing mechanism and the flow QoS specification scheme, the router must contain an admission control process, to determine if sufficient resources are available to make the reservation, and a policy control process, to determine if the user has permission to make the reservation. If the RSVP process gets an acceptance indication from both the admission control and policy control processes, it sends the appropriate parameter values to the packet classifier and packet scheduler. The packet classifier determines the QoS class of packets according to the requirements, and the packet scheduler manages various queues to guarantee the required QoS. To guarantee the bandwidth and delay characteristics reserved by RSVP, a fair packet-scheduling scheme, such as weighted fair queuing, can be employed. Fair scheduling isolates data streams and gives each stream a percentage of the bandwidth on a link. This percentage can be varied by applying weights derived from RSVP's reservations. The admission control process, packet classifier, and packet scheduler are collectively called traffic control.

1.2.2   RSVP DESIGN GOALS

We explain the RSVP design goals by giving the problem and RSVP solution associated with each goal[18] in Table 1.1.

1.3   RSVP FEATURES

Figure 1.3 shows the router model employed by RSVP. Data arrives on the incoming interfaces from the previous hops and is routed to one of the next hops through the outgoing interfaces. An RSVP sender uses the PATH message to communicate with receivers informing them of flow characteristics. RSVP provides receiver-initiated reservation of resources, using different reservation styles to fit a variety of applications. RSVP receivers periodically alert networks to their interest in a data flow, using RESV messages that contain the source IP address of the requester and the destination IP address, usually coupled with flow details. The network allocates the needed bandwidth and defines priorities. RSVP decouples the packet classification and scheduling from the reservation operation, transporting the messages from the

TABLE 1.1
RSVP Goals

Goal	Problem	RSVP Solution
Accommodate heterogeneous receivers	Receivers in the same (multicast) session, and paths to these receivers, can have different capacities and require different QoS	RSVP allows receivers to make different reservations
Adapt to changing multicast group membership	New members can join a multicast group at any time, and existing members can leave at any time	RSVP gracefully handles group member changes to scale to large groups
Adapt to route changes	Routing protocols adapt to changes in topology and load by establishing new routes	RSVP handles route changes by automatically reestablishing resource reservations along new paths if adequate resources are available
Control protocol overhead	Refreshing reservations over the multicast routing tree can create a high overhead	RSVP overhead does not grow proportional to the group size, and parameters are used to control the overhead
Use network resources efficiently	Sometimes resources are wasted if each sender in the same session makes a separate reservation along its multicast tree	RSVP allows users to specify their needs so that the aggregate resources reserved for the group reflect the resources actually needed. Receivers can specify the specific senders that can use the reserved resources
Accommodate heterogeneous underlying technologies	The protocol design should interoperate and coordinate with different routing algorithms and other components	RSVP design is relatively independent of the flow specification, routing, admission control and packet scheduling functions

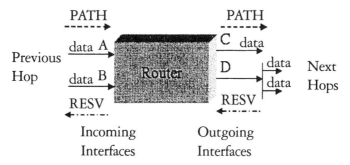

Figure 1.3 RSVP router model

source and destination as opaque data. Periodic renewal of state allows networks to be self correcting despite routing changes and loss of service. This enables routers to understand their current topologies and interfaces, as well as the amount of network bandwidth currently supported. These features are discussed below.

1.3.1 RECEIVER-INITIATED SETUP

The RSVP protocol provides receiver-initiated setup of resource reservations for both unicast and multicast data flows. For multicast data flows, reservation requests merge when they progress up the multicast tree. The reservation for a single receiver travels only until it reaches a reserved branch of the tree. Receiver-initiated reservation works better than sender-initiated reservation because it is the receivers that know their possibly different (in this case we call them heterogeneous receivers) and possibly changing requirements and limitations. Hence, receiver-controlled setup is more scalable. RSVP reserves resources in only one direction, so the sender is logically separate from the receiver.

1.3.2 PACKET CLASSIFICATION AND SCHEDULING

RSVP does not determine which packets can use the resources; it specifies only the amount of resources reserved for each flow. RSVP interacts with the packet classifier and the packet scheduler to determine the classes (and perhaps routes) and achieve the required QoS. Thus, RSVP transports reservation data as opaque data. An RSVP reservation request consists of a FlowSpec, specifying the desired QoS, as well as a FilterSpec, defining the flow to receive the desired QoS. The FlowSpec is used to set parameters in the packet scheduler, while the FilterSpec is used in the packet classifier.

The FlowSpec in a reservation request will generally include a service class and two sets of numeric parameters: (1) an RSpec (R for reserve) that defines the desired QoS, and (2) a TSpec (T for traffic) that describes the data flow. The basic FilterSpec format defined in the present RSVP specification has a very restricted form: sender IP address and, optionally, the UDP/TCP port number SrcPort.

1.3.3 SOFT STATE

The soft state feature of RSVP increases the robustness of the protocol. RSVP adapts to changing group memberships and changing routes throughout the application lifetime in a manner that is transparent to applications. This is accomplished through *soft state*, which means that state maintained at network switches expires after a certain period of time (called *cleanup timeout interval*), unless it gets reinstated. Reinstatement is performed through periodic "refresh" messages sent by the end users (both senders and receivers) every "refresh timeout" period, to automatically maintain state in the switches along the reserved paths. In the absence of such refresh messages, reservation state in the routers times out. This is also one way of releasing resources in reservations that are shared by multiple receivers, as explained next.

1.3.4 RSVP Reservation Styles

The RSVP protocol supports several reservation styles to fit a variety of application requirements. A reservation style allows the applications to specify how reservations for the same session are aggregated at the intermediate switches. The basic distinction among different reservation styles is whether a separate reservation should be established for each upstream sender in the same session, or if a single reservation can be shared among the packets of selected senders. The selection of senders in such a case can be done through an explicit list of senders or through a wildcard that selects all the senders in the session (see Table 1.2). Reservation styles supported by RSVP include wildcard filters, fixed filters, and shared explicit filters.

TABLE 1.2
Reservation Styles

	Reservations	
	Distinct	Shared
Sender selection		
Explicit	Fixed filter	Shared explicit
Wildcard	None	Wildcard filter

The fixed filter creates a distinct reservation per specified sender (without installing separate reservations for each receiver to the same sender). The total reservation on a link for a given session is the sum of the flow specifications for all requested senders. An example of applications that can use the fixed filter is video conferencing applications for which enough resources must be reserved for the number of video streams a receiver wishes to watch simultaneously.

The wildcard filter shares the same reservation among all upstream senders, reserving resources to satisfy the largest resource request (regardless of the number of senders). A wildcard reservation automatically extends to new senders in the session as they appear. An example of applications that can use the wildcard filter is audio conferencing applications where all the senders can share the same set of reserved resources, since multiple senders are unlikely to transmit at the same time.

The last type of reservation style is the shared explicit filter, where a single reservation is created but can be shared only by selected upstream senders. An example of applications that use the shared explicit filter is audio conferencing applications where the receivers want to block traffic from specific senders. Note that the shared explicit filter incurs more overhead than the wildcard filter. In addition, reservations of different styles cannot be merged.

Example:
Figure 1.4 illustrates a router with two incoming interfaces, A and B, and two outgoing interfaces, C and D. There are three upstream senders and three downstream receivers. Outgoing interface D is connected to a broadcast LAN. Thus, R2 and R3 are reached

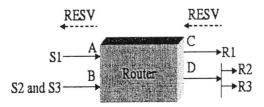

Figure 1.4 Router configuration

via different next-hop routers (not shown). Data packets from each sender shown in Figure 1.4 are routed to both outgoing interfaces.

The FlowSpec is given as a multiple of some base resource quantity R. In Tables 1.3–1.5, the Receives column shows the RSVP reservation requests received over outgoing interfaces C and D (for interface D, the requests received from R2 and R3 are separated by the word *and*). The Reserves column shows the resulting reservation state for each interface. The Sends column shows the reservation requests that are sent upstream to previous hops A and B.

Table 1.3 shows an example of merging wildcard filter style reservations. Merging is required twice in this example. First, each of the two next hops on interface D requests a reservation, and these two requests must be merged into the FlowSpec 2R used to make the reservation on interface D. Second, the reservations on the interfaces C and D must be merged in order to forward the reservation requests upstream. The larger FlowSpec 3R is forwarded upstream to each previous hop.

TABLE 1.3
Wildcard Filter Example

Receives		Reserves		Sends	
Interface	(Sender, Rate)	Interface	(Sender, Rate)	Interface	(Sender, Rate)
C	(*,3R)	C	(*,3R)	A	(*,3R)
D	(*,2R) and (*,R)	D	(*,2R)	B	(*,3R)

Table 1.4 shows fixed filter style reservations. For each outgoing interface, there is a separate reservation for each source that has been requested, but this reservation will be shared among all the receivers that made the request. The flow descriptors for senders S2 and S3, received through outgoing interfaces C and D, are placed into the request forwarded to previous hop B. The three different flow descriptors (from R1, R2, and R3) specifying sender S1 are merged into the single request (S1,4R) sent to previous hop A.

Table 1.5 shows an example of shared explicit style reservations. When such reservations are merged, the resulting FilterSpec is the union of the original Filter-Specs, and the resulting FlowSpec is the largest FlowSpec.

TABLE 1.4
Fixed Filter Example

Receives		Reserves		Sends	
Interface	(Sender, Rate)	Interface	(Sender, Rate)	Interface	(Sender, Rate)
C	(S1,4R),(S2,5R)	C	(S1,4R),(S2,5R)	A	(S1,4R)
D	(S1,3R),(S3,R) and (S1,R)	D	(S1,3R),(S3,R)	B	(S2,5R),(S3,R)

TABLE 1.5
Shared Explicit Example

Receives		Reserves		Sends	
Interface	(Sender, Rate)	Interface	(Sender, Rate)	Interface	(Sender, Rate)
C	((S1,S2),R)	C	((S1,S2),R)	A	(S1,4R)
D	((S1,S3),4R) and (S2,2R)	D	((S1,S2,S3),4R)	B	((S2,S3),4R)

1.4 RSVP MESSAGES

This section explains RSVP message formats and message processing, routing, and merging. As previously mentioned, RSVP receivers use the reserve (RESV) message to periodically advertise to the network their interest in a flow, specifying the flow and filter specifications. RSVP senders, on the other hand, use the PATH message to indicate that they are senders and give information such as the previous hop IP address, the multicast group address, templates for identifying traffic from that sender, and sender traffic specifications. The message is sent to all receivers in the multicast tree using the forwarding table maintained by the multicast routing protocol. The RESV message is forwarded back to the sources by reversing the paths of PATH messages (using the previous hop IP address stored from PATH messages). RSVP supports one pass with advertising (OPWA): the PATH messages contain a field (AdSpec) to gather information that may be used to predict the end-to-end QoS. The results are delivered by RSVP to the receivers which construct, or dynamically adjust, an appropriate reservation request.

1.4.1 MESSAGE FORMATS AND MESSAGE PROCESSING

An RSVP message consists of a common header, followed by a body consisting of a variable number of variable length, typed *objects*. The main RSVP messages are PATH and RESV messages. In addition, there is a reservation confirmation message (ResvConf), two types of error messages (PathErr and ResvErr), and two types of teardown messages (PathTear and ResvTear). These messages are shown in Figure 1.5 and briefly explained in Table 1.6. This section examines them in more detail.

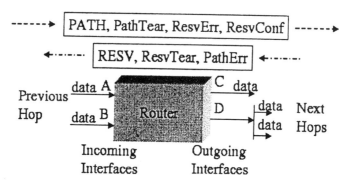

Figure 1.5 RSVP messages

TABLE 1.6
RSVP Messages

Message	Meaning	Purpose
PATH	Path establishment	Used by senders to specify their traffic characteristics
RESV	Reservation request	Used by receivers to advertise to the network their interest in a flow
ResvConf	Reservation confirmation	Indicates to the receiver successful installation of a reservation at an upstream node
PathErr	Path error	Indicates to the sender an error in the path message
ResvErr	Reservation error	Indicates to the receivers that a reservation request has failed or an active reservation has been preempted
PathTear	Path teardown	Deletes path state and dependent reservation state
ResvTear	Reservation teardown	Deletes reservation state

1.4.1.1 PATH Message

The PATH message contains the following:

- *Session identifier* and *timeout values* to control refresh frequencies.
- *Previous hop address*. This is maintained at every node to route RESV messages in the reverse path taken by PATH messages.
- *Sender Template*. The sender template describes the data packets that the sender wi¹l originate. This is given as a FilterSpec to distinguish the sender packets from others in the same session on the same link. The sender template may specify only the sender IP address and optionally the sender port, assuming the same protocol identifier specified for the session.

- *Sender TSpec.* This defines the traffic characteristics of the data flow that the sender will generate. The TSpec is used by traffic control to prevent over-reservation and unnecessary admission control failures.
- *AdSpec.* The AdSpec (optional) includes parameters describing the properties of the data path (including the availability of specific QoS control services) and parameters required by specific QoS control services to operate correctly. An AdSpec received in a PATH message is passed to the local traffic control, which returns an updated AdSpec. The updated version is then forwarded in PATH messages sent downstream.

The PATH message may also contain integrity objects and policy data objects.

Each RSVP-capable node along the path captures the PATH message and processes it to create path state for the sender defined by the sender template and session objects (the next section defines the state information maintained at each node). The sender TSpec and objects such as policy data and AdSpec are also stored in the path state. The RSVP process forwards PATH messages and replicates them as required by multicast sessions (modifying the previous hop and AdSpec fields). This operation uses routing information RSVP obtains from the appropriate routing process (see Figure 1.6). Periodically, the RSVP process scans the path state to create new PATH messages to forward towards the receivers.

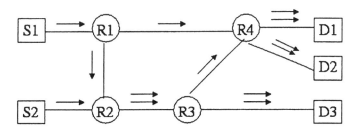

Figure 1.6 RSVP PATH messages

1.4.1.2   RESV Message

The RESV message contains the following:

- *Session identifier* and *timeout values* to control refresh frequencies.
- *Hop address* which contains the address of the interface through which the message was sent and the interface on which reservation is required.
- *Confirmation required or not* and the *receiver address* to which to send the confirmation.
- *Reservation style, FilterSpec,* and *FlowSpec.* In case of wildcard filters, only a FlowSpec needs to be given. For a fixed filter, the FilterSpec and FlowSpec for each sender should be given. For shared explicit reservations, one FlowSpec and a set of FilterSpecs must be given.
- *Set of senders* to which to forward the RESV message (scope).

The RESV message may also contain integrity and policy data objects.

The RSVP process at each intermediate node first passes the reservation request to admission control and policy control. If either test fails, the reservation is rejected and the RSVP process returns an error message to the appropriate receivers. If both succeed, the node sets the packet classifier to select the data packets defined by the FilterSpec. RSVP then interacts with the appropriate link layer to obtain the desired QoS defined by the FlowSpec. The action to control QoS occurs at the upstream end of the link, although the RSVP reservation request originates from receivers downstream. Once the reservation is made at the node, the reservation request is propagated upstream towards the appropriate senders. The FlowSpec in the RESV message may have been modified by the traffic control mechanism, and reservations from different downstream branches of the multicast tree for the same sender (or set of senders) are merged as reservations travel upstream. The RESV messages will be propagated immediately to the next node *only if there will be a net change after merging.* Otherwise, the messages are refreshed periodically. RESV messages must finally be delivered to the sender hosts themselves, so the hosts can set up appropriate traffic control parameters for the first hop.

1.4.1.3   Confirmation Messages

When a receiver originates a reservation request, it can also request a confirmation message by including in the RESV message a confirmation request object containing its IP address. The reservation confirmation (ResvConf) message is sent by an upstream node to indicate that the request was probably, but not necessarily due to merging, installed in the network. If the reservation request is merged with a larger one at an intermediate node, the intermediate node sends the confirmation message, because the reservation request is not propagated upstream in this case. This reservation might then fail if the merged request fails.

1.4.1.4   Error Messages

There are two RSVP error messages: ResvErr and PathErr.

PathErr messages are sent upstream to the sender that created the error, and they do not change path state in nodes through which they pass.

As for ResvErr messages, there are many ways for a syntactically valid reservation request to fail at a node along the path. Nodes may also preempt an established reservation. Because the failed request may be a combination of a number of requests, a ResvErr message must be sent to all of the appropriate receivers. In addition, merging heterogeneous requests creates a potential problem (called the *killer reservation* problem), in which a request could deny service to another.

The problem is simple when a reservation R1 is already in place. If another receiver makes a larger reservation R2, the result of merging R1 and R2 may be rejected by admission control in an upstream node. The service to R1 will not be denied, however, because when admission control fails for a reservation request existing reservations are left in place.

When the receiver making a reservation R2 is persistent even though admission control is failing for R2 at a certain node, another receiver should be able to establish a smaller reservation R1 that would succeed if not merged with R2. To enable this, a ResvErr message establishes an additional state, called *blockade state*, in each node through which it passes. Blockade state in a node modifies the merging procedure to omit the offending FlowSpec from the merge, allowing a smaller request to be established. A reservation request that fails admission control creates blockade state but is left in place in nodes downstream of the failure point.

1.4.1.5 Teardown Messages

RSVP teardown messages remove path or reservation state immediately. There are two types of RSVP teardown messages: PathTear and ResvTear. A PathTear message travels towards all receivers downstream from its point of initiation and deletes path state, as well as all dependent reservation state. A ResvTear message travels towards all senders upstream from its point of initiation and deletes reservation state. It is possible to tear down any subset of the established state; for path state, the granularity for teardown is a single sender, while for reservation state the granularity is an individual FilterSpec.

Teardown requests can be initiated by end systems or by routers as a result of state timeout or service preemption. The state deletion is immediately propagated to the next node only if there will be a net change after merging (same as with the reservation state). Hence, a ResvTear message will prune the reservation state back as far as possible.

1.4.2 STATE DATA

The following data structures are maintained by the RSVP protocol at each node[7]:

- *Path state block* stores state from the PATH message for each session and sender template.
- *Reservation state block* holds a reservation request that arrived in a particular RESV message, corresponding to the triple: session, next hop, and FilterSpec list.
- *Traffic control state block* holds the reservation specification that has been handed to traffic control for a specific outgoing interface. In general, this information is derived from the reservation state block for the same outgoing interface. The traffic control state block defines a single reservation for the triple: session, outgoing interface, and FilterSpec list.
- *Blockade state block* contains an element of blockade state (see Section 1.4.1). Depending upon the reservation style in use, this information may be per session and sender template pair, or per session and previous hop pair.

1.4.3   Message Routing

PATH messages are sent with the same source and destination addresses as the data so they will be routed correctly through non-RSVP clouds. On the other hand, RESV messages are sent hop-by-hop; each RSVP-capable node forwards a RESV message to the unicast address of a previous RSVP hop.

RSVP acquires routing entries by sending route queries to the routing protocol. As previously mentioned, RSVP needs to send a different copy of the PATH message on each outgoing interface. Hence, RSVP simulates its own multicast forwarding so it can specify a single interface to send a multicast packet without any loop back.

RSVP may ask routing to notify it when a particular route changes. Route change notification enables RSVP to quickly adapt its reservations to changes in the route between a source and destination. For multicast destinations, a route change consists of any local change in the multicast tree for a source-group pair (including prunes and grafts), as well as routing changes due to failed or recovered links. RSVP adapts to route changes by resending PATH or RESV messages where needed. If routing cannot support route change notification, RSVP must poll routing for route entries in order to adapt to route changes.

1.4.4   Message Merging

RSVP merges RESV messages in the network for the *same* reservation style (see Figure 1.7). The RESV messages are forwarded only until the point where they

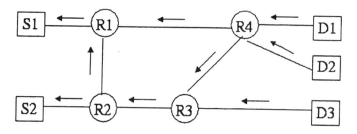

Figure 1.7 RSVP RESV messages

merge with a larger request. A RESV message forwarded to a previous hop carries a FlowSpec that is the largest of the FlowSpecs requested by the next hops to which the data flow will be sent. Since FlowSpecs are opaque to RSVP, the rules for comparing FlowSpecs are defined in the integrated services specifications. RSVP must call service-specific routines to perform FlowSpec comparison and merging.

FlowSpecs are generally multidimensional vectors, and each of their TSpec and RSpec components may itself be multidimensional. It may not be possible to strictly order two FlowSpecs. For example, if one request specifies a lower bandwidth than the other, but the other specifies a looser delay bound, neither is larger than the other. In this case, instead of taking the larger, the service-specific merging routines must be able to return a third FlowSpec that is at least as large as each of them. This is the least upper bound (LUB) of the flows. In some cases, a FlowSpec at least as small is needed, and this is the greatest lower bound (GLB).

1.5 RSVP INTERFACES

RSVP interacts with other components in the router through well-defined interfaces. The RSVP/policy control interface will be discussed in Section 1.11 because it is still being specified. The remaining interfaces are briefly described below.

- *Application/RSVP Interface.* The application/RSVP interface should allow the application to register a session, define a sender, reserve resources, and release resources. It should also inform the application of the receipt of all types of RSVP messages.
- *RSVP/Traffic Control Interface.* This interface enables establishing a reservation, modifying a reservation, deleting a FlowSpec, deleting or adding a FilterSpec, updating the AdSpec, and preempting a reservation.
- *RSVP/Routing Interface.* This interface supports route querying, route change notification, and interface list discovery.
- *RSVP/Packet I/O Interface.* RSVP must be able to use the promiscuous receive mode for RSVP messages, force a packet to be sent to a specific interface, specify the source address and time-to-live in PATH messages, and send messages with the router alert option.
- *Service-Dependent Manipulations.* RSVP must be able to compare Flow-Specs, compute their LUBs and GLBs, and compare and sum TSpecs (as explained in the last section of this chapter).

1.6 RSVP MANAGEMENT

The simple network management protocol (SNMP) version 2 defines an architecture for network management for the Internet protocol suite and a framework for accessing, describing, and naming objects to be managed. Managed objects are accessed via a virtual information store, which is called the management information base (MIB). Each object type is named by an administratively assigned name, called the object identifier.

The RSVP MIB is composed of the following sections defined in RFC 2206[2]:

- General objects
- Session statistics table
- Session sender table
- Reservation requests received table
- Reservation requests forwarded table
- RSVP interface attributes table
- RSVP neighbor table

1.7   RSVP SECURITY

In 1995, an architecture for providing security in IP versions 4 and 6 (IPSEC) was developed. Two methods for IP security were defined. The first method introduces an authentication header (AH) in IP packets after the IP header, but before the information being authenticated. The authentication header can provide authentication, integrity, and possibly non-repudiation, depending on the cryptographic algorithm employed. The second method is the IP encapsulating security payload (ESP). ESP can provide confidentiality and integrity, and possibly authentication to IP packets.

RSVP as specified in Braden et al.[8] can support the two above mentioned IP security protocols, but only on a per-address, per-protocol basis, not on a per-flow basis. This is because RSVP relies on transport protocol port numbers (e.g., TCP or UDP ports). For flows without such port numbers, such as IPSEC packet flows where such information is encrypted, flow definition is solely dependent on the IP address and protocol.

In Berger and O'Malley,[3] RSVP is extended to permit per-flow use of the AH and ESP techniques. This is accomplished through using the IPSEC security parameter index (SPI) instead of the transport protocol port numbers. This, however, necessitates that the FilterSpec object contain the SPI. The session object will also require the addition of a virtual destination port to be able to demultiplex sessions beyond the IP destination address. Therefore, the processing of the RESV and PATH messages is modified. One limitation of this method, however, is that when the wildcard filter is used, all flows to the same IP destination address and with the same IP protocol identifier will share the same reservation.

Hop-by-hop integrity and authentication of RSVP messages and sessions can be provided through an integrity object. This is especially important in order to protect the integrity of the admission control mechanism against corruption and spoofing. A scheme is proposed in Baker[1] to transmit the result of applying a cryptographic algorithm to a one-way function or digest of the message together with a secret authentication key.

1.8   USE OF RSVP WITH INTEGRATED SERVICES

We first give an overview of the integrated services model then describe how RSVP can be used to set up reservations for integrated services.

1.8.1   INTEGRATED SERVICES

The integrated services model was based on the premise that applications can be either inelastic (real-time) which requires end-to-end delay bounds, or elastic, which can wait for data to arrive. Real-time applications can be further subdivided into those that are intolerant to delay and those that are more tolerant, called delay adaptive.[6] These three application types were directly mapped onto three service categories to be provided to IP traffic: the guaranteed service for delay-intolerant applications, the controlled load service for delay-adaptive applications, and the currently available best-effort service for elastic applications. The guaranteed service gives firm bounds on the throughput and delay, while the controlled load service tries to approximate the performance of an unloaded packet network. In this section, we will provide a brief overview of these two services. Each service is specified by a TSpec and an RSpec, as previously discussed.

1.8.1.1   Guaranteed Quality of Service

The guaranteed service gives firm end-to-end delay bounds as well as bandwidth guarantees. If the traffic of the flow obeys the TSpec, the packets are guaranteed to be delivered within the requested delay bound. The service does not give any guarantees on the delay variation (jitter). The TSpec of the flow is given in the form of a token bucket (bucket rate and bucket depth), a peak rate, a minimum policed unit, and a maximum packet size. The RSpec is described using a rate and a slack term.

Given the token bucket parameters and the data rate given to the flow, it is possible to compute a bound on the maximum queuing delay (thus the maximum delay) experienced by packets. This is because the network elements are required to approximate the fluid model of service. The network element must also export two error-characterization terms which represent how the network element implementation deviates from the fluid model.

At the edge of the network, arriving traffic is compared against the TSpec and policed. In addition, traffic is reshaped at heterogeneous branch points (when TSpec for all branches in a multicast tree is not the same) and at merge points. Reshaping means that the traffic is reconstructed to conform to the TSpec. Reshaping needs to be done only if the TSpec on the outgoing link is less than the TSpec reserved on the immediate upstream link.[13]

1.8.1.2   Controlled Load Service

The controlled load service approximates the behavior of best-effort service with underload conditions. It uses admission control to ensure that adequate resources are available to provide the requested level of quality with overload conditions. Applications can assume that (1) a high percentage of the transmitted packets are successfully delivered to the destinations, and (2) the transit delay experienced by a high percentage of the delivered packets will not highly exceed the minimum transit delay. The controlled load service does not give specific delay or loss guarantees (thus there is no RSpec). Over all timescales significantly larger than the burst time, a controlled load service flow should experience little or no average packet queuing delay and little or no congestion loss.[14]

The controlled load service can borrow bandwidth needed to clear bursts from the network, using an explicit borrowing scheme within the traffic scheduler or an implicit scheme based on statistical multiplexing and measurement-based admission control. Information from measurement of the aggregate traffic flow or specific knowledge of traffic statistics can be used by the admission control algorithm for a multiplexing gain.

As with the guaranteed service, the TSpec for the controlled load service is given by a token bucket, a peak rate, a minimum policed unit, and a maximum packet size. Over all time periods T, the length of the burst should never exceed $rT+b$, where r is the token bucket rate and b is the bucket depth. Nonconformant controlled load traffic is forwarded on a best-effort basis only under overload conditions.

1.8.2 Using RSVP to Set Up Reservations for Integrated Services

As previously mentioned, the RSVP specification does not define the internal format of the RSVP objects related to invoking QoS control services. Interfaces to the QoS control services are also defined in a general format. RFC 2210[15] defines the usage and contents of three RSVP protocol objects:

- FlowSpec: includes the QoS service desired by the receivers, a description of the traffic flow to which the resource reservation should apply, and the parameters required to invoke the service
- AdSpec: includes parameters describing the properties of the data path (including the availability of specific QoS services), and parameters required by the QoS services to operate correctly
- Sender TSpec: includes a description of the data traffic generated by the sender

RFC 2210 also specifies a procedure for applications using RSVP facilities to compute the minimum MTU (maximum transmission unit) over a multicast tree and return the result to the senders to avoid fragmentation.

1.9   SUPPORT OF RSVP BY LINK LAYERS

IP can operate over a number of link layers, including ATM, 802 technologies such as Ethernet, and point to point links such as PPP. This section discusses the support of RSVP by ATM and IEEE 802.

1.9.1   ATM Networks

Table 1.7 illustrates the different principles that underlie the design of IP and ATM signaling, especially multicast. Different receivers in an IP multicast group can specify different QoS requirements through RSVP. In addition, receivers are allowed to dynamically change their QoS requirements throughout the connection lifetime (since reservations are periodically refreshed). Group membership also changes throughout connection lifetime.

Since ATM networks provide QoS guarantees, it is natural to map RSVP QoS specifications to ATM QoS specifications and establish the appropriate ATM switched virtual connections (VCs) to support the RSVP requirements. The issue, however, is complicated by the factors that were previously mentioned: RSVP allows heterogeneous receivers and reservation parameter renegotiation, while ATM does

TABLE 1.7
IP/RSVP Multicast versus ATM Multipoint

Category	IP/RSVP	ATM UNI 3.1
Connection type	Connectionless	Virtual connections
Cell ordering	Not guaranteed	Guaranteed
QoS	New services are being added to best effort	CBR, rt-VBR, nrt-VBR, ABR, UBR and more (GFR)
QoS setup time	Separate from route establishment	Concurrent with route establishment
Renegotiation	Allowed	Not allowed
Heterogeneity	Receiver heterogeneity	Uniform QoS to all receivers
Tree Orientation	Receiver-based	Sender-based (UNI 4.0 adds leaf-initiated join)
State	Soft (periodic renewal)	Hard
Directionality	Unidirectional	Unidirectional point-to-multipoint VCs
Tree construction	Different algorithms	Multicast servers or meshes

not. The solution for providing RSVP over ATM must tackle these problems, ensuring scalability. It must also support both UNI 3.1 and UNI 4.0, which support only point-to-multipoint connections.

The problem of supporting RSVP over ATM consists of two main subproblems: first, mapping the IP integrated services to ATM services and, second, using ATM VCs as part of the integrated services Internet. The IP guaranteed service is mapped to constant bit rate (CBR) or real-time variable bit rate (VBR-rt); the controlled load service is mapped to non real-time VBR (VBR-nrt) or available bit rate (ABR) with a minimum cell rate; and the best-effort service is mapped to unspecified bit rate (UBR) or ABR. The second subproblem, managing ATM VCs with QoS as part of the integrated services Internet, entails computing the number of VCs needed and designating the traffic flows that are routed over each VC. Two types of VCs are required: data VCs that handle the actual data traffic, and control VCs which handle the RSVP signaling traffic. The control messages can be carried on the data VCs or on separate VCs.

The best scheme for VC management should use a minimal number of VCs, waste minimal bandwidth due to duplicate packets, and handle heterogeneity and renegotiation in a flexible manner. Proposals that significantly alter RSVP should be avoided. Furthermore, using special servers might introduce additional delays, so cut-through forwarding approaches are preferred. The problem of mapping RSVP to ATM is simplified by the fact that while RSVP reservation requests are generated at the receiver, actual allocation of resources occurs at the sub-net sender. Thus senders establish all QoS VCs, and receivers must be able to accept incoming QoS VCs. The key issues to tackle are data distribution, receiver transitions, end-point identification, and heterogeneity. Several heterogeneity models are defined by Crawley et al.[9] that provide different capabilities to handle the heterogeneity problem. The dynamic QoS problem can be solved by establishing new VCs with minimal signaling, but a timer should guarantee that the rate at which VCs are established is not excessively high.

1.9.2   IEEE 802 Networks

The IETF is currently working on supporting integrated services and RSVP over IEEE 802 networks. In particular, the subnet bandwidth manager (SBM) protocol is being defined as a signaling protocol for RSVP-based admission control over IEEE 802-style networks. SBM defines the operation of RSVP-enabled hosts/routers and link layer devices (switches, bridges) to support reservation of LAN resources for RSVP-enabled data flows. In the absence of any link-layer traffic control or priority queuing mechanisms in the underlying LAN (such as a shared LAN segment), the SBM-based admission control mechanism limits only the total amount of traffic load imposed by RSVP-enabled flows. A protocol entity called designated SBM (DSBM) exists for each managed segment and is responsible for admission control over the resource reservation requests originating from the DSBM clients in that segment. DSBM obtains information such as the limits on fraction of available resources that can be reserved on each managed segment under its control. Then, for each interface attached, a DSBM client determines whether a DSBM exists on the interface. When a DSBM client sends or forwards an RSVP PATH message over an interface attached to a managed segment, it sends the PATH message to the segment DSBM instead of sending it to the RSVP session destination address (as is done in conventional RSVP processing). The DSBM processes the RSVP RESV message based on the bandwidth available and returns a ResvErr message to the requester if the request cannot be granted. If sufficient resources are available and the reservation request is granted, the DSBM forwards the RESV message towards the previous hop, based on its local PATH state for the session. The addition of a DSBM for admission control over managed segments results in some additions to the RSVP message-processing rules at a DSBM client.[16]

1.10 RSVP INTEROPERABILITY

RSVP must operate correctly even when two RSVP-capable routers are joined by an arbitrary "cloud" of non-RSVP routers. This is because RSVP will be deployed gradually in the Internet and might never be implemented in some parts. An intermediate cloud that does not support RSVP will be unable to perform resource reservation. If that cloud has sufficient capacity, however, it may still provide useful real-time service.

Both RSVP and non-RSVP routers forward PATH messages towards the destination address using their local routing table. The PATH message carries to each RSVP-capable node the IP address of the last RSVP-capable router. RESV messages are thus forwarded directly to the next RSVP-capable router on the path(s) back towards the source, as shown in Figure 1.8. Non-RSVP-capable nodes can affect the QoS provided to receivers. Thus, RSVP passes a non-RSVP flag bit (also called global break bit) to the local traffic control mechanism when there are non-RSVP-capable hops in the path. Traffic control forwards such information on the service capability to receivers using the AdSpecs.

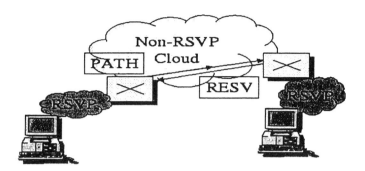

Figure 1.8 Interoperation with non-RSVP capable routers

Some topologies of RSVP and non-RSVP routers can cause RESV messages to arrive at the wrong RSVP-capable node or to arrive at the wrong interface of the correct node. To handle the wrong interface case, a Logical Interface Handle (LIH) is used. In addition to the address of the previous node, the PATH message includes a LIH defining the logical outgoing interface, and this is stored in the path state block. A RESV message arriving at the addressed node carries both the IP address and the LIH of the correct outgoing interface.

1.11 OPEN ISSUES AND CURRENT WORK

RSVP is recommended to be deployed gradually and with caution, first in intranets, then limited Internet Service Provider environments, before being used on a large scale in the Internet. This is because RSVP is still immature and may be changed if problems are found. In addition, some issues pertaining to RSVP remain unresolved. For example, scalability of RSVP is a major concern, since resource requirements for running RSVP on a router increase proportionally to the number of RSVP reservations. To overcome this problem, it is foreseen that the "edge" of the backbone will aggregate groups of streams requiring special service, as is discussed below.

Security considerations are also a subject of current study. It is essential to protect against modified reservation requests used to obtain service to unauthorized parties or lock up network resources. Hop-by-hop checksums and encryption were proposed to detect reservation requests that are modified between RSVP neighbor routers. Key management and distribution for such encryption is not yet in place. Policies for making or limiting reservations are also still being studied. Caution is warranted because of limited experience with setting and controlling such policies.[12]

Much work is currently being done on RSVP security and support by specific link layers, as well as on policy control, aggregation of flows, routing interface, label switching, and diagnostics. Some of these issues, such as security and support by specific link layers, have been previously discussed. The remaining issues will be discussed in this section.

1.11.1 AGGREGATION AND DIFFERENTIATED SERVICES

An extremely large number of flows travels on network backbone links. Many researchers have expressed concerns about how well RSVP will scale to such situations, since RSVP exhibits overhead in terms of state, bandwidth, and computation required for each flow. One of the solutions proposed to this problem is the aggregation of flows. Aggregation, however, must be provided without affecting the end-to-end QoS guarantees of individual flows.

QoS aggregation in RSVP has two major components. The first is the extension of RSVP to support aggregate QoS requests made by a set of flows rather than individual flows. Aggregate requests are not currently supported by RSVP and require the definition of new filter specifications. The second component is the aggregation of a large number of individual RSVP requests without precluding support for individual QoS guarantees where feasible. This serves to ensure individual end-to-end QoS guarantees, without requiring the awareness of individual flows on every segment of their path.

Routers at the edge of a region doing aggregation keep detailed state, while in the interior of the region routers keep a greatly reduced amount of state. Tunneling of RSVP requests and the use of the router alert option have been proposed to reduce RSVP overhead in the backbones. Packets can be tagged at the edge with scheduling information that will be used in place of the detailed state.[5] One way of doing that is through the use of the type of service (TOS) field in the IP header. This idea has been adopted by the Differentiated Services Working Group, which first met in March 1998.

The aim of the group is to find an alternative to per-flow processing and per-flow state to enable the deployment of differentiated services in large carrier networks. A number of proposals have emerged which suggest that RSVP can be used with the differentiated services model to meet the needs of large Internet service providers. Most of these proposals envision the use of the differentiated services model in large core networks and the use of RSVP in peripheral stub networks (see Figure 1.9). This model enables the scalability of the differentiated services model to be combined with the fine granularity, minimal management requirements, admission control, and policy support of the integrated services and RSVP model.[4]

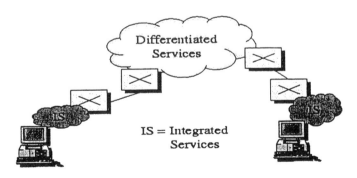

Figure 1.9 Interoperation of integrated services and differentiated services

1.11.2 POLICY CONTROL

Since RSVP allows users to reserve network resources, it is important to have firm control over which users are allowed to reserve resources, and how much resources they can reserve. Network managers and service providers must be able to monitor, control, and enforce use of network resources and services based on policies derived from criteria such as the identity of users and applications, traffic or bandwidth requirements, security considerations, and time-of-day or week.

A policy is defined to be the combination of rules and services, where rules define the criteria for resource access and usage. Policy control determines if a user has the permission to make a reservation. This can be as simple as access approval or as complex as sophisticated accounting and debiting mechanisms. The RSVP specification in RFC 2205[8] contains a place holder for policy support in the form of policy data objects that can be carried in RSVP messages. A policy object contains policy-related information, such as policy elements, which are units of information necessary for the evaluation of policy rules, such as the identity of the user. The mechanisms and message formats for policy enforcement are not yet specified. The interface between RSVP and the local policy modules (LPMs) at the network nodes is also not specified. LPMs are responsible for receiving, processing, and forwarding policy data objects. LPMs may also rewrite and modify the objects as they pass through policy nodes.

The policy enforcement point (PEP) is the point where the policy decisions are actually enforced. This usually resides in a network node (for example, a router). The policy decision point (PDP) is the point where policy decisions are made. This may or may not be local to the network node, and if not local it may or may not be located in a policy server. In some cases, local policy decisions need to be made. Such preliminary policy decisions are made by a local decision point (LDP). The final decision is still made by the PDP.[17] Figure 1.10 illustrates these components.

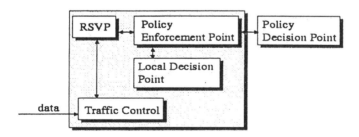

Figure 1.10 Policy control components

When the PEP receives a message requiring a policy decision, it first consults the LDP, if available. Then the PEP sends a policy decision request to the PDP. That request may contain one or more policy objects, in addition to the admission control information in the original message that triggered the policy decision request. It will also contain the LDP decision, if any. The PDP returns the policy decision, and the PEP enforces the policy decision by appropriately accepting or denying the request.

The PDP may also return additional information to the PEP. The interaction between the PEP and PDP can use a protocol such as the Common Open Policy Service (COPS) protocol.

1.11.3 ROUTING AND LABEL SWITCHING

Routing

A number of proposals for the RSVP interface to the routing protocol are currently being discussed. Some proposals suggest that the interface should not only allow RSVP to request information and services from routing, but also to pass relevant information to the routing protocol. For example, it may be desirable to use different routes for flows belonging to different applications, having different QoS requirements, or different specified explicit routes, even if the packets of these two flows have the same source and destination addresses. If RSVP is to support this type of routing, its interface to routing must allow it to pass information such as port numbers, QoS requirements, and explicit routes, in addition to the addresses. Thus, the RSVP module should pass to the routing module all relevant information that may be useful in making a routing decision.[11]

Label Switching

Multiprotocol Label Switching (MPLS) allows labels to be bound to various granularities of forwarding information, including application flows. Labels can be allocated and bound to RSVP flows, and RSVP messages can be used to distribute the appropriate binding information. Hosts and routers that support both label switching and RSVP can associate labels with RSVP flows. This enables label-switching routers to identify the appropriate reservation state for a packet based on its label value. Two new objects are defined for this purpose: *RSVP label* carries a label in an RSVP message, and *hop count* enables time-to-live processing for RSVP flows that pass through ATM label-switching routers. There are several alternatives to mapping RSVP flows to labels, one of which specifies a model in which, on a given link, each sender to a single RSVP session is associated with one label.[10]

1.11.4 DIAGNOSTICS

Diagnostic messages for RSVP are useful for collecting information about the RSVP state along the path. Such information includes information on different hops when a path or reservation request has failed, as well as feedback regarding the details of a reservation that has been made, such as whether, where, or how, the reservation request was merged with those of others. This information can be useful for debugging purposes and for network resource management. Diagnostic messages are independent from any other RSVP control messages and do not change RSVP state. This diagnostic tool can be invoked by a client from any host that may or may not be a participant of the RSVP session to be diagnosed.

Two types of RSVP diagnostic packets are defined: diagnostic request (DREQ) and reply (DREP). A client invokes RSVP diagnostic functions by generating a

DREQ packet and sending it along the RSVP path to be diagnosed. This DREQ packet specifies the RSVP session and a sender host to that session. The DREQ packet starts collecting information at the last node and proceeds backwards towards the sender. Each RSVP-capable router receiving the DREQ packet adds to the packet a response data object containing the router RSVP state for the specified RSVP session, then it forwards the request via unicast to the router that it believes to be the previous hop for the given sender. When the DREQ packet reaches the sender, the sender changes the packet type to DREP and sends the completed response to the original requester.[19]

GLOSSARY

Admission Control Process determines if sufficient resources are available to make the reservation.

AdSpec includes parameters describing the properties of the data path and parameters required by specific QoS control services to operate correctly. The AdSpec is generated by data sources or intermediate network elements and may be used and updated inside the network before being delivered to receiving applications. The receivers can use the AdSpec to predict the end-to-end service.

FilterSpec defines the flow to receive the desired QoS and is contained in an RSVP reservation request. The basic FilterSpec format defined in the present RSVP specification has a very restricted form: sender IP address and optionally the UDP/TCP port number SrcPort.

Fixed Filter creates a distinct reservation per specified sender (without installing separate reservations for each receiver to the same sender).

FlowSpec specifies the QoS desired by the receivers and is contained in an RSVP reservation request. The FlowSpec will generally include a service class and two sets of numeric parameters: an RSpec that defines the desired QoS, and a TSpec that describes the data flow.

Packet Classifier determines the quality of service class of packets according to the requirements.

Packet Scheduler manages the various queues to guarantee the required quality of service.

Policy Control Process determines if the user has permission to make the reservation.

Reservation Protocol (RSVP) is the means by which applications communicate their requirements to the network in an efficient and robust manner. RSVP does not provide any network service; it simply communicates any end-system requirement to the network. Thus RSVP can be viewed as a switch state establishment protocol, rather than just a resource reservation protocol.

Reservation Styles refers to the method by which reservation requests from various receivers in the same session are aggregated inside the network: whether a separate reservation should be established for each upstream sender in the same session, or if a single reservation can be shared among the packets of selected senders (or all senders in that session).

RSpec (R for reserve) gives the quality of service requested from the network.

Session refers to a simplex data flow with a particular (unicast or multicast) destination, transport-layer protocol, and an optional (generalized) destination port.

Shared Explicit Filter is a filter where a single reservation is created, but can only be shared by selected upstream senders.

Soft State is the state maintained at network switches which is periodically and automatically refreshed by RSVP.

Traffic Control refers to the admission control process, packet classifier and packet scheduler.
TSpec (T for traffic) describes the flow traffic pattern to the network.
Wildcard Filter shares the same reservation among all upstream senders, reserving resources to satisfy the largest resource request (regardless of the number of senders).

REFERENCES

1. F. Baker. RSVP cryptographic authentication. Work in progress, draft-ietf-rsvp-md5-05.txt, August 1997.

2. F. Baker, J. Krawczyk, and A. Sastry. RSVP management information base using SMIv2. RFC 2206, September 1997. http://info.internet.isi.edu:80/in-notes/rfc/files/rfc2206.txt.

3. L. Berger and T. O'Malley. RSVP extensions for IPSEC data flows. RFC 2207, September 1997. http://info.internet.isi.edu:80/in-notes/rfc/files/rfc2207.txt.

.4. Y. Bernet, R. Yavatkar, P. Ford, F. Baker, and L. Zhang. A framework for end-to-end QoS combining RSVP/Intserv and differentiated services. Work in progress, draft-bernet-intdiff-00.tx, March 1998.

5. S. Berson and S. Vincent. Aggregation of internet integrated services state. Work in progress, draft-berson-classy-approach-01.txt, November 1997.

6. R. Braden, D. Clark, and S. Shenker. Integrated services in the internet architecture: an overview. RFC 1633, June 1994. http://info.internet.isi.edu:80/in-notes/rfc/files/rfc1633.txt.

7. R. Braden and L. Zhang. Resource reservation protocol (RSVP) - version 1 message processing rules. RFC 2209, September 1997. http://info.internet.isi.edu:80/in-notes/rfc/files/rfc2209.txt.

8. R. Braden, L. Zhang, S. Berson, S. Herzog, and S. Jamin. Resource ReSerVation Protocol (RSVP). RFC 2205, September 1997. http://info.internet.isi.edu:80/in-notes/rfc/files/rfc2205.txt.

9. E. Crawley, L. Berger, S. Berson, F. Baker, M. Borden, and J. Krawczyk. A framework for integrated services and RSVP over ATM. Work in progress, draft-ietf-issll-atm-framework-03.txt, April 1998.

10. B. Davie, Y. Rekhter, E. Rosen, A. Viswanathan, V. Srinivasan, and S. Blake. Use of label switching with RSVP. Work in progress, draft-dietf-mpls-rsvp-00.txt, March 1998.

11. R. Guerin, S. Kamat, and E. Rosen. Extended RSVP-routing interface. Work in progress, draft-guerin-ext-rsvp-routing-intf-00.txt, July 1997.

12. A. Mankin, F. Baker, R. Braden, S. Bradner, M. O'Dell, A. Romanow, A. Weinrib, and L. Zhang. Resource reservation protocol (RSVP) - version 1 applicability statement: Some guidelines on deployment. RFC 2208, September 1997. http://info.internet.isi.edu:80/in-notes/rfc/files/rfc2208.txt.

13. S. Shenker, C. Partridge, and R. Guerin. Specification of guaranteed quality of service. RFC 2212, September 1997. http://info.internet.isi.edu:80/in-notes/rfc/files/rfc2212.txt.

14. J. Wroclawski. Specification of the controlled-load network element service. RFC 2211, September 1997. http://info.internet.isi.edu:80/in-notes/rfc/files/rfc2211.txt.

15. J. Wroclawski. The use of RSVP with IETF integrated services. RFC 2210, September 1997. http://info.internet.isi.edu:80/in-notes/rfc/files/rfc2210.txt.

16. R. Yavatkar, D. Hoffman, Y. Bernet, and F. Baker. SBM (subnet bandwidth manager): Protocol for RSVP-based admission control over IEEE 802-style networks. Work in progress, draft-ietf-issll-is802-bm-05.txt, November 1997.

17. R. Yavatkar, D. Pendarakis, and R. Guerin. A framework for policy-based admission control. Work in progress, draft-ietf-rap-framework-00.txt, November 1997.

18. L. Zhang, S. Deering, D. Estrin, S. Shenker, and D. Zappala. RSVP: A new Resource ReSerVation Protocol. *IEEE Network Magazine*, September 1993. ftp://parcftp.xerox.com/pub/net-research/rsvp.ps.Z.

19. L. Zhang and A. Terzis. RSVP diagnostic messages. Work in progress, draft-ietf-rsvp-diagnostic-msgs-03.txt, November 1997.

2 Multimedia Over IP: RSVP, RTP, RTCP, RTSP

Chunlei Liu

CONTENTS

The future Integrated Services Internet will provide means to transmit real-time multimedia data across networks. RSVP, RTP, RTCP and RTSP are the foundation of real-time services. This paper is a detailed survey of the four related protocols.

2.1 MULTIMEDIA NETWORKING: GOAL AND CHALLENGE

Computer networks were designed to connect computers at different locations so that they can share data and communicate. In the old days, most of the data carried on networks was textual data. Today, with the rise of multimedia and network technologies, multimedia has become an indispensable feature on the Internet. Animation, voice, and video clips become more and more popular on the Internet. Multimedia networking products such as Internet telephony, Internet TV, and videoconferencing have appeared on the market. In the future, people will enjoy other multimedia products in distance learning, distributed simulation, distributed work groups, and other areas.

For networkers, multimedia networking is to build the hardware and software infrastructure and application tools to support multimedia transport on networks so that users can communicate in multimedia. Multimedia networking will greatly boost the use the of computers as a communication tool. We believe that someday multimedia networks will replace telephone, television, and other inventions that have dramatically changed our lives.

2.1.1 THE REAL-TIME CHALLENGE

Multimedia networking is not a trivial task. We can expect at least three difficulties. First, compared with traditional textual applications, multimedia applications usually require much higher bandwidth. A typical piece of 25-s 320 × 240 QuickTime movie could take 2.3MB, which is equivalent to about 1000 screens of textual data. This would have been unimaginable in the old days when only textual data was transmitted on the net.

Second, most multimedia applications require real-time traffic. Audio and video data must be played back continuously at the rate they were sampled. If the data does not arrive in time, the playback will stop and human ears and eyes can easily pick up the artifact. In Internet telephony, human beings can tolerate a latency of about 250 msec. If the latency exceeds this limit, the voice will sound like a call routed over a long satellite circuit and users will complain about the quality of the call. In addition to the delay, network congestion also has more serious effects on real time traffic. If the network is congested, the only effect on nonreal-time traffic is that the transfer takes longer to complete, but real-time data becomes obsolete and will be dropped if it doesn't arrive in time. If no proper action is taken, the retransmission of lost packets would aggravate the situation and jam the network.

Third, a multimedia data stream is usually bursty. Just increasing the bandwidth will not solve the burstiness problem. For most multimedia applications, the receiver has a limited buffer. If no measure is taken to smooth the data stream, it may overflow or underflow the application buffer. When data arrives too quickly, the buffer will overflow and some data packets will be lost, resulting in poor quality. When data arrives too slowly, the buffer will underflow and the application will starve.

Contrary to the high bandwidth, real-time and bursty traffic of multimedia data in real life networks are shared by thousands and millions of users and have limited bandwidth and unpredictable delay and availability. How to solve these conflicts is a challenge multimedia networking must face.

The possibility of answering this challenge comes from the existing network software architecture and emerging hardware. The basis of the Internet, TCP/IP and UDP/IP, provides a range of services that multimedia applications can use. Fast networks like Gigabit Ethernet, FDDI, and ATM provide the high bandwidth required by digital audio and video.

Consequently the design of real-time protocols for multimedia networking becomes imperative before the multimedia age comes.

2.1.2 MULTIMEDIA OVER THE INTERNET

There are other ways to transmit multimedia data, such as over dedicated links, cables, and ATM. However, the idea of running multimedia over the Internet is extremely attractive.

Dedicated links and cables are not practical because they require special installation and new software. Without an existing technology like LAN or WAN, the software development will be extremely expensive. ATM was said to be the ultimate solution for multimedia because it supports very high bandwidth, is connection-oriented, and can tailor different levels of quality of service to different types of applications. But at this moment, very few users have ATM networks reaching their organization, and even fewer have an ATM connection to their desktops.

On the other hand, the Internet is growing exponentially. The well-established LAN and WAN technologies based on the IP protocol suite connect bigger and bigger networks all over the world to the Internet. In fact, the Internet has become the platform of most networking activities. This is the primary reason to develop multimedia protocols over the Internet. Another benefit of running multimedia over IP is that users can have integrated data and multimedia service over one single network, without investing in another network and building the interface between two networks.

At the current time, IP and Ethernet seem to be more favored in desktops and LANs, with ATM in WANs.

As a shared datagram network, the Internet is not naturally suitable for real-time traffic. To run multimedia over the Internet, several issues must be solved. First, multimedia means extremely dense data and heavy traffic. The hardware has to provide enough bandwidth.

Second, multimedia applications are usually related to multicast, i.e., the same data stream, not multiple copies, is sent a group of receivers. For example, in videoconferencing, the video data needs to be sent to all participants at the same time. Live video can be sent to thousands of recipients. The protocols designed for multimedia applications must take into account multicast in order to reduce the traffic.

Third, the cost attached to shared network resources is the unpredictable availability. However, real-time applications require guaranteed bandwidth when the transmission takes place, so there must be some mechanism for real-time applications to reserve resources along the transmission path.

Fourth, the Internet is a packet-switching datagram network where packets are routed independently across shared networks. The current technologies cannot guarantee that real-time data will reach the destination without being jumbled and jerky. Some new transport protocols must be used to take care of the timing issues so that audio and video data can be played back continuously with correct timing and synchronization.

Fifth, there should be some standard operations for applications to manage the delivery and present the multimedia data.

The answers to the above issues are the protocols that are discussed in this chapter.

2.1.3 THE SOLUTION

The Internet carries all types of traffic, each of which has different characteristics and requirements. For example, a file transfer application requires that some quantity of data be transferred in an acceptable amount of time, while Internet telephony requires that most packets get to the receiver in less than 0.3 seconds. If enough bandwidth is available, best-effort service fulfills all of these requirements. When resources are scarce, however, real-time traffic will suffer from the congestion.

The solution for multimedia over IP is to classify all traffic, allocate priority for different applications, and make reservations. The Integrated Services working group in the IETF (Internet Engineering Task Force) developed an enhanced Internet service model called Integrated Services Internet that includes best-effort service and real-time service.[2] The real-time service will enable IP networks to provide quality of service to multimedia applications. Resource ReSerVation Protocol (RSVP), together with Real-time Transport Protocol (RTP), Real-Time Control Protocol (RTCP), and Real-Time Streaming Protocol (RTSP), provides a working foundation for real-time services. Integrated Services allows applications to configure and manage a single infrastructure for multimedia applications and traditional applications. It is a comprehensive approach to providing applications with the type of service they need and the quality they choose. This chapter is a detailed review of these four protocols.

2.2 RSVP — RESOURCE RESERVATION PROTOCOL

RSVP is the network control protocol that allows the data receiver to request a special end-to-end quality of service for its data flows. Real-time applications use RSVP to reserve necessary resources at routers along the transmission paths so that the requested bandwidth can be available when the transmission actually takes place. RSVP is a main component of the future Integrated Services Internet which can provide both best-effort and real-time service.[2]

2.2.1 DEVELOPMENT

RSVP design has been a joint effort of Xerox Corporation's Palo Alto Research Center (PARC), MIT, and Information Sciences Institute of the University of California (ISI).[1] The RSVP specification was submitted to the Internet Engineering

Steering Group (IESG) for consideration as a Proposed RFC in November 1994. In September 1997, RSVP Version 1 Functional Specification and the following other, related Internet drafts were approved as Proposed Standards:

- RFC 2205, Resource ReSerVation Protocol (RSVP) — Version 1 Functional Specification
- RFC 2206, RSVP Management Information Base using SMIv2
- RFC 2207, RSVP Extensions for IPSEC Data Flows
- RFC 2208, RSVP Version 1 Applicability Statement: Some Guidelines on Deployment
- RFC 2209, RSVP Version 1 Message Processing Rules

The RSVP working group of the IETF is developing other protocols to be used with RSVP.

2.2.2 How Does RSVP Work?

RSVP is used to set up reservations for network resources. When an application in a host (the data stream receiver) requests a specific quality of service (QoS) for its data stream, it uses RSVP to deliver its request to routers along the data stream paths. RSVP is responsible for the negotiation of connection parameters with these routers. If the reservation is set up, RSVP is also responsible for maintaining router and host states to provide the requested service.

Each node capable of resource reservation has several local procedures for reservation setup and enforcement (see Figure 2.1). *Policy control* determines whether the user has administrative permission to make the reservation. In the future, authentication, access control, and reservation accounting will also be implemented by policy control. *Admission control* keeps track of the system resources and determines whether the node has sufficient resources to supply the requested QoS.

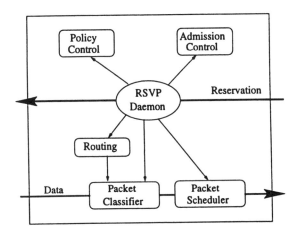

Figure 2.1 Reservation at a node on the data flow path

The RSVP daemon checks with both procedures. If either check fails, the RSVP program returns an error notification to the application that originated the request. If both checks succeed, the RSVP daemon sets parameters in the packet classifier and packet scheduler to obtain the requested QoS. The *packet classifier* determines the QoS class for each packet and the *packet scheduler* orders packet transmission to achieve the promised QoS for each stream. The RSVP daemon also communicates with the routing process to determine the path to send its reservation requests and to handle changing memberships and routes.

This reservation procedure is repeated at routers along the reverse data stream path until the reservation merges with another reservation for the same source stream.

Reservations are implemented through two types of RSVP messages: PATH and RESV. The PATH messages are sent periodically from the sender to the multicast address. A PATH message contains *flow spec* to describe sender template (data format, source address, source port) and traffic characteristics. This information is used by receivers to find the reverse path to the sender and to determine what resources should be reserved. Receivers must join the multicast group in order to receive PATH messages. RESV messages are generated by the receivers and contain reservation parameters including *flow spec* and *filter spec*. The filter spec defines what packets in the flow should be used by the packet classifier. The flow spec is used in the packet scheduler and its content depends on the service. RESV messages follow the exact reverse path of PATH messages, setting up reservations for one or more senders at every node.

The reservation states RSVP builds at the routers are *soft states*. The RSVP daemon needs to send refresh messages periodically to maintain the reservation states. The absence of a refresh message within a certain time will destroy the reservation state. By using soft states, RSVP can easily handle changing memberships and routes.

The reservation requests are initiated by the receivers. They do not need to travel all the way to the source of the sender. Instead, they travel upstream until they meet another reservation request for the same source stream then merge with that reservation. Figure 2.2 shows how the reservation requests merge as they progress up the multicast tree.

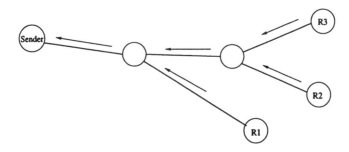

Figure 2.2 Reservation merging

This reservation merging leads to the primary advantage of RSVP: scalability A large number of users can be added to a multicast group without increasing the data traffic significantly. Consequently, RSVP can scale to large multicast groups, and the average protocol overhead decreases as the number of participants increases.

The reservation process does not actually transmit the data and provide the requested QoS. However, through reservation RSVP guarantees the network resources will be available when the transmission actually takes place.

Although RSVP sits on top of IP in the protocol stack, it is not a routing protocol, but rather an Internet control protocol. Actually, RSVP relies on the underlying routing protocols to find where it should deliver the reservation requests. RSVP is also intended to cooperate with unicast and multicast routing protocols. When the RSVP-managed flow changes its path, the routing module will notify the RSVP module of the route changes. Therefore, RSVP can quickly adjust the resource reservation to new routes.

The delivery of reservation parameters is different from the determination of these parameters. How to set the connection parameters to achieve the requested QoS is the task of QoS control devices; the role of RSVP is just a general facility to distribute these parameters. Since different applications may have different QoS control devices, RSVP is designed to treat these QoS parameters as opaque data to be delivered to and interpreted by the control modules at the routers. This logical separation of QoS control devices and distribution facility simplifies the design of RSVP and makes it more adaptive to new network technologies and applications.

2.2.3 RSVP FEATURES

- RSVP flows are simplex.

RSVP distinguishes senders and receivers. Although in many cases a host can act both as a sender and as a receiver, one RSVP reservation only reserves resources for data streams in one direction.

- RSVP supports both multicast and unicast and adapts to changing memberships and routes.

RSVP is designed for both multicast and unicast. Since the reservations are initiated by the receivers and the reservation states are soft, RSVP can easily handle changing memberships and routes. A host can send IGMP (Internet Group Management Protocol) messages to join a multicast group. Reservation merging enables RSVP to scale to large multicast groups without causing heavy overhead for the sender.

- RSVP is receiver-oriented and handles heterogeneous receivers.

In heterogeneous multicast groups, receivers have different capacities and levels of QoS. The receiver-oriented RSVP reservation requests facilitate the handling of heterogeneous multicast groups. Receivers are responsible for choosing their own level of QoS, initiating the reservation, and keeping it active as long as it wants. The senders divide traffic in several flows, each of which is a separate RSVP flow with a different level of QoS. Each RSVP flow is homogeneous and receivers can choose to join one or more flows. This approach makes it possible for heterogeneous receivers to request different QoS tailored to their particular capacities and requirements.

- RSVP has good compatibility.

Efforts have been made to run RSVP over both IPv4 and IPv6. It provides opaque transport of traffic-control and policy-control messages in order to be more adaptive to new technologies. It also provides transparent operation through nonsupporting regions.

2.2.4   RSVP INTERFACES

RSVP communicates with both applications at the end hosts and various components inside network routers. For application programmers, the most important part is the client application programming interface. In an implementation developed by the ISI and Sun Microsystems, the following fundamental functions make up an RSVP client library called RAPI (RSVP Application Programming Interface):

- rapi_session() creates and initiates an API session and returns a session handle for further reference.
- rapi_sender() is called by a sender application to specify the parameters of its data flow. The RSVP daemon at the sender will use this flow spec to create PATH messages for this flow.
- rapi_reserve() is used to make, modify, or delete a reservation.
- rapi_release() asks the RSVP daemon to tear down the reservation.

For more information about RAPI, see the Internet Draft[4] by Braden and Hoffman.

2.3   RTP — REAL-TIME TRANSPORT PROTOCOL

RTP is an IP-based protocol providing support for the transport of real-time data such as video and audio streams. The services provided by RTP include time reconstruction, loss detection, security, and content identification. RTP is primarily designed for multicast of real-time data, but it can be also used in unicast. It can be used for one-way transport, such as video-on-demand, as well as interactive services such as Internet telephony.

RTP is designed to work in conjunction with the auxiliary control protocol RTCP to get feedback on quality of data transmission and information about participants in the ongoing session.

2.3.1   DEVELOPMENT

Attempts to send voice over networks began in the early seventies. Several patents on packet transmission of speech, time stamp, and sequence numbering were granted in the seventies and eighties. In 1991, a series of voice experiments were completed on DARTnet. In August 1991, the Network Research Group of Lawrence Berkeley

National Laboratory released an audio conference tool vat for DARTnet use. The protocol used was referred later as RTP version 0.

In December 1992, Henning Schulzrinne, GMD Berlin, published RTP version 1. It went through several states of Internet drafts and was finally approved as a proposed standard on November 22, 1995 by the IESG. This version[5,6] was called RTP version 2 and was published as RFC 1889, *RTP: A Transport Protocol for Real-Time Applications* and RFC 1890, *RTP Profile for Audio and Video Conferences with Minimal Control*.

On January 31, 1996, Netscape announced Netscape LiveMedia based on RTP and other standards. Microsoft claims that their NetMeeting Conferencing Software supports RTP. The latest extensions have been made by an industry alliance around Netscape Inc., which uses RTP as the basis of the RTSP.

2.3.2 HOW DOES RTP WORK?

As discussed in the first section, the Internet is a shared datagram network, and packets sent on the Internet have unpredictable delay and jitter. However, multimedia applications require appropriate timing in data transmission and playback. RTP provides timestamping, sequence numbering, and other mechanisms to take care of the timing issues. Through these mechanisms, RTP provides end-to-end transport for real-time data over a datagram network.

Timestamping is the most important information for real-time applications. The sender sets the timestamp according to the instant the first octet in the packet was sampled. Timestamps increase by the time covered by a packet. After receiving the data packets, the receiver uses the timestamp to reconstruct the original timing in order to play out the data at the correct rate. Timestamping is also used to synchronize different streams with timing properties, such as audio and video data in MPEG. However, RTP itself is not responsible for the synchronization. This has to be done in the application level.

UDP does not deliver packets in timely order, so sequence numbers are used to place the incoming data packets in the correct order. They are also used for packet loss detection. Notice that in some video formats, when a video frame is split into several RTP packets all of them can have the same timestamp. Consequently, times-tamp alone is not enough to put the packets in order.

The payload type identifier specifies the payload format as well as the encoding/compression schemes. From this payload type identifier the receiving application knows how to interpret and play out the payload data. Default payload types are defined in RFC 1890.[6] Example specifications include PCM, MPEG1/MPEG2 audio and video, JPEG video, Sun CellB video, and H.261 video streams. More payload types can be added by providing a profile and payload format specification. At any given time of transmission, an RTP sender can send only one type of payload although the payload type may change during transmission, for example to adjust to network congestion.

Another function is source identification. It allows the receiving application to know where the data is coming from. For example, in an audio conference a user could tell who is talking from the source identifier.

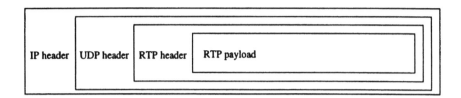

Figure 2.3 RTP data in a UDP/IP packet

The above mechanisms are implemented through the RTP header. Figure 2.3 shows an RTP packet encapsulated in a UDP/IP packet.

RTP is typically run on top of UDP to make use of its multiplexing and checksum functions. TCP and UDP are the two most commonly used transport protocols on the Internet. TCP provides a connection-oriented and reliable flow between two hosts, while UDP provides a connectionless but unreliable datagram service over the network. UDP was chosen as the target transport protocol for RTP for two reasons. First, RTP is primarily designed for multicast; the connection-oriented TCP does not scale well and therefore is not suitable. Second, for real-time data, reliability is not as important as timely delivery. Even more, reliable transmission provided by retransmission as in TCP is not desirable. For example, in network congestion, some packets might get lost and the application would result in lower but acceptable quality. If the protocol insists on reliable transmission, the retransmitted packets could possibly increase the delay, jam the network, and eventually starve the receiving application.

RTP and RTCP packets are usually transmitted using UDP/IP service. However, efforts have been made to make them transport-independent so they can also be run on CLNP(Connectionless Network Protocol), IPX (Internetwork Packet Exchange), AAL5/ATM, or other protocols.

In practice, RTP is usually implemented within the application. Many issues, such as lost packet recovery and congestion control, have to be implemented in the application level.

To set up an RTP session, the application defines a particular pair of destination transport addresses (one network address plus a pair of ports for RTP and RTCP). In a multimedia session, each medium is carried in a separate RTP session, with its own RTCP packets reporting the reception quality for that session. For example, audio and video would travel on separate RTP sessions, enabling a receiver to select whether or not to receive a particular medium.

An audio-conferencing scenario presented in RFC 1889[5] illustrates the use of RTP. Suppose each participant sends audio data in segments of 20-ms duration. Each segment of audio data is preceded by an RTP header, and the resulting RTP message is placed in a UDP packet. The RTP header indicates the type of audio encoding that is used, e.g., PCM. Users can opt to change the encoding during a conference in reaction to network congestion or, for example, to accommodate low-bandwidth requirements of a new conference participant. Timing information and a sequence number in the RTP header are used by the receivers to reconstruct the timing produced by the source, so, in this example, audio segments are contiguously played out at the receiver every 20 ms.

2.3.3   RTP Fixed Header Fields

The RTP header has the format shown in Figure 2.4

The first twelve octets are present in every RTP packet, while the list of CSRC (contributing source) identifiers is present only when inserted by a mixer. The fields have the following meaning:

version (V): 2 bits. Version of RTP. The newest version is 2.

padding (P): 1 bit. If set, the packet contains one or more additional padding octets at the end which are not part of the payload. The last octet of the padding contains a count of how many padding octets should be ignored.

extension (X): 1 bit. If set, the fixed header is followed by exactly one header extension.

CSRC count (CC): 4 bits. Indicates the number of CSRC identifiers that follow the fixed header. This number is more than one if the payload of the RTP packet contains data from several sources.

marker (M): 1 bit. Defined by a profile, the marker is intended to allow significant events such as frame boundaries to be marked in the packet stream.

payload type (PT): 7 bits. Identifies the format of the RTP payload and determines its interpretation by the application.

sequence number: 16 bits. Increments by one for each RTP data packet sent. It may be used by the receiver to detect packet loss and to restore packet sequence. The initial value is randomly set.

timestamp: 32 bits. The sampling instant of the first octet in the RTP data packet. It may be used for synchronization and jitter calculations. The initial value is randomly set.

SSRC: 32 bits. SSRC is a randomly chosen number to distinguish synchronization sources within the same RTP session. It indicates where the data was combined or the source of the data if there is only one source.

CSRC list: 0 to 15 items, 32 bits each. Contributing sources for the payload contained in this packet. The number of identifiers is given by the CC field.

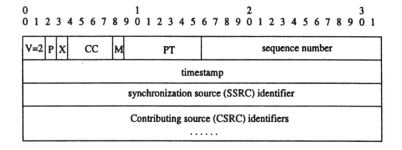

Figure 2.4 RTP packet header

2.3.4   RTCP — Real-time Control Protocol

RTCP is the control protocol designed to work in conjunction with RTP. It is standardized in RFC 1889 and RFC 1890. In an RTP session, participants periodically send RTCP packets to convey feedback on quality of data delivery and information on membership. RFC 1889 defines five RTCP packet types to carry control information. These five types are:

- **RR** (receiver report) — Receiver reports are generated by participants that are not active senders. They contain reception quality feedback about data delivery, including the highest packet number received, the number of packets lost, interarrival jitter, and timestamps to calculate the round-trip delay between the sender and the receiver.
- **SR** (sender report) — Sender reports are generated by active senders. In addition to the reception quality feedback as in RR, they contain a sender information section, providing information on intermedia synchronization, cumulative packet counters, and number of bytes sent.
- **SDES** (source description items) — They contain information to describe the sources.
- **BYE** — Indicates end of participation.
- **APP** (application specific functions) — Is now intended for experimental use as new applications and new features are developed.

Through these control information packets, RTCP provides the following services:

- **QoS monitoring and congestion control**

This is the primary function of RTCP. RTCP provides feedback to an application about the quality of data distribution. The control information is useful to the senders, the receivers, and third-party monitors. The sender can adjust its transmission based on the receiver report feedback. The receivers can determine whether a congestion is local, regional, or global. Network managers can evaluate the network performance for multicast distribution.

- **Source identification**

In RTP data packets, sources are identified by randomly generated 32-bit identifiers. These identifiers are not convenient for human users. RTCP SDES packets contain textual information, called *canonical names*, as globally unique identifiers of the session participants. They may include the user's name, telephone number, e-mail address and other information.

- **Inter-media synchronization**

RTCP sender reports contain an indication of real time and the corresponding RTP timestamp. This can be used in intermedia synchronization, such as lip synchronization in video.

- **Control information scaling**

RTCP packets are sent periodically among participants. When the number of participants increases, it is necessary to balance between getting up-to-date control information and limiting the control traffic. In order to scale up to large multicast groups, RTCP has to prevent the control traffic from overwhelming network resources. RTP limits the control traffic to at most 5% of the overall session traffic. This is enforced by adjusting the RTCP generating rate according to the number of participants.

2.3.5 RTP FEATURES

- RTP provides end-to-end delivery services for data with real-time characteristics, such as interactive audio and video. But RTP itself does not provide any mechanism to ensure timely delivery. It needs support from lower layers that actually have control over resources in switches and routers. RTP depends on RSVP to reserve resources and to provide the requested quality of service.
- RTP does not assume anything about the underlying network, except that it provides framing. RTP is typically run on the top of UDP to make use of its multiplexing and checksum service, but efforts have been made to make RTP compatible with other transport protocols, such as ATM AAL5, and IPv6.
- Unlike usual data transmission, RTP does not offer any form of reliability or flow/congestion control. It provides timestamps and sequence numbers as hooks for adding reliability and flow/congestion control, but how to implement is totally left to the application.
- RTP is a protocol framework that is deliberately not complete. It is open to new payload formats and new multimedia software. By adding new profile and payload format specifications, one can tailor RTP to new data formats and new applications.
- RTP/RTCP provides functionality and control mechanisms necessary for carrying real-time content, but it is not responsible for the higher-level tasks such as assembly and synchronization. These have to be done at the application level.
- The flow and congestion control information of RTP is provided by RTCP sender and receiver reports.

2.3.6 RTP IMPLEMENTATION RESOURCES

RTP is an open protocol that does not provide preimplemented system calls. Implementation is tightly coupled to the application itself. Application developers must add the complete functionality in the application layer by themselves. However, it is always more efficient to share and reuse code rather than start from scratch. The RFC 1889 specification[5] itself contains numerous code segments that can be borrowed directly by the applications. Some implementations with source code are available on the web for evaluation and educational purposes. Many modules in the source code can be usable with minor modifications. The following is a list of useful resources:

- **vat** — http://www-nrg.ee.lbl.gov/vat/
- **tptools** — ftp://ftp.cs.columbia.edu/pub/schulzrinne/rtptools/
- **NeVoT** — http://www.cs.columbia.edu/~{}hgs/rtp/nevot.html
- **RTP Page** — http://www/cs.columbia.edu/~{}hgs/rtp — maintained by Henning Schulzrinne and a very complete reference
- **RTP Library** — http://www.iasi.rm.cnr.it/iasi/netlab/gettingSoftware.html — by E. A. Mastromartino and offers convenient ways to incorporate RTP functionality into C++ Internet applications

2.4 RTSP — REAL-TIME STREAMING PROTOCOL

Instead of being sent as a single large multimedia file that must be stored then played back, multimedia data is usually sent across the network in streams. Streaming breaks data into packets with a size suitable for transmission between the servers and clients. The real-time data flows through the transmission, decompresses and plays back pipeline just like a water stream. A client can play the first packet and decompress the second while receiving the third. Thus the user can start enjoying the multimedia without waiting until the end of transmission.

RTSP is a client-server multimedia presentation protocol to enable controlled delivery of streamed multimedia data over an IP network. It provides "VCR-style" remote control functionality for audio and video streams, such as pause, fast forward, reverse, and absolute positioning. Sources of data include both live data feeds and stored clips.

RTSP is an application-level protocol designed to work with lower-level protocols like RTP and RSVP to provide a complete streaming service over the Internet. It provides a means for choosing delivery channels (such as UDP, multicast UDP, and TCP) and delivery mechanisms based upon RTP. It works for large audience multicast as well as single-viewer unicast.

2.4.1 DEVELOPMENT

RTSP was jointly developed by RealNetworks, Netscape Communications, and Columbia University. It was developed from the streaming practice and experience of RealNetworks' RealAudio and Netscape's LiveMedia. The first draft of the RTSP protocol was submitted to IETF on October 9, 1996 for consideration as an Internet Standard. Since then, it has gone through significant changes. The latest version is draft-ietf-mmusic-rtsp-07.[7]

Although RTSP is still an Internet Draft at IETF and is likely to have significant changes, products using RTSP are available today. The major online players, Netscape, Apple, IBM, Silicon Graphics, VXtreme, Sun, and other companies, have claimed their support to RTSP, although Microsoft's absence is conspicuous.

2.4.2 RTSP Operations and Methods

RTSP establishes and controls streams of continuous audio and video media between the media servers and the clients. A media server provides playback or recording services for the media streams while a client requests continuous media data from the media server. RTSP is the "network remote control" between the server and the client. It provides the following operations:

- Retrieval of media from the media server: The client can request a presentation description and ask the server to set up a session to send the requested data.
- Invitation of a media server to a conference: The media server can be invited to the conference to play back media or to record a presentation.
- Adding media to an existing presentation: The server or the client can notify each other about any additional media becoming available.

RTSP aims to provide the same services on streamed audio and video just as HTTP does for text and graphics. It is designed intentionally to have similar syntax and operations so that most extension mechanisms to HTTP can be added to RTSP.

In RTSP, each presentation and media stream is identified by an RTSP URL. The overall presentation and the properties of the media are defined in a presentation description file, which may include the encoding, language, RTSP URLs, destination address, port, and other parameters. The presentation description file can be obtained by the client using HTTP, e-mail, or other means.

RTSP differs from HTTP in several aspects. First, while HTTP is a stateless protocol, an RTSP server has to maintain session states in order to correlate RTSP requests with a stream. Second, HTTP is basically an asymmetric protocol where the client issues requests and the server responds, but in RTSP both the media server and the client can issue requests. For example, the server can issue a request to set playback parameters of a stream.

In the current version, the services and operations are supported through the following methods:

- **OPTIONS** — The client or the server tells the other party the options it can accept.
- **DESCRIBE** — The client retrieves the description of a presentation or media object identified by the request URL from the server.
- **ANNOUNCE** — When sent from client to server, ANNOUNCE posts the description of a presentation or media object identified by the request URL to a server. When sent from server to client, ANNOUNCE updates the session description in realtime.
- **SETUP** — The client asks the server to allocate resources for a stream and start an RTSP session.
- **PLAY** — The client asks the server to start sending data on a stream allocated via SETUP.

- **PAUSE** — The client temporarily halts the stream delivery without freeing server resources.
- **TEARDOWN** — The client asks the server to stop delivery of the specified stream and free the resources associated with it.
- **GET_PARAMETER** — Retrieves the value of a parameter of a presentation or a stream specified in the URI.
- **SET_PARAMETER** — Sets the value of a parameter for a presentation or stream specified by the URI.
- **REDIRECT** — The server informs the clients that it must connect to another server location. The mandatory location header indicates the URL the client should connect to.
- **RECORD** — The client initiates recording a range of media data according to the presentation description.

Note that some of these methods can be sent either from the server to the client or from the client to the server, but others can be sent in only one direction. Not all these methods are necessary in a fully functional server. For example, a media server with live feeds may not support the PAUSE method.

RTSP requests are usually sent on a channel independent of the data channel. They can be transmitted in persistent transport connections, as a one-connection-per-request/response transaction, or in connectionless mode.

2.4.3 RTSP Features

- RTSP is an application-level protocol with syntax and operations similar to HTTP but for audio and video. It uses URLs like those in HTTP.
- An RTSP server needs to maintain states, using SETUP, TEARDOWN, and other methods.
- RTSP messages are carried out-of-band. The protocol for RTSP may be different from the data delivery protocol.
- Unlike HTTP, in RTSP both servers and clients can issue requests.
- RTSP is implemented on multiple operating system platforms; it allows interoperability between clients and servers from different manufacturers.

2.4.4 RTSP Implementation Resources

Although RTSP is still an IETF draft, there are a few implementations already available on the web. The following is a collection of useful implementation resources:

- RTSP Reference Implementation — http://www6.real.com/rtsp/reference.html

This is a source code testbed for the standards community to experiment with RTSP compatibility.

- RealMedia SDK — http://www6.real.com/realmedia/index.html

This is an open, cross platform, client-server system where implementors can create RTSP-based streaming applications. It includes a working RTSP client and server, as well as the components to quickly create RTSP-based applications which stream arbitrary data types and file formats.

- W3C's Jigsaw — http://www.w3.org/Jigsaw/

A Java-based web server. The RTSP server in the latest beta version was written in Java.

- IBM's RTSP Toolkit — http://www.research.ibm.com/rtsptoolkit/

IBM's toolkits derived from tools developed for ATM/video research and other applications in 1995-1996. Its shell-based implementation illustrates the usefulness of the RTSP protocol for nonmultimedia applications.

2.5 SUMMARY

This chapter discusses the four related protocols for real-time multimedia data in the future Integrated Services Internet.

RSVP is the protocol that deals with lower layers that have direct control over network resources to reserve resources for real-time applications at the routers on the path. It does not deliver the data.

RTP is the transport protocol for real-time data. It provides timestamp, sequence number, and other means to handle the timing issues in real-time data transport. It relies on RSVP for resource reservation to provide quality of service. RTCP is the control part of RTP that helps with quality of service and membership management.

RTSP is a control protocol that initiates and directs delivery of streaming multimedia data from media servers. It is the "Internet VCR remote control protocol." Its role is to provide the remote control; the actual data delivery is done separately, most likely by RTP.

ACKNOWLEDGMENT

This chapter was originally written as a paper in Professor Raj Jain's "Recent Advances in Networking" class in summer, 1997 at Ohio State University. The author sincerely thanks Professor Jain for his helpful guidance.

REFERENCES

1. L. Zhang, S. Deering, D. Estrin, S. Shenker, and D. Zappala, RSVP: A New Resource ReSerVation Protocol, *IEEE Network*, Vol 7, no 5, pp 8-18, September 1993.
2. R. Braden, D. Clark, and S. Shenker, Integrated Services in the Internet Architecture: an Overview, ftp://ds.internic.net/rfc/rfc1633.txt, RFC1633, ISI, MIT, and PARC, June 1994.

3. L. Zhang, S. Berson, S. Herzog, and S. Jamin, Resource ReSerVation Protocol(RSVP)--Version 1 Functional Specification, ftp://ds.internic.net/rfc/rfc2205.txt, RFC2205, September 1997.

4. R. Braden and D. Hoffman, RAPI — An RSVP Application Programming Interface, ftp://ftp.ietf.org/internet-drafts/draft-ietf-rsvp-rapi-00.txt, Internet Draft, June 1997.

5. H. Schulzrinne and S. Casner, R. Frederick, and V. Jacobson, RTP: A Transport Protocol for Real-Time Applications, ftp://ds.internic.net/rfc/rfc1889.txt, RFC1889, January, 1996.

6. H. Schulzrinne, RTP Profile for Audio and Video Conferences with Minimal Control, ftp://ds.internic.net/rfc/rfc1890.txt, RFC1890, January 1996.

7. H. Schulzrinne, A. Rao, and R. Lanphier, Real Time Streaming Protocol (RTSP), ftp://ds.internic.net/internet-drafts/draft-ietf-mmusic-rtsp-07.txt, Internet Draft, January 1998.

3 Video Transmission over Wireless Links: State of the Art and Future Challenges

Antonio Ortega and Masoud Khansari

CONTENTS

In this chapter we give an overview of potential applications of wireless video and discuss the challenges that will have to be overcome for these systems to become reality. We focus on video coding issues and outline how the requirements of (i) low bandwidth and low power consumption and (ii) robustness to variable channel conditions are being addressed in state of the art video compression research. We provide an overview of these areas, emphasizing in particular the potential benefits of designing video coding algorithms that are aware of the channel transmission conditions and can adjust their performance in real time to increase the overall robustness of the system.

3.1   INTRODUCTION

The increased popularity of mobile phones among consumers has made the wireless communications industry one of the fastest growing industries ever. At the beginning, the industry concentrated on providing users with telephone access from their cars. The technology, however, was at its early stages, as was evident by the need to use bulky handset radios, which were mounted onto the cars. This, in comparison with the current small handsets, shows the significant progress that has been made in both the system and the radio transmission technologies.

As wireless access becomes more commonplace, it will be necessary to support services other than voice and data. In particular, image and video communications over wireless links are likely to grow in importance over the coming years. This chapter addresses wireless video, which we believe to be one of the most promising emerging technologies for future mobile communications. Our goal is to report on the state of the art in this technology, especially from the perspective of video compression, and to highlight some of the remaining challenges that will have to be addressed to make wireless video a reality.

3.1.1   Potential Applications of Wireless Video

While there are numerous applications in which wireless delivery of video is likely to play a role, we distinguish between two major scenarios, namely, those where interactive and noninteractive communication takes place. Each scenario has different implications when it comes to system-level trade-offs.

In an interactive environment the goal will be to support two-way video or to be able to respond in a very fast manner to user commands (for example, fast forward in a one-way playback application). In this case, there will be strict constraints on delay, which will have to be kept low so that interactivity can be preserved. For example, a good rule of thumb is that system delays of the order of 100ms are completely transparent to the users. Although longer delays may be tolerable, they can have a negative impact on the perceived quality of the service. In addition to delay constraints, interactive applications involving two-way video also require that the portable device provide video encoding in addition to decoding, which will place severe demands on power consumption (not only for the purpose of signal processing, but also for transmission, display, capture, etc.)

In a noninteractive application we can assume one-way video being delivered without as strict a constraint on the delay. This could be the case if we considered the local distribution of video signals (for example displaying a particular video feed to several receivers in a household or office). Here, we are likely to have access to higher bandwidth and to deal with more predictable channel characteristics. This chapter will concentrate more on the low rate, high channel-variability case, as it presents the most challenges for efficient video transmission.

To further illustrate the significant differences between these two types of situations, let us consider two concrete examples of wireless video transmission.

Local video distribution

The success of the cordless phone in the consumer market has demonstrated the usefulness of providing mobility even within a limited area such as a home. It is thus likely that this type of short-range mobility will be supported for other types of applications, in addition to traditional telephony. For example, given that personal computers (PCs) have found their way into many households within the United States and Europe, it is likely that they will be used more frequently to control household appliances. PCs are thus envisioned to expand their role as central intelligence of the household, to control appliances within the home, and to provide connectivity to the outside world. Based on this vision, one has to provide radio connectivity between the PC and the many networked appliances. Providing such connectivity, one such application is to send video signals representing the screen from the PC to a light portable monitor. The user can then access and control the PC remotely (e.g., to check e-mail). Another application is to use the DVD player of the PC (or the video-on-demand feed received over the network) and use a TV monitor to display the video signal. These applications will likely require a high-bandwidth channel from the PC to the receiver and a lower bandwidth one from mobile to PC. The relatively limited mobility and range make it possible to have a more controlled transmission environment, thus higher bandwidth and video quality may be possible.

Car videophone

As an example of an interactive environment consider the provision of video conferencing services to mobile receivers as a direct extension of the existing wireless telephony. Here one can assume that low bandwidth and the effects of mobility (time-varying channels) are going to be the most significant factors. Unless the system has to be fully portable (so that it can be used outside of the car), power will not be a major issue. In recent years, there has been a significant increase in the amount of data that is delivered to moving vehicles; for example, in addition to telephony, some cars are equipped to be linked to geopositioning systems (GPS). These kinds of services are likely to grow, and one can foresee maps, traffic information, video news updates, and even two-way video being proposed as extra features for high-end vehicles in coming years. The most significant characteristics of these applications are the low bandwidth available, the potentially extreme variations in channel conditions, and the delay sensitivity of the system, in the case where interactivity is required.

3.1.2 PRESENT AND FUTURE WIRELESS COMMUNICATIONS SYSTEMS

While our main interest here is the support of video services, we start by briefly discussing the evolution of wireless communication systems in recent years. The major trend to be observed is the move towards all digital service, with data services being supported in addition to traditional voice service.

With the first generation of wireless communications systems, service providers have developed a new infrastructure parallel to the traditional phone network (Public System Telephone Network or PSTN).[1] This infrastructure uses the cellular concept to provide capacity (via frequency reuse) and the necessary range to cover a wide geographical area. In first generation systems, radio transmission is analog and the necessary signaling protocols have been developed to provide seamless connectivity despite mobility.[2] Parallel to this activity, cordless phone standards such as CT2 and DECT were developed to provide wireless connectivity to the telephone network within home and office environments.[3] In comparison with the cellular system, these systems are of lower complexity and use smaller transmission power because of the limited required range.

The more advanced second generation wireless systems use digital transmission to improve the overall capacity of the system, and also to provide security and other added functionalities. Different incompatible systems have been developed and implemented in Japan, Europe, and North America for the radio transmission.[4-8] Sophisticated speech compression methods such as Code-Excited Linear Prediction (CELP) are used to compress digital speech while almost achieving the toll-quality speech of PSTN.[9] Also, new higher frequency bands have been allocated and auctioned by the Federal Communications Commission (FCC) for the rollout of the new Personal Communication System (PCS). The PCS uses the existing telephony signaling infrastructure to provide nationwide coverage and uses smaller transmission power, which translates into smaller handsets and longer battery life.

In the meantime, the emergence of the Internet with data applications such as e-mail or web-browsing and the popularity of lap-tops have resulted in increased interest in providing wireless connectivity for data networks. Different wireless LAN protocols have been proposed, and the IEEE 802.11 working group has defined a new MAC protocol with three different physical layer possibilities.[10] At the same time, in Europe a new protocol known as High Performance Radio Local Area Network (HIPERLAN) has been proposed.[11] These proposals tend to use the public Industrial Scientific Medical (IMS) bands and can provide an aggregated bandwidth of up to 2 Mbps. They, however, support only a limited mobility and coverage.

The current cellular system is primarily targeted at the transmission of speech, and even though it can support extensive user mobility and coverage, it provides only a limited bandwidth (around 10 Kbps).[12] This is clearly inadequate for many multimedia applications. Therefore, a new initiative, known as third-generation cellular systems (IMT-2000), has been started; it emphasizes providing multimedia services and applications. The proposed systems are based mostly on Code Division Multiple Access (CDMA) and are able to support applications with a variety of rate requirements. Two main candidates are CDMA-2000 (proposed by Qualcomm and supported by North American standard groups) and Wideband CDMA (WCDMA) (proposed jointly by Japan and Europe).[13-18] The main improvements over second-generation systems are more capacity, greater coverage, and high degree of service flexibility. Third-generation systems also provide a unified approach to both macro-and microcellular systems by introducing a hierarchical cell organization. Third-generation systems provide enough bandwidth and flexibility (e.g., possibility of adaptive rate transmission) to bring multimedia information transmission (specifically video) closer to reality.[18]

3.1.3   VIDEO OVER WIRELESS: ALGORITHM DESIGN CHALLENGES

While significant progress is being made in developing a digital transmission infrastructure, there is by no means a guarantee that advanced real-time services, such as video transmission, will be widely deployed in the near future. This is in part because of the demanding requirements placed on video transmission to achieve efficiency over the challenging wireless channels. These requirements can be derived by considering the characteristics of typical transmission environments, namely low bandwidth, low power, and time-varying behavior.[19] In this chapter we address these issues by describing first the video coding algorithms then discussing how channel characteristics need to be taken into account in the video transmission design.

First, the low bandwidth of the transmission link calls for low or very low rate compression algorithms. We outline some of the progress made in this area in recent years, in particular through algorithms like MPEG-4[20] or H.263.[21] In addition, it is obvious that the devices to be used for video display and capture have to be particularly efficient because they have to be portable. This places constraints on the weight and power consumption of these devices thereby calling for compression algorithms that are optimized for power consumption, memory, etc.

Given the variable nature of the channels we consider, it is necessary to consider video compression algorithms that are scalable, that can compress the same video input at various rates (consequently with different decoded output qualities). In other words, if the channel can provide different qualities of service (QoS) the video encoder should be able to likewise provide different rates and video qualities. We present some of the approaches that have been proposed for scalable video and indicate how these can be incorporated into practical video transmission environment. An alternative approach to deal with channel variability is to let the source coder adjust its rate to match the expected channel behavior (i.e., transmit fewer bits in instances of poor channel quality). We also discuss these rate control approaches.

Finally, even if efficient channel coding is used, the variations in channel conditions due to roaming will result, in general, in losses of information. Thus, the video applications will have to provide sufficient built-in robustness to ensure that the quality of the decoded video is not overly affected by the channel unreliability.

3.1.4   OUTLINE OF THE CHAPTER

This chapter is organized as follows. Section 3.2 provides a brief introduction to the typical architecture of the wireless access network and describes the channel behavior to be expected under such transmission conditions. Section 3.3 introduces the basic components in currently used video compression environments and discusses the delay constraints imposed by the transmission environment and how these are translated into conditions on the video encoding process. It also briefly describes how some of the requirements of efficient wireless video (very low rate and power consumption) are met in state of the art systems. Section 3.4 motivates that error correction techniques are not sufficient to provide the required robustness and that, in fact, video coding algorithms have to be especially designed to support transmission over a wireless link. For each of the techniques that can be used to increase the video coding robustness (e.g., redundancy, packetization, etc.) we describe how specific video techniques can be found to improve the performance over purely channel-coding approaches.

3.2 WIRELESS ACCESS NETWORK ARCHITECTURE AND CHANNEL MODELS

We start by providing a quick overview of the channel characteristics that are most important for video transmission. First we describe a typical wireless access network architecture, then the methods used to achieve robustness in mobile environments. Finally, we describe some of the models that are used to characterize overall transmission performance in terms of packet losses.

3.2.1 ARCHITECTURE

In a typical cellular system a mobile station (MS) can communicate only with the nearest base station (BS). Each BS will be connected to a network that allows it to communicate with other BS,[1-2] so that communication between MSs and between a MS and the fixed telephone network is possible. Obviously, power is much more constrained at the MS than at the BS, and a BS has much more processing power than a MS. Also, since every connection within a given cell is established through the corresponding BS, each BS has knowledge of the status of all the connections within its cell. As a result, there is a significant amount of network intelligence at the BS which does not exist at the MS.

Therefore the two links or connections in a wireless access network (the one from a base station to a mobile station, the downlink, and the one in the reverse direction, the uplink) have very different characteristics. For example, for the downlink channel, the transmitter has an almost unlimited amount of processing power, whereas the conservation of power at the receiver is of utmost importance. Consequently, for this link the transmitter can be significantly more complex than the receiver. The situation is reversed for the uplink channel; the transmitter would tend to be simpler than the receiver.

Note that the same will be true when designing a video compression algorithm targeted for transmission over a wireless link. For example, consider the case when a video signal is transmitted to many mobile stations simultaneously. In this scenario, one can tradeoff a more complex video compression encoder for a simple decoder (see Meng et al.[22] for examples of this approach). Alternatively, video transmission from a MS to another user may be power-limited to such an extent that a low complexity (and low compression performance) algorithm may be preferred, even if bandwidth efficiency is sacrificed.

It is also worth noting that there is another popular architecture, the so-called *Ad hoc* network, in which mobile stations can communicate with each other without the need of a base station. In this network all the communication links are symmetric and there is no difference between the transmitter and the receiver. Transmitting video signals over this network favors the use of symmetric video coding algorithms in which both the encoder and the decoder are of about the same complexity. *Ad hoc* networks are more likely to be used in a local configuration to provide communication between a number of users in a reduced geographical area.

3.2.2 CHANNEL CHARACTERISTICS AND ERROR CORRECTION

Let us consider now the characteristics of typical wireless links. We first discuss the various impairments that affect the physical channel behavior, then we discuss how these can be addressed using various channel-coding techniques.

3.2.2.1 Physical Channel Characteristics

The first type of channel impairment is due to the loss of signal as the distance between the transmitter and the receiver increases or as shadowing occurs. Clearly this loss of signal will depend heavily on the surrounding geographical environment, with significant differences in behavior between, for instance, urban and rural environments. These channels are also subject to the effect of various interferences, which produce an ambient noise usually modeled as an additive white Gaussian noise.

A major contributor to the degradation of the received signal is what is known as multipath fading. This occurs when different duplicates of the same transmitted signal reach the receiver, and each version of the signal has a different phase and signal level. This situation is common in any transmission environment where the signal is reflected off buildings and other objects. Reflections can result in changes in phase and in attenuation of the signal, such that signals arriving through different paths may combine to generate a destructive addition at the receiver. If this is the case, the result can be a considerable drop in the signal-to-noise ratio (SNR). As the SNR fluctuates, the bit error rate will vary as well, and if the fluctuations are sufficiently severe the connection itself can be dropped. Typically the magnitude of the received signal under fading conditions is modeled using the Rayleigh distribution, and its phase is assumed to follow a uniform distribution. Other models, based on two or four paths, are also used in practice. Since our concern here is not the behavior of the physical link but the performance of the link after channel coding, suffice it to say that fading conditions are characterized by the fact that the distributions of low SNR periods is "bursty" in nature. Thus, rather than observing randomly distributed errors, we may be seeing that during fading periods hardly any information gets across correctly. For fading channels, measuring the average performance (i.e., the error rate averaged over the duration of a long transmission period) may not be as meaningful as characterizing the worst case performance; for example, if severe fading occurs does the connection get lost? Given that fading is determined by location, speed of the user, etc., there is a great degree of variability in channel conditions.

Typically, the fading experienced by each frequency component in the transmitted signal is statistically independent. However, in situations where the mobile station is not stationary this is no longer true, due to Doppler frequency shift. Because of the Doppler shift, the frequency of the received signal is shifted from that of the transmitted signal based on a random cosine distribution. The effect of Doppler shift becomes more pronounced when higher frequency bands (e.g., tens of Ghz) are used or the mobile stations are moving at a faster speed.

Radio transceivers use a combination of channel coding, different methods of diversity, channel equalization, and power control to combat the channel impairments. However, wireless channels are not only a hostile environment because of multipath and Doppler effects, but in many cases they are also a shared environment where many users compete for the same available bandwidth. While users can cooperate to avoid having a negative impact on each other's transmission quality, there is no guarantee that conflicts can be avoided when, for example, two users try to use the same transmission resources. Because one user's signal can have destructive effects on the other users, the challenge is therefore not to improve the performance of a specific user or connection but to simultaneously improve the overall performance of the system for all the users and the links.

3.2.2.2 Channel Coding for Wireless Channels

There are two main types of error control techniques that can be used in wireless environments, specifically open-loop and closed-loop techniques. We describe first open-loop systems, where the channel encoder provides redundancy which can be used by the decoder to correct potential channel errors.

The most popular approach to achieve robustness in wireless channels is to introduce diversity. Diversity is achieved when multiple coded versions of the same information arrive at the receiver, with each replica degraded by different *independent* noise. Spatial, frequency, and time diversity are the main modalities, and in each the underlying assumption is that while multipath fading may degrade the SNR, its effect is "local."

For example, in a broadband signal only a few frequencies will be attenuated, given that multipath fading is typically frequency selective. Thus, by spreading the information over a broad range of frequencies it is possible to achieve robustness. This frequency diversity approach is the one taken in CDMA spread spectrum techniques and explains their wide popularity. Alternative methods using frequency multiplexing are also used in practice.

The fading period (as the user moves) will not last a long period of time. Then if one can time interleave the data produced by the channel coder for transmission, bits transmitted at different times are likely to be subject to different noise. Thus, while the error patterns affecting the transmitted stream will be bursty, after time de-interleaving, the error patterns observed by the channel decoder will look more "random." This time diversity approach is also popular. In fact, time interleaving combined with channel coding is useful for nonfading channels, as demonstrated by the recent success of "turbo" codes.[23]

Finally, while the user remains stationary, fading is also spatially localized; that is, small changes in the position of the antenna can result in large increases and decreases of SNR. Thus spatial diversity approaches operate by providing multiple antennas at either transmitter or receiver, either to ensure that at least one of the antennas receives a strong signal, or to combine the information received over all the antennas to achieve better overall SNR performance.

With any of these diversity scenarios, the aggregated received noise becomes an Additive White Gaussian Noise (AWGN), and the performance of the system asymptotically approaches that of the AWGN channel.

Closed-loop approaches are such that the channel coder is allowed to adjust to the channel conditions. These approaches are more complicated to implement because the transmitter must estimate the state of the channel in order to adjust its transmission parameters. Channel state estimation is achieved with the help of the receiver, which can send information back to the transmitter about observed channel conditions. For example, if information has been corrupted due to excessive channel errors, the receiver can request retransmission. Conversely, if the receiver detects that few channel errors occur (so the channel conditions are good), it can request the transmitter to lower its transmit power or to reduce the amount of redundancy in the bitstream. Examples of closed-loop power control can be found in the design of the uplink of the CDMA-based IS-95 systems.[6] Closed-loop recovery methods based on Automatic Repeat reQuest (ARQ) have also been proposed[24] where the receiver can request retransmission of the erroneously received data. This method is especially efficient for channels with burst errors because it requires only the redundancy needed to provide error correction during fade periods. In contrast, open-loop approaches operate with basically the same level of redundancy under all channel conditions. This means that during periods of good channel behavior (e.g., when the user is close to a BS) the amount of channel protection is excessive.

Closed-loop approaches are better suited for data applications with no stringent delay requirement. While it may appear a priori questionable that such closed-loop methods be applicable to video (given the delay-sensitive nature of the video data), we will see that in fact closed-loop methods can also prove useful for video.

3.2.3 CHANNEL MODELING

While channel coding aims at supporting error-free transmission, this becomes a nearly impossible goal as users roam about. Thus, given that channel errors are unavoidable, we provide now an overview of models that characterize the effective channel rate variations that the end-user application observes. These models aim at characterizing the effective channel performance after any applicable channel decoding has been used. Thus, in typical video transmission environments video data will be packetized and error detection techniques will be used to determine if a particular packet has been corrupted, even after error correction has been applied. If the video encoder is to adjust its coding parameters when channel variations occur, the parameter of interest will then be the probability that packets of video data will be lost. Therefore the models we describe provide statistical characterization of the probability of packets' being corrupted.

As indicated above, fading will tend to produce localized changes in SNR so that overall packet loss will also tend to be bursty. Consequently, if a packet is lost, subsequent packets are likely to be lost too. Various models have been proposed that attempt to capture the burstiness of errors in wireless channels. Of particular interest are Markov models[24–29] that are characterized by a series of states, each having associated different channel behavior. Transitions between states occur with given probabilities and are used to capture the variability of the channel. Figure 3.1 shows an example of a two-state Markov model,[25,28] where we assume that the channel can be in only one of two states (*good* and *bad* channel behavior). The modeling process consists of finding estimates for the transition probabilities. Models such as these can be used in conjunction with rate control techniques to let the source coder adjust to channel variations, as will be discussed further in Section 3.4.5.

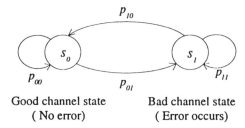

Good channel state   Bad channel state
(No error)   (Error occurs)

Figure 3.1 Two state Markov channel model

3.3 VIDEO COMPRESSION FOR WIRELESS COMMUNICATIONS

While a complete introduction to state of the art video coding techniques goes far beyond the scope of this chapter, we present here a brief description of the main principles underpinning video compression design. We then consider the specific changes required to adapt these algorithms to the demanding wireless environment, and in particular how low rate, low power, and low delay can be supported. In Section 3.4 we revisit some of these design issues but will consider them only from the perspective of adding robustness to the transmission.

3.3.1 TYPICAL VIDEO COMPRESSION ALGORITHMS

Most practical video compression algorithms achieve high compression ratios with good quality decoded images by exploiting the various types of redundancy (spatial, temporal, and statistical) existing in video sequences. The main trade-off in video coding is that between the fidelity of the approximation to the original images (i.e., the distortion introduced by coding) and the number of bits required to represent the images (i.e., the rate). Compression performance can then be evaluated by considering the rate required to achieve a specific quality level. What is important is how redundancy is exploited to achieve good compression performance.

First, in most natural images neighboring image pixels exhibit significant similarity. For example, in typical scenes the images contain many flat regions (i.e., where low frequencies predominate) or regions with regular patterns or textures (i.e., where a specific range of frequencies contains most of the information). This spatial redundancy can be exploited so that only a few bits are needed to represent these regions. Most practical systems employ a transform, such as the discrete cosine transform (DCT), which provides estimates of the frequency contents of each image region (e.g., each block of 8 × 8 pixels, in the DCT case). Given that most of the energy is concentrated in a few coefficients, only a few bits will be needed to provide a good approximation to the original image block.

Second, consecutive frames in a video sequence will also exhibit similarity, especially in video scenes with low motion. This temporal redundancy is exploited through motion estimation, a technique that breaks video frames into nonoverlapping blocks and then searches the previous frame for blocks that provide the most similarity to each block in the current frame. For example, this technique predicts a block belonging to an object in the current frame from a block in the same object in the previous frame.

Finally, when data has been quantized we are left with a number of symbols to be transmitted, and these symbols are drawn from a discrete set, i.e., they can take a finite number values after quantization. Given that these symbols, e.g., the quantized DCT coefficients, are not equally likely, further compression gains can be achieved by taking advantage of this statistical redundancy. Gains are achieved through *entropy coding*: the most frequently observed symbols are represented by shorter codewords, while the least frequently used are represented by longer codes. Variable length coding techniques are very useful in terms of compression but, as will be seen later, are risky when used in a lossy environment. A single bit error can trigger a loss of synchronization in the bitstream and lead to loss of most of the information in the sequence.

Our purpose here is not the basics of image and video coding. Detailed discussions of these topics can be found in numerous textbooks.[30,31] Instead, our emphasis is that compression techniques that exploit temporal or spatial redundancy will result in different rates for different frames at a given quality level. In fact, the degree of redundancy, therefore the resulting rate for a given distortion, can fluctuate widely from scene to scene. For example, scenes with high motion content will require more bits than more stationary ones.

Figure 3.2 provides a diagram of the basic building blocks of an MPEG coder,[31] which include a DCT transform, block-based motion estimation, and quantization. It should be noted that the selection of quantization step size decided by the encoder determines the rate-distortion trade-off, and it allows the encoder to adjust the number of bits/frame to suit various constraints, such as those imposed by the channel. Note also that this video coder operates essentially as a closed-loop predictor, that is, the encoder and the decoder both use previously quantized frames to generate predictions for the current frame (the decoder uses this predicted value to reconstruct the frame once it receives the coded residue from the encoder). This is a particular concern when operating over a lossy channel: if, after the occurrence of an undetected channel error, the encoder and decoder start to operate with different frames in the prediction loop, the result can be a complete loss of synchronization between the two, with potentially disastrous results for the video quality at the receiver.

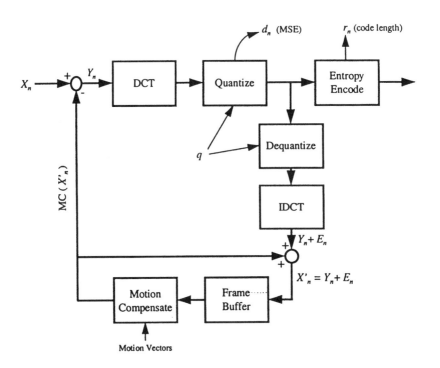

Figure 3.2 Block diagram of a typical MPEG coder. The quantization parameter can be adjusted to make the rate comply with the channel constraints. X_m represents the current video frame, and $MC(X_m^1)$ is the motion-compensated prediction frame, based on the previously transmitted frame.

3.3.2 DELAY CONSTRAINTS IN VIDEO COMMUNICATIONS

Let us now consider a typical real-time transmission as illustrated in Figure 3.3. As just described, video frames require a variable bit rate, thus it is necessary to have buffers at encoder and decoder to smooth the bit rate variations. Assuming the video input and output devices capture and display frames at a constant rate, and no frames are dropped during transmission, the end-to-end delay in the system will remain constant.[32]

Let us call ΔT the end-to-end delay: a frame coded at time t has to be decoded at time $t + \Delta T$. That imposes a constraint on the rate that can be used for each frame (it has to be low enough that transmission can be guaranteed within the delay). For example, let a coding unit be coded at time t and assume that it will have to be available at the decoder at time $t + \Delta T$. If each coding unit lasts t_u seconds, then the end-to-end delay can be expressed as $\Delta N = \Delta T/t_u$ in coding units. For example, if a

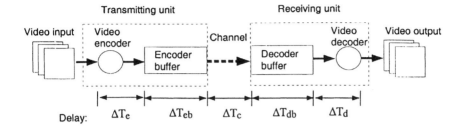

Figure 3.3 Delay components in a communication system

video encoder compresses 30 frames/sec and the system operates with an end-to-end delay of $\Delta T = 2$ sec, then the decoder will wait 2 sec to decompress and display the first frame (assuming no channel transmission delay) and at any given time there will be $\Delta N = 2/(1/30) = 60$ video frames in the system (stored in the encoder or decoder buffers or being transmitted). The video encoder will have to ensure that the rate selection for each frame is such that no frames arrive too late at the decoder.

Consider the case when transmission takes place over a constant bit rate (CBR) channel. Of the delay components of Figure 3.3 only ΔT_{eb} and ΔT_{ed}, which is the time spent in encoder and decoder buffer, respectively, will now be variable. Consider, for example, ΔT_{eb}. This delay will be at most B_{max}/C, where B_{max} is the physical buffer size at the encoder and C is the channel rate in bits/sec. It is clear that B_{max} has to be smaller than $\Delta T \cdot C$; otherwise we could store in the buffer frames which will then experience too much delay.

If we consider the transmission of a variable rate sequence we will either (1) have to use very large buffers (and correspondingly long end-to-end delays), or (2) have to adjust the source rate, thus the delivered quality, to make it possible to use a smaller buffer (shorter delay) without losing any data. The delays required for such a sequence would be exceedingly long; therefore, in practical applications it is necessary to perform rate control to adjust the coding parameters and meet the delay constraints.

As shown in Figure 3.2, it is possible to adjust the video rate (and the quality) by modifying the quantization stepsizes used for each frame. A natural way to approach these problems is to consider the Rate-Distortion trade-offs in the allocation. R-D techniques for rate control have been described in the literature.[33–36] See Ortega and Ramchandran[37] for a more detailed treatment of these techniques.

Note that even in cases where transmission is performed over a variable bit rate (VBR) channel, or where the sequence is preencoded and stored (e.g., in a Digital Versatile Disk, DVD), it is also necessary to perform rate allocation. For example, to store a full-length movie in a DVD it may be necessary to preanalyze the movie, then allocate appropriate target rates to the various parts of the movie. In this way, allocating more bits to the more challenging scenes and fewer to the easier ones will result in a globally uniform quality. R-D based approaches for rate control for VBR channels have also been studied.[38–41] Approaches more suitable for storage in disk-based video servers are considered in Miao and Ortega.[42]

In summary, video coders produce a variable number of bits/frame and are subject to strict delay constraints. Thus, any rate adjustment at the video encoder to accommodate changing channel conditions will have to take into account the delay constraints to avoid having data arrive at the decoder too late to be decoded.

The overall delay in the system ΔT is an important parameter. For noninteractive applications its sole significance is the latency it introduces (i.e., decoding does not start until ΔT seconds after data was first received at the decoder). For interactive applications ΔT is even more critical, since large values of end-to-end delay degrade the perceived interactivity. On the other hand, since channel variations are bursty, a longer delay allows increased robustness to changes (if the delay is long enough the transmitter can wait until the fading conditions have subsided to transmit data without errors). Thus, the selection of an end-to-end delay for a particular application will have to balance conflicting demands for high robustness, low latency, and high interactivity.

3.3.3 CODING FOR LOW RATE CHANNELS

The basic principles outlined in Section 3.3.1 still apply to video coding at low rates, i.e., the main sources of compression gains come from exploiting temporal, spatial, and statistical redundancies.

The ITU-T standards for video conferencing and video phone (H.261[43–44] and its successor H.263[21]) have traditionally been representative of the state of the art for low bit rate video, targeting bit rate ranges in the order of 64-384kbps. While these standards initially were meant for a fairly narrow range of applications (e.g., they assumed "head and shoulders" sequences as their expected source material), technology has improved enough that the last generation of low bit rate coders in the ITU-T family, the H.263 series,[21] can handle more challenging sequences with good results. Still, to achieve good quality at very low transmission rates (64kbps and below) one has to start with video source material captured at a reduced frame rate and with limited frame sizes. Typically, frame sizes of 176×144 pixels (QCIF) are typical of these applications, with frame rates as low as ten frames/sec or less.

Most of the gains achieved in going from H.261 to H.263 can be attributed to more efficient motion estimation, rather than to specific optimizations of the systems for low rates. In the mid nineties, as the process to define the ISO MPEG-4 standard[20] got underway, there was substantial interest in defining very low bit rate coding algorithms that would go beyond the transform coding paradigm and introduce so called second-generation coding techniques,[45] such as segmentation, or even more advanced object-based approaches.[46–47] However, the investigation of these techniques seems to have led to the conclusion that second-generation image-coding approaches are not sufficiently mature yet, and their complexity may still place them beyond the reach of affordable and reliable implementations with today's technology. As a result the emerging MPEG-4 standard[20] has not abandoned the transform coding techniques. However, one of its main novelties has been to provide added functionality for multimedia applications, notably the possibility of choosing as the basic coding unit video objects that have arbitrary shapes, instead of being limited to rectangular frames.

In summary, recent technology has contributed to increased quality for low bit rate video, along with, in the case of MPEG-4, some interesting additional functionality. However, these techniques address generic low bit rate applications (e.g., including video streaming over the Internet). As will be seen in Section 3.4, additional features are required to provide robustness in an error-prone environment such as a wireless link. As will be seen, many useful robustness tools have been proposed within the H.263 Annexes.[21,48]

3.3.4 POWER CONSTRAINTS AND SYSTEM LEVEL TRADE-OFFS

As anybody who has used a cell phone can easily attest, power efficiency and the related issue of weight are two of the most important considerations in a useful wireless communications system. There is an excellent overview of recent work on these issues.[49] Obviously there are many components that affect the power consumption of a wireless communications device, including the RF circuitry, the communication-specific processing, and of course the image/video compression and corresponding display, if video support is included. See the *IEEE Personal Communications Magazine*[49] for a detailed analysis of some of these issues.

It is also clear that, compared to the required power for display or transmission, video processing power may not always rank among the top power-consumption components in a portable video communicator. Still, video processing will be a significant component in power consumption budget. Recent research has demonstrated that the power consumption can be dramatically reduced by intelligent trade-offs in the image coding/decoding process. For example, Meng et al.[22] and references show reductions of two orders of magnitude in power consumption with respect to a JPEG[50] image decoder. The coding algorithm used is designed such that decoding is particularly simple, with the DCT replaced by simple (i.e., integer coefficient) subband filtering. In addition, coding efficiency is based on a pyramid vector quantization (PVQ) scheme which has further advantages in terms of complexity, as decoding can be implemented based on a look-up table. The basic lesson from this and related work is that it is possible to implement algorithms with low power in mind, but even larger gains can be achieved if one is willing to *design* the compression algorithms with power efficiency, and not just coding efficiency, in mind.

There is an increased awareness in the coding community that implementation issues should be at the forefront of the coding algorithm design, rather than being addressed as an afterthought. For example, in the ongoing JPEG2000[51] standardization process, which is set to define the next generation of wavelet-based image coding techniques, special care has been given to taking into account both the complexity and the memory usage of competing algorithms.[52]

Since in cellular and most other wireless access networks the communication among mobile stations is done through a base station, there are always two different links to consider: downlink (forward link) and uplink (reserve link). If the video application is interactive and the same video codec is used in both directions, then the performance of the overall system is determined by the worst-case link. Note that the decision of which link has poorer performance is dependent on the channel recovery method being used. In one-way applications or when it is possible to use different video encoders for each direction, by using asymmetric video codec, one can improve the performance. For example, for the downlink we use a more complex video encoder to simplify the operation of the decoder. The situation is reversed for the uplink where one would like to use a simple encoder, but since there is more processing power at the base station the constraint on the complexity of the decoder is more relaxed.

As in Meng et al.[22] it may be a good idea to consider very simple decoders to minimize the power consumption at the receiver, since the MS receivers are clearly much more power limited than the BS. For the same reasons, it may be easier to have higher power (thus rate) in transmitting from the BS to the MS. In general, therefore, power is the fundamental source of asymmetry in the overall system design and this may lead to interesting system-level trade-offs. Power efficiency can thus be addressed at three levels: (1) efficient implementation of existing systems, (2) algorithm design driven by power efficiency, and (3) intelligent system-level trade-offs. The latter two levels have the most potential for significant power reductions but are also the least studied and understood.

3.4   ACHIEVING ROBUSTNESS THROUGH SOURCE AND CHANNEL CODING

The goal of this section is to motivate that video coding techniques are required to support robust communication over a wireless channel and to complement channel coding approaches such as those described in Section 3.2. While in other transmission environments it is possible to effectively decouple source and channel coding, time-varying channels such as a wireless link make it more difficult to guarantee error-free transmission. If errors do occur then video coding techniques are needed in order to achieve graceful degradation, i.e., to provide lower, but acceptable, quality video as the channel conditions worsen.

3.4.1   THE SEPARATION PRINCIPLE AND THE NEED FOR JOINT SOURCE CHANNEL CODING

In his seminal work on transmission of information over a noisy channel, Shannon[53] showed that the two fundamental functions of source coding (compressing a source so that it requires the minimum number of bits to represent it, given a fidelity criterion) and channel coding (adding redundancy to a bitstream to protect it against channel errors) can be done separately without loss of optimality. This is known as the *separation principle* and has been a fundamental guideline in the design of communication systems. Based on this principle, a source encoder tries to remove as much redundancy as possible from the source, irrespective of the channel and the channel encoder being used. At the same time, the channel encoder adds enough redundancy to provide reliable, i.e. error-free, transmission of the information over the noisy channel, and again this can be done irrespective of what the information being transmitted is. Shannon showed that if the entropy of the transmitted information is smaller than the channel capacity, then reliable communication is possible and can be achieved through separate source and channel coders. If source entropy exceeds channel capacity, then reliable communication is impossible. An implicit assumption in developing the separation principle is that there is no delay constraint on the operation of both the source and the channel encoder/decoders. Also, it is assumed that the channel is ergodic (hence stationary).

It is fair to say that in a wide variety of applications, in particular point-to-point transmission over channels that exhibit mostly time-invariant behavior (e.g., phone lines), these assumptions are valid and provide a perfect framework to design communication systems. However, both broadcast channels[54] and time-varying channels, such as those considered in this chapter, no longer fit into the framework assumed by the separation principle.

In a system design based on the separation principle the channel encoder takes the output of the source encoder (bits) and adds redundancy so that all bits are equally protected. In fact, not all bits produced by the encoder have equal importance. For example, when transform coding is used to compresss images, coefficients representing low frequencies have more importance in terms of perceptual quality. In general, for each type of bits produced by a source coder, it is possible to measure the effect errors would have on the reconstructed signal, and these can then be used to provide unequal levels of protection to the different types of data.[55,56] Obviously, unequal error protection is needed only if channel errors cannot be avoided; otherwise one would be better off by simply providing a high level of protection to all the information so that it is all received error-free.

 The separation principle guided us in the design of a system that would be virtually error free and this would be true in general, assuming the statistical characteristics of the channel do not change. However, typical radio channels in a wireless access network, as discussed in Section 3.2, are characterized by their time-variability. Thus a standard design based on separation would have to assume a specific model for the channel: if conditions were worse than predicted, the service would be interrupted. If the channel conditions are worse than modeled only a small fraction of the time, this might be acceptable. However, to ensure that a service interruption happens only rarely, the system may have to be designed under very pessimistic assumptions. This in turn means that during periods of favorable channel conditions the system may be operating at a higher level of redundancy than would be required by the channel at that time, resulting in inefficient use of resources.

In summary, for both time-varying and broadcast channels, trade-offs tend to be more complex than for the more traditional point-to-point channel with time-invariant characteristics. This leads to the development of *joint source channel-coding* techniques, which aim at alleviating the effect of time variability by enabling quality of service that can be adjusted to match different types of channel behavior.

Although there is a more detailed discussion of joint source channel coding,[57] we outline here some of the basic principles. In a joint source source-channel coder design, the two objectives are (1) to design source coders that clearly separate source information into different classes according to their importance as far as the decoded image quality is concerned (these are called multiresolution, scalable, or layered codecs), and (2) to provide channel coding mechanisms that allow unequal error protection, so important information can be better protected than less important information. Note that we provide an overview of a broad range of techniques which we put under the umbrella of joint source channel coding. Some are truly joint in that the source and channel coding techniques are designed together. In other cases the designs are carried out independently, but channel coding is made aware of the relative importance of the information that is being carried. In fact, the latter set of techniques, which do not involve a joint design, are by far the most widely used in practice.

3.4.2 PACKETIZATION AND SYNCHRONIZATION

A simple way of achieving some degree of protection, without necessarily introducing modifications in the source coder, is to packetize the data and provide mechanisms that limit error propagation. While these are not strictly speaking joint source channel coding techniques (since the channel coder is not modified) they do show how having knowledge of data can help reduce the effect of errors. In fact, most of the current activities on error-resilient source coding have focused on methods to prevent error propagation or to facilitate resynchronization after errors have occurred.

Packetization consists of breaking up the video data into segments and transmitting each of these segments or packets, with individual error detection being performed on each of them to determine whether the information, or payload, of the packet was corrupted. If the packet was corrupted the corresponding information can be discarded by the decoder. A priori, the packetization process could take place without any knowledge of the contents of the bit stream. However, careful packetization is essential to improve the system's performance. For example, if packet boundaries always coincide with codeword boundaries it is easier to prevent the propagation of errors from a corrupted packet to the next.

More generally it is desirable to design compression algorithms with reduced error propagation, such that a particular error affects only local (spatial or temporal) video data. This can be achieved not only through packetization, but also by introducing explicit synchronization markers which are unlikely to be corrupted (for example, to be corrupted a synchronization marker formed by a fixed pattern of 16 bits would require the occurrence of many consecutive bit errors.)

There are two major reasons for the existence of error propagation in compressed video bitstreams: the use of variable length codes for entropy coding and the reliance on predictive techniques to increase the coding performance.

Entropy coders used in practice generate *variable length* codes, i.e., the number of bits transmitted can be different for each symbol. Unique decoding of their output is possible because they satisfy the prefix condition; no valid codeword is a prefix of another valid codeword. When channel errors occur and a value of even one single bit is altered, it may no longer be possible to correctly decode not only the current symbol but, in some cases, the complete sequence of symbols following the location of the error. This may happen if the current symbol is incorrectly decoded with a different length from that of the original transmitted symbol. When this happens, the decoder may not be able to determine the output correctly until decoded symbols fall again at the correct codeword boundaries.

Resilience can be increased by introducing synchronization markers, which guarantee that the error will not propagate beyond the marker. However, explicit synchronization has a cost since the markers themselves do not convey any useful video information. In some cases it is possible to replace explicit synchronization markers by introducing constraints in the bit stream, such that groups of codewords have a total fixed length, even if individual codewords themselves have variable length. See the reference section[58–61] for examples of synchronization techniques used in image coding.

Another approach for robust entropy coding consists of using fixed-length codewords. For example, it is possible to group several symbols so that together they are assigned a fixed-length representation. This approach is used in the design of the scalar vector quantizers,[62] for example. It has the drawback of added complexity, as the encoder has to assign a single (fixed length) code to a block of inputs.

Finally, even variable-length entropy codes can be designed with inherent error resilience properties. For example, reversible variable length codes (RVLC)[63] are such that it is possible to decode them both in the forward and the backward directions (i.e., starting from the beginning of the sequence and working to the end, or vice versa). When, due to a bit error, synchronization is lost while decoding in the forward direction, one can start decoding in the backward direction. Using this method, it is possible to isolate the erroneously received codeword and to limit the error propagation. In some instances it is even possible to correct some of the errors.

The second source of error propagation is the utilization of predictive techniques, in particular motion-compensated prediction as used in most state of the art video coders, including both the ITU-T H.263 coders and the ISO MPEG family. As discussed earlier, for a given block within the current frame, the motion estimation algorithm finds the best match in the previous frame. The difference between the block and its best match is what is known as residual signal. The information transmitted over the channel consists of the residual signal and the corresponding motion vector field. Clearly, since prediction is being used, any erroneously received motion vector and residual information can affect not only the current but many subsequent reconstructed frames: the current (and erroneously reconstructed) frame is used to predict future frames and because the predictors in encoder and decoder will be different, so will the reconstructed images. This situation will last until the next intracoded frame (a frame coded without motion estimation) is transmitted. Increased robustness can be achieved by forcing intraframe coding to be used at regular intervals or by refreshing all the blocks in a frame following some pseudo-random pattern.[64]

Other approaches are possible if a closed-loop system is used. For example, in H.263+ coders[21] it is possible to use arbitrary frames as references in the prediction loop. We can then use as a reference frame not the previous frame (which may not have been correctly received) but rather the *the latest acknowledged frame* — the latest frame known to have been received correctly by the decoder. In this way, we guarantee that encoder and decoder will use the same information in their prediction loops. Other approaches use the feedback channel to increase the speed at which the system can be resynchronized after an error by forcing a refresh of regions that are known to have been corrupted.[65] The decoder can request the refresh through the back channel after detecting that errors have occurred.

3.4.3   REDUNDANCY VS ROBUSTNESS TRADE-OFF

As described in Section 3.3.1 video compression algorithms operate by removing redundancy from the original source. But, considering the robustness issues discussed above as an example, the price for robustness is a reduction in the coding efficiency. This is a fundamental trade-off in designing error-resilient source coding: resilience can be achieved at the cost of increased redundancy, thus lower coding efficiency. This section considers approaches that explicitly avoid removing all the redundancy in order to increase the robustness.

Illustrating this trade-off is a very simplistic example where an image is transmitted and a few random bit errors occur. Assume first that the image is transmitted without compression. Then these bit errors (1) may be only perceptually noticeable if they affect the most significant bits of a particular pixel and (2) will never affect more than one pixel at a time, since each pixel is coded independently (no prediction or variable-length coding is used). Conversely, if the image has been efficiently compressed, variable-length codes are likely to be used; therefore a single bit error can cause complete loss of synchronism and render the decoded image useless. However it is not clear that it is better to transmit a very large (uncompressed) image to make the system more error resilient.

Of course typical trade-offs are not quite as drastic, and one can achieve reasonable compression performance while being robust. For example, it is possible to compress the data so that even though individual codewords have variable length, groups of codewords have constant length. In this way, full advantage of the statistical redundancy of the data is not taken, but some additional robustness is achieved. Another example is a system where motion estimation can be turned off to avoid errors propagating through several frames, i.e., the temporal redundancy is not fully removed.

It is important to note that redundancy is useful not just to limit the error propagation but also to allow better estimation of the information that was lost due to errors. For example, if not all spatial redundancy has been removed one can interpolate lost information from the received information in the local neighborhood. Similarly, as Hagenauer[66] has shown, if there is residual statistical redundancy in the transmitted codewords this can be used to aid in decoding. For example, if transmitted signals are not equally likely, it is possible to take this into account in weighting the likelihood of the missing information.

Thus, we can in general state that an image/video compression algorithm which has been optimized to squeeze out all redundancy will tend to be less robust than one where redundancy (statistical or temporal redundancy as in the two examples above) is carefully preserved to prevent error propagation or enable easier reconstruction.

3.4.3.1   Diversity and Multiple Description Coding

A particularly interesting approach for robust coding is to explicitly design a source encoder that can preserve some redundancy. Note that this is contrary to the traditional communication system design where redundancy is introduced by the channel coder only and is independent of the type of data transmitted.

Redundant source encoders can introduce structured correlation and redundancy at the signal level, before the entropy coding is applied. One such approach is Multiple Description Coding (MDC), where the same signal is encoded using two (or possibly more) encoders. An appropriate amount of structured correlation is constructed between these streams. Each stream is then transmitted along a separate path, or, in the case of wireless transmission, using a different path in the diversity structure. For example, each version of the signal could be transmitted on a different frequency, or at different times, or from a different antenna. Thus, assuming that not all the "channels" (frequencies, times, etc.) fade at the same time, the system is conceptually sending information over several virtual links, each subject to independent loss probabilities. Note that in a standard diversity system the redundancy is introduced only through the channel coding, while in a MDC approach redundancy is built into the source as well.

At the receiver, it is possible to reconstruct the original signal using either of the streams. Moreover, because of the inherent correlation between the two transmitted streams, one can estimate to some extent the lost information and improve the reconstructed signal. This is in contrast with the layered coding approaches to be discussed next, where no reconstruction is possible at the absence of the base layer (containing the most important signal information). When both streams are received without error, a better replica of the transmitted signal can be reproduced. In comparison with repetitive transmission of the same signal, MDC provides better performance both when there is no information loss and when one of the streams is received erroneously. MDC is particularly attractive in conjunction with transmission diversity.

Interest in MDC has been ongoing with progress being made since the early eighties in both the theory,[67,68] including the design of optimal quantizers[69–70] and its applications to speech coding under various transmission environments.[71–72] Some of the recent developments have included the development of Internet-based applications which use MDC to provide robustness[73–74] and the extension of MDC to transform coding[75–76] and wavelet coding.[77–78] The recent burst of activity in this area indicates that it may be one of the most promising approaches for robust image/video transmission.

3.4.4 Unequal Error Protection and Scalable Coding

So far, we have assumed that all robustness is achieved exclusively through source coding techniques, under the assumption that channel errors could happen, and that errors would occur at random and affect with equal likelihood the whole bitstream.

However, in practical source encoders each bit of the compressed bit sequence contains different amounts of information, and the effect of a loss caused by erroneously receiving a bit will be different in each case. It is therefore natural to try to protect different portions of a bit stream differently, so that important information is protected using stronger channel codes. This approach, known as Unequal Error Protection (UEP), assumes that channel coders can adequately provide different levels of protection.

3.4.4.1   Channel Coding Techniques Providing UEP

UEP can be achieved in many different ways in a channel coding environment. One basic principle of UEP codes is that each channel codeword carries several types of information, with different levels of protection for each. For example, assume two levels of protection are provided; *coarse* data is the information that is heavily protected, and *detail* is the information that is less protected. UEP can be achieved by *embedding* the coarse information in the fine information.

This can be best illustrated by the concept of clouds and satellites as introduced by Cover.[54] The fine information is transmitted by selecting one among all the possible channel codewords. However, these codewords are also grouped into *clouds*, i.e., sets of codewords that are relatively close to each other while being somewhat more distant from codewords in other clouds. Thus a selection of a cloud conveys information that is better protected than the information conveyed by the satellites within the cloud.

When one of the satellite points is received the information about which cloud was transmitted can be decoded; this can be done even if the error is large, since confusing different points within the cloud does not affect the cloud information. The least protected information is carried by the specific codeword within the cloud, which is itself more susceptible to errors. Examples of codes designed with this property can be found in Kasami et al.[79] and Lin, Lin, and Lin.[80] This approach can also be used with the selection of modulation points[81] rather than with specific codes. Generally in a given modulation scheme it is possible to control the degree of protection provided by each modulation point by selecting the distances between the modulation points.

Once one allows the selection of modulation points in arbitrary positions it is possible to combine the design of source coders and channel modulation schemes. This truly joint source channel coding design then no longer has the goal to minimize distortion for a given rate, but rather to minimize overall (received) distortion for a given average transmission power budget.[82–84]

Probably the most popular example of UEP codes are the Rate-Compatible Punctured Convolutional (RCPC) codes introduced by Hagenauer.[85–87] RCPCs are essentially convolutional codes that can be *punctured* at predetermined positions (i.e., certain output bits are not transmitted). In this way the transmitter can control the rate. If the puncturing is increased fewer bits are transmitted, thus a lower degree of protection is offered; conversely, if less puncturing is performed then more redundancy is present and more protection will be offered.

3.4.4.2   Scalable Video Coding Techniques

While any compression algorithm will tend to produce bits with unequal importance, it may be desirable to design algorithms where the encoder has control of the relative importance of the information. This will lead us to the design of scalable or multi-resolution algorithms.

In layered coding, a source signal is separated into a number of streams (or layers). This separation is usually based on the importance of the information of each layer in the reconstruction of the signal. Therefore, different layers contain information with different levels of significance. The most important layer, called base layer, contains information without which the reconstruction of the signal would not be possible. By adding more layers to this base layer, a better replica of the original signal can be reconstructed. The base layer corresponds to high priority information whereas the enhancement layer is of lower priority and its loss does not have catastrophic consequences.

Different methods can be used to produce multiple layers from a single source signal. One such method is to split the bit stream generated by the source encoder into multiple streams. For example, for the motion-compensated based source encoders (such as MPEG and H.263), the motion vectors are of more significance for the reconstruction than is the residual information. Therefore, one can construct a base layer containing bits corresponding to these motion vectors and other control information, and another stream, the enhancement layer, can then contain the bits corresponding to the residual information. A different splitting strategy than the above can be used.

The MPEG-2 standard[31] provides several methods to achieve scalability: temporal, SNR, and spatial scalability. In a temporal scalability framework, the base layer may consist of video data encoded at a reduced frame rate, for example 10 frames/sec instead of the original 30 frames/sec of the source material. The enhancement layer provides the remaining frames. SNR scalability consists of transmitting the low frequency DCT coefficients in the base layer (as they tend to convey most of the energy in the sequence) and the higher frequency coefficients in the enhancement layer. Finally spatial scalability consists of coding as the base layer a sequence that is produced by reducing the size of the frames in the original sequence. The enhancement layer contains the difference between the the base layer frames interpolated to the larger size and the original frames.

Other examples of scalable coding techniques include wavelet or subband coding,[88–89] where the resulting image subbands can be, if needed, separately coded and transmitted. An alternative method which offers additional flexibility is the so-called Pyramid coding.[90–91] In this framework a coarse version of the signal (generated, for example, through spatial decimation) is compressed as the base layer, then the difference between the decoded base layer interpolated to the full resolution and the original image is coded as the enhancement layer.

In summary, with a scalable coder and efficient UEP channel codes it is possible to match important source information to high levels of protection. For example in wavelet coders such as those described by Shapiro[92] and Said and Pearlman,[93] the bits produced by the encoder are generated in order of importance, but if a single bit is lost all remaining bits in the file are useless. Thus, as discussed in Sherwood and Zeger,[56] one can use an RCPC as a channel code with the maximum redundancy (and protection) available at the beginning of the file, and with little or no protection available towards the last bits in the file.

3.4.5 ADAPTATION TO CHANNEL CONDITIONS AND RATE CONTROL

So far we have described approaches that are mostly suited for open-loop error control environments, i.e., where no feedback from the receiver/decoder is provided to the source encoder/transmitter. We now show how feedback information can be used to further improve the performance. The basic principle as illustrated by Figure 3.4 is simple: we should transmit less information during periods of poor channel performance. This is achieved by feeding back information about the channel state to the rate control algorithm at the encoder. The channel state information can be, as in Figure 3.4, simply the result of the latest acknowledgment in an ARQ scheme: a negative acknowledgment indicates poor channel behavior.

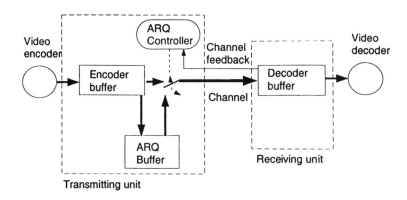

Figure 3.4 Diagram of buffers in the system

Let us formalize this by considering once again the delay constraints affecting video transmission that were disussed in Section 3.3.2. In both CBR and VBR transmission cases data is stored in buffers at encoder and decoder. Assume a variable channel rate of $C(i)$ during the i-th coding unit interval, then, the encoder buffer state at time i is

$$B(i) = \max B(i-1) + r_{ix(i)} - C(i), 0$$

with $B(0) = 0$ being the initial state of the buffer, and $r_{ix(i)}$ is the rate required by frame i when quantizer $x(i)$ is used.

Let us consider now what constraints need to be applied on the encoder buffer state (controlling the encoder buffer suffices to guarantee that the delay constraints are met[32,38]). First, the buffer state $B(i)$ cannot grow indefinitely because the physical buffer memory will be limited. If B_{max} is the physical memory available then we need to guarantee that $B(i) \leq B_{max}$ at all times. In addition, in order for the delay constraints not to be violated we need to guarantee that the data corresponding to coding unit i is transmitted before $t_i + \Delta T$, that is, transmission has to be completed during the next ΔN coding unit intervals. Intuitively, in order for this constraint to be met, all we need to ensure is that the future channel rates, over the next ΔN units, are sufficient to transmit all the data in the buffer.

Let us define the effective buffer size, $B_{eff}(i)$, as the sum of future channel rates over the next ΔN intervals:

$$B_{eff}(i) = \sum_{k=i+1}^{i+\Delta N} C(k),$$

Then it is easy to see, as demonstrated in Hsu et al.[38] and Reibman and Haskell,[32] that correct transmission is guaranteed if

$$B(i) \leq B_{eff}(i), \forall i.$$

As an example, consider the case where $C(i) = \bar{C} = R_T/N$ is constant. Then if the system operates with an end-to-end delay ΔN the buffer can store no more than $\Delta N \cdot \bar{C}$ bits at time i. For a detailed analysis of the relationship between buffering and delay constraints, see Reibman and Haskell[32] and Hsu et al.[38]

We call this the *effective* size because it defines a constraint imposed regardless of the physical buffer size. In general the applicable constraint will be imposed by the smallest of $B_{eff}(i)$ and B_{max}. Assuming that sufficient physical buffer storage is available, that B_{max} is always larger than $B_{eff}(i)$, our problem becomes, to find the optimal set of quantizers $x(i)$ for each i such that the buffer occupancy

$$B(i) = \max(B(i-1) + r_{ix(i)} - C(i), 0),$$

is such that

$$B(i) \leq B_{eff}(i)$$

and some metric $f(d_{1x(1)}, d_{2x(2)}, \ldots, d_{Nx(N)})$ is minimized. The metric to be used could be, for example, the mean squared error.

It is interesting that the constraints depend on the channel rates. When the channel rates can be chosen by the user (e.g., transmission over a network) this leads to interesting questions on the best combination of source and channel rates given constraints on the channel rates that can be utilized.[38-40]

In the wireless environment we are considering we have a variable channel rate because packets of video information can be randomly lost. However, we have no control over the behavior of the summer and consequently on the channel rates. Thus, deterministically ensuring that transmission will be successful requires us to know $B_{eff}(i)$ at all times i. Because the behavior of the channel is not deterministic and $B_{eff}(i)$ is a function of the *future* channel rates, we will never be able to know $B_{eff}(i)$ exactly. Instead what we can do is use some of the models described in Section 3.2.3 in order to *predict* $B_{eff}(i)$. This approach is described in more detail elsewhere[41,94-96] and it can be shown to perform better than approaches where there is no feedback about the channel state. In general, if a closed-loop system is used we can take advantage of the available channel information to adapt the information transmitted by the encoder and improve overall performance. Other examples of this approach can be found in Hafez and Rajugopal[97] and Liu and Zarki.[98]

In summary, given the constraints imposed by the delay and our observation of the channel state, we can optimize the selection of source coding rates. Other approaches may allow us to modify the level of redundancy or the power used for transmission. The principle remains the same: in a closed-loop error control scheme it is possible to adjust the parameters (both source and channel coding) to best match the observed state of the channel.

3.5   CONCLUSIONS

In this chapter we have introduced some of the major issues in the design of video coding algorithms that are suitable for wireless communications. The requirements for these algorithms to become reality go beyond achieving reasonable quality at low rate and with low power consumption. In fact one of the key requirements for these systems to become reality is to provide a robustness to errors that, as discussed in Section 3.4, is best achieved when the video encoder is aware of the characteristics of the underlying channel. This section has also illustrated a number of techniques to increase the robustness, ranging from simple packetization to sophisticated rate adaptation algorithms based on channel state feedback. We expect much of the research involved in making wireless video a reality to concentrate on some of these issues.

REFERENCES

1. C. Lee, *Mobile Cellular Telecommunications.* McGraw-Hill Inc., New York, 1995.
2. R. Steele (Ed), *Mobile Radio Communications.* Pentech, London, 1992.
3. W. Tuttlebee (Ed), *Cordless Telecommunication in Europe.* Springer-Verlag, 1990.
4. *GSM Recommendation, European Telecommunications Standardization Institute, Sophia Antipolis, France,* 1988.
5. *Dual-Mode Subscriber Equipment –Network Equipment Compatibility Specification* Telecommunications Industry Association, 1989.
6. *Mobile-station-base-station compatibility standard for dual-mode wideband spread spectrum cellular system.* Telecommunications Industry Association, July 1993.
7. *Recommended minimum performance standards for dual-mode wideband spread spectrum mobile stations.* Telecommunications Industry Association, Jan. 1994.
8. *Mobile Station-Base Station Compatibility Standard for Dual-Mode Wideband Spread Spectrum Systems.* Telecommunications Industry Association, Washington, D.C.
9. *Coding of Speech at 8 Kbit/s using conjugate-structure algebraic code-excited linear prediction (CS-CELP).* ITU draft recommendation G.729, Feb. 1996.
10. *Wireless LAN Medium Access Control (MAC) and Physical Layer (PHY) specification,* 1997.
11. *Radio Equipment and Systems (RES); High Performance Radio Local Area Network (HIPERLAN), Type 1, functional specifications, ETSI Final Draft prETS 300652*
12. L. Hanzo and J. Stefanov, "The pan-european digital cellular mobile radio system-known as GSM." In *Mobile Radio Communications.* R. Steele, Ed. : Pentech, London, pp. 677–773, 1992.

13. http://www.tiaonline.org/standards/sfg/imt2k/
14. E. G. Tiedemann Jr., Y.-C. Jou, and P. Odenwalder, "The evolution of IS-95 to a third generation system and to the IMT-2000 era." *Proc. ACTS Summit*, pp. 924–929, 1997.
15. E. G. Tiedemann Jr., "CDMA2000: The evolution of is-95 to third generation systems." *Proc. of the 3rd International Symposium on Multi-Dimensional Mobile Communications*, (Menlo Park, CA), 1998.
16. E. Dahlman, B. Gudmundson, M. Nilsson, and J. Skold, "UMTS/IMT 2000 based wideband CDMA." *IEEE Commun. Mag.*, vol. 39, pp. 70–80, Sept. 1998.
17. F. Adachi, M. Sawahashi, and H. Suda, "Wideband DS-CDMA for next-generation mobile communications systems." *IEEE Commun. Mag.*, vol. 39, pp. 56–69, Sept. 1998.
18. T. Ojanpera and R. Prasad, "An overview of air interface multiple access for IMT-2000/UMTS." *IEEE Commun. Mag.*, vol. 39, pp. 82–95, Sept. 1998.
19. L. Hanzo, "Bandwidth-efficient wireless multimedia communications." Proc. IEEE, vol. 86, pp. 1342–1382, 1998.
20. "Special issue on MPEG-4." *IEEE Trans. on Circ. and Syst. for Video Technol.*, Feb. 1997.
21. ITU-T, "Video coding for low bitrate communication." ITU-T Recommendation H.263; version 1, Nov. 1995, version 2, Jan. 1998.
22. T. H. Meng, A. C. Hung, E. K. Tsern, and B. M. Gordon, "Low-power signal processing system design for wireless applications." *IEEE Personal Commun.*, vol. 5, pp. 20–31, June 1998.
23. S. Benedetto, D. Divsalar, G. Montorsi, and F. Pollara, "Serial concatenated of interleaved codes: performance analysis, design, and iterative decoding." *IEEE Trans. Inform. Theory*, vol. 44, pp. 909–926, 1998.
24. M. Khansari, A. Jalali, E. Dubois, and P. Mermelstein, "Low bit-rate video transmission over fading channels for wireless microcellular systems." *IEEE Trans. on Circ. and Sys. for Video Technol.*, pp. 1–11, Feb. 1996.
25. E. N. Gilbert, "Capacity of burst-noise channel." Bell Syst. Tech. J., vol. 39, pp. 1253–1265, Sept. 1960.
26. M. Zorzi, R. R. Rao, and L. B. Milstein, "ARQ error control for fading mobile radio channels." IEEE Trans. on Veh. Tech., vol. 46, pp. 445–455, May 1997.
27. M. Zorzi and R. R. Rao, "ARQ error control for delay-constrained communications on short-range burst-error channels." Proc. IEEE VTC'97, 1997.
28. M. Zorzi, R. R. Rao, and L. Milstein, "On the accurary of a first-order Markov model for data transmission on fading channels." *Proc. IEEE ICUPC'95*, 1995.
29. M. Zorzi, R. R. Rao, and L. Milstein, "A Markov model for block errors on fading channels." *Proc. PIMRC96*, 1996.
30. R. J. Clarke, *Digital Compression of Still Images and Video,* Academic Press, 1995.
31. J. Mitchell, W. Pennebaker, C. E. Fogg, and D. J. LeGall, *MPEG Video Compression Standard.* Chapman & Hall, New York 1997.
32. A. R. Reibman and B. G. Haskell, "Constraints on variable bit-rate video for ATM networks." *IEEE Trans. on CAS for Video Technol.*, vol. 2, pp. 361–372, Dec. 1992.
33. S.-W. Wu and A. Gersho, "Rate-constrained optimal block-adaptive coding for digital tape recording of HDTV." *IEEE Trans. on Circ. and Syst. for Video Technol.*, vol. 1, pp. 100–112, Mar. 1991.
34. A. Ortega, K. Ramchandran, and M. Vetterli, "Optimal trellis-based buffered compression and fast approximation." *IEEE Trans. on Image Proc.*, vol. 3, pp. 26–40, Jan. 1994.

35. W. Ding and B. Liu, "Rate control of MPEG video coding and recording by rate-quantization modeling." *IEEE Transactions on Circ. and Syst. for Video Technol.*, vol. 6, pp. 12–20, Feb. 1996.

36. L.-J. Lin and A. Ortega, "Bit-rate control using piecewise approximated rate-distortion characteristics." *IEEE Trans. on Circ. and Sys. for Video Technol.*, vol. 8, pp. 446–459, Aug. 1998.

37. A. Ortega and K. Ramchandran, "Rate-distortion techniques in image and video compression." *IEEE Signal Processing Mag.*, Nov. 1998.

38. C.-Y. Hsu, A. Ortega, and A. Reibman, "Joint selection of source and channel rate for VBR video transmission under ATM policing constraints." *IEEE J. on Sel. Areas in Commun.*, vol. 15, pp. 1016–1028, Aug. 1997.

39. J.-J. Chen and D. W. Lin, "Optimal bit allocation for coding of video signals over ATM networks." *IEEE J. on Sel. Areas in Commun.*, vol. 15, pp. 1002–1015, Aug. 1997.

40. W. Ding, "Joint encoder and channel rate control of VBR video over ATM networks." *IEEE Trans. on Circ. and Syst. for Video Technol.*, vol. 7, pp. 266–278, Apr. 1997.

41. C.-Y. Hsu, A. Ortega, and M. Khansari, "Rate control for robust video transmission over burst-error wireless channels," *IEEE J. on Sel. Areas in Commun.*, 1999. To appear.

42. Z. Miao and A. Ortega, "Rate control algorithms for video storage on disk based video servers in *Proc. of Asilomar Confer. on Signals Syst. and Comp.*, Nov. 1998.

43. *ITU-T Recommendation H.261: Video codec for audiovisual services at p×64 Kbits*, Mar. 1993.

44. M. Liou, "Overview of the px64 kbit/s video coding standard." *Comm. of the ACM*, vol. 34, pp. 59–63, Apr. 1991.

45. M. Kunt, A. Ikonomopoulos, and M. Kocher, "Second generation image coding techniques." *Proc. of the IEEE*, vol. 73, pp. 549–579, Apr. 1985.

46. H. G. Musmann, M. Hötter, and J. Ostermann, "Object-oriented analysis-synthesis coding of moving images." *Signal Processing: Image Commun.*, vol. 1, pp. 117–138, Oct. 1989.

47. J. Ostermann, "Object-oriented analysis-synthesis coding based on the source model of moving flexible 3D objects." *IEEE Trans. on Image Proc.*, vol. 3, pp. 705–710, Sept. 1994.

48. N. Färber, E. Steinbach, and B. Girod, "Robust H.263 compatible video transmission over wireless channels." In *Proc. PCS'96*, pp. 575–578, 1996.

49. *IEEE Personal Commun. Mag.*, vol. 5, no. 3. Special Issue on Energy Management in Personal Communications and Mobile Computing.

50. W. Pennebaker and J. Mitchell, *JPEG Still Image Data Compression Standard.* Van Nostrand Reinhold, 1994.

51. D. Lee, "New work item proposal: JPEG 2000 image coding system." ISO/IEC JTC1/SC29/WG1 N390, 1996.

52. C. Chrysafis and A. Ortega, "Line Based Reduced Memory Wavelet Image Compression." in *Proc. IEEE Data Compression Conf.*, pp. 308–407, IEEE Computer Society Press, Los Alamitos, California, 1998.

53. C. E. Shannon, "A mathematical theory of communication." *Bell Syst. Tech. J.*, vol. 27, pp. 379–423, 1948.

54. T. Cover, "Broadcast channels." *IEEE Trans. on Inform. Theory*, vol. IT-18, pp. 2–14, Jan. 1972.

55. K. Fazel and J. L. Huillier, "Application of unequal error protection codes on combined source-channel coding." In *IEEE Int. Conf. Commun.*, pp. 320.5.1–6, Apr. 1990.

56. G. Sherwood and K. Zeger, "Progressive image coding for noisy channels." *IEEE Signal Processing Lett.*, vol. 4, pp. 189–191, Jul. 1997.

57. K. Ramchandran and M. Vetterli, Multiresolution Joint Source-Channel Coding for Wireless Channels. *Wireless Communications: A Signal Processing Perspective*, Editors: V. Poor and G. Wornell. Prentice-Hall, 1998.

58. D. Yu and M. W. Marcellin, "A fixed-rate quantizer using block-based entropy-constrained quantization and run-length coding." *IEEE Trans. on Image Proc.*, Aug. 1996. Submitted, Available in http://www-spacl.ece.arizona.edu/.

59. D. W. Redmill and N. G. Kingsbury, "The EREC: An error-resilient technique for coding variable-length blocks of data." *IEEE Trans. on Image Proc.*, vol. 5, pp. 565–574, Apr. 1996.

60. C. D. Creusere, "A new method of robust image compression based on the embedded zerotree wavelet algorithm." *IEEE Trans. on Image Proc.*, vol. 6, pp. 1436–1442, Oct. 1997.

61. Y. Yoo and A. Ortega, "Constrained bit allocation for error resilient JPEG coding." In *Proc. of Asilomar Confer. on Signals, Syst. and Comp.*, 1997.

62. R. Laroia and N. Farvardin, "A structured fixed rate vector quantizer derived from a variable-length scalar quantizers: Part I - Memoryless sources." *IEEE Trans. Inform. Theory*, vol. 39, pp. 851–867, May 1993.

63. J. Wen and J. Villasenor, "Reversible variable length codes for efficient and robust image and video coding." In *Proc 1998 IEEE Data Compression Confer.*, pp. 471–480, 1998.

64. S. McCanne, M. Vetterli, and V. Jacobson, "Low-complexity video coding for receiver-driven layered multicast." *IEEE J. on Sel. Areas in Commun.*, vol. 15, pp. 983–1001, Aug. 1997.

65. E. Steinbach, N. Färber, and B. Girod, "Standard compatible extension of H.263 for robust video transmission in mobile environments." *IEEE Trans. on Circ. and Sys. for Video Technol.*, vol. 7, pp. 872–881, Dec. 1997.

66. J. Hagenauer, "Source-controlled channel decoding." *IEEE Trans. on Commun.*, vol. 43, pp. 2449–2457, Sept. 1995.

67. A. A. El-Gamal and T. M. Cover, "Achievable rates for multiple descriptions," *IEEE Trans. Inform. Theory*, vol. IT-28, pp. 851–857, Nov. 1982.

68. J. C. Batllo and V. Vaishampayan, "Asymptotic performance of multiple description transform codes." *IEEE Trans. Inform. Theory*, vol. 43, no. 2, pp. 703–707, 1997.

69. V. A. Vaishampayan, "Design of multiple description scalar quantizers." *IEEE Trans. Inform. Theory*, vol. 39, no. 3, pp. 821–834, 1993.

70. A. Ingle and V. A. Vaishampayan, "DPCM system design for diversity systems with applications to packetized speech." *IEEE Trans. Speech and Audio Process.*, vol. 3, no. 1, pp. 48–58, 1995.

71. N. S. Jayant and S. W. Christensen, "Effects of packet losses in waveform coded speech and improvements due to an odd-even sample-interpolation procedure." *IEEE Trans. Commun.*, vol. COM-29, pp. 101–109, Feb. 1981.

72. S.-M. Yang and V. A. Vaishampayan, "Low-delay communication for rayleigh fading channels: An application of the multiple description quantizer." *IEEE Trans. Commun. Theory*, vol. 43, pp. 2771–2783, Nov. 1995.

73. V. H. M. A. Sasse, M. Handley, and A. Watson, "Reliable audio for use over the internet." In *Proc. INET*, 1995.

74. M. Podolsky, C. Romer, and S. McCanne, "Simulation of FEC-based error control for packet audio on the internet." *INFOCOM'98*, 1998.

75. M. T. Orchard, Y. Wang, V. Vaishampayan, and A. R. Reibman, "Redundancy rate-distortion analysis of multiple description coding using pairwise correlating transforms." *ICIP'97*, 1997.

76. V. K. Goyal and J. Kovacevic, "Optimal multiple description transform coding of large sc Gaussian vectors." *Proc. of IEEE Data Compression Confer.*, 1998.

77. S. D. Servetto, K. Ramchandran, V. Vaishampayan, and K. Nahrstedt, "Multiple description wavelet based image coding." *ICIP'98*, 1998.

78. W. Jiang and A. Ortega, "Multiple description coding via polyphase transform and selective quantization." In *Proc. of Visual Commun. and Image Process.*, 1999.

79. T.Kasami, S.Lin, V. Wei, and S. Yamamura, "Coding for the binary symmetric broadcast channel with two receivers." *IEEE Trans. on Inform. Theory*, vol. IT-31, pp. 616–625, Sept. 1985.

80. M.-C. Lin, C.-C. Lin, and S. Lin, "Computer search for binary cyclic UEP codes of odd length up to 65." *IEEE Trans. on Inform. Theory*, vol. 36, pp. 924–935, July 1990.

81. K. Ramchandran, A. Ortega, K. M. Uz, and M. Vetterli, "Multiresolution broadcast for digital HDTV using joint source-channel coding." *IEEE J. on Sel. Areas in Commun.*, vol. 11, pp. 6–23, Jan. 1993.

82. N. Phamdo, N. Farvardin, and T. Moriya, "A unified approach to tree-structured and multistage vector quantization for noisy channels." *IEEE Trans. on Inform. Theory*, vol. IT-39, pp. 835–850, May 1993.

83. N. Farvardin and V. Vaishampayan, "Optimal quantizer design for noisy channels: An approach to combined source-channel coding." *IEEE Trans. on Inform. Theory*, vol. IT-33, pp. 827–838, Nov. 1987.

84. I. Kozintsev and K. Ramchandran, "Robust image transmission over energy-constrained time-varying channels using multiresolution joint source-channel coding." *IEEE Trans. on Signal Process., Special Issue on Wavelets and Filter Banks*, vol. 46, pp. 1012–1026, Apr. 1998.

85. J. Hagenauer, "Rate-compatible punctured convolutional codes (rcpc codes) and their applications." *IEEE Trans. on Commun.*, vol. COM 36, pp. 389–400, Apr. 1988.

86. J. Hagenauer, N. Seshadri, and C.-E. W. Sundberg, "The performance of rate-compatible punctured convolutional codes for digital mobile radio." *IEEE Trans. on Commun.*, vol. 38, pp. 966–980, July 1990.

87. R. V. Cox, J. Hagenauer, N. Seshadri, and C. Sundberg, "Variable rate sub-band speech coding and matched convolutional channel coding for mobile radio channels." *IEEE Trans. on Signal Process.*, vol. 39, pp. 1717–1731, 1991.

88. M. Vetterli and J. Kovacevic, *Wavelets and Subband Coding*. Prentice-Hall, 1995.

89. D. Taubman and A. Zakhor, "Multirate 3-D subband coding of video." *IEEE Trans. on Image Process.*, vol. 3, Sept. 1994.

90. P. J. Burt and E. H. Adelson, "The laplacian pyramid as a compact image code." *IEEE Trans. on Commun.*, vol. 31, pp. 532–540, Apr. 1983.

91. K. M. Uz, M. Vetterli, and D. LeGall, "Interpolative multiresolution coding of advanced television with compatible sub-channels." *IEEE Trans. on CAS for Video Technol., Special Issue on Signal Process. for Advanced Television*, vol. 1, pp. 86–99, Mar. 1991.

92. J. M. Shapiro, "Embedded image coding using zerotrees of wavelet coefficients." *IEEE Trans. on Signal Process.*, vol. 41, pp. 3445–3462, Dec. 1993.

93. A. Said and W. A. Pearlman, A new fast and efficient image coder based on set partitioning in hierarchical trees. *IEEE Trans. Circ. and Syst. for Video Technol.*, pp. 243–250, June 1996.

94. A. Ortega and M. Khansari, Rate control for video coding over variable bit rate channels with applications to wireless transmission. *Proc. of the 2nd. Intl. Confer. on Image Process.*, 1995.

95. C.-Y. Hsu, A. Ortega, and M. Khansari, Rate control for robust video transmission over wireless channels. *Proc. of Visual Commun. and Image Process*, 1997.

96. C.-Y. Hsu, *Rate control for video transmission over variable rate channels.* Ph.D. thesis, University of Southern California, Aug. 1998.

97. R. H. M. Hafez and G. R. Rajugopal, Adaptive rate controlled robust video communications over packet wireless networks. *ACM/Baltzer Mobile Networks and Appl. J.*, vol. 3, no. 1, pp. 33–47, 1998.

98. H. Liu and M. El Zarki, Adaptive source rate control real-time wireless video transmission. *ACM/Baltzer Mobile Networks and Appl. J.*, vol. 3, no. 1, pp. 49–60, 1998.

4 JMAPI - Java Management Application Programming Interface

Mario Francois Jauvin

CONTENTS

4.1 WHAT IT IS

The Java Management Application Programming Interface (JMAPI) is a collection of Java classes, provided by Sun Microsystems, that allow network management software developers to write management applications using a standardized platform-independent application programming interface. The initial API was based on the Java Developer Kit (JDK) 1.0 API specification. JMAPI was subsequently modified to support the 1.1 JDK specification in order to make use of its security and the Remote Method Invocation (RMI).

JMAPI is not a network or system management product and by itself cannot provide any management functionality. Rather, it is designed to be used by developers of network management systems and also by network element hardware vendors. A look at the JMAPI architecture will clarify these concepts.

4.2   ARCHITECTURE

The architecture of JMAPI consists of the components at the highest-level as shown in Figure 4.1.

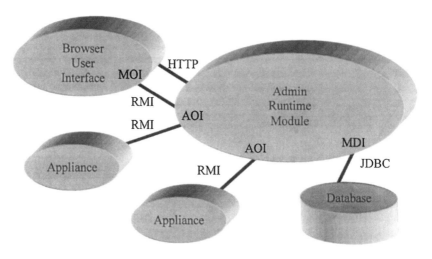

Figure 4.1 JMAPI Architecture

Everything except for the database portion runs in a Java environment.

4.2.1   BROWSER USER INTERFACE

The Browser User Interface (BUI) is somewhat a misnomer. The BUI actually is the JMAPI component from which an administrator can issue management operations or get management status from the objects being managed. The BUI can run either as an applet in a browser or as a standalone Java application with or without a graphical user interface. The BUI interacts entirely with one or more Admin Runtime Module and a HyperText Transfer Protocol (HTTP) server.

The BUI is comprised of the following modules:

- Admin View Module
- Managed Object Interfaces
- Java-enabled browser

4.2.1.1 Admin View Module

The Admin View Module (AVM) contains the key client-side classes for developers of JMAPI-applets whose primary role is to provide user interface and application-level functionality. The user interface is built completely on top of JDK Abstract Window Toolkit (AWT). The AVM classes are broken down into three packages: AVM Help to provide a general purpose help environment, AVM Base to provide AWT extensions for integrated management solution building and AVM Integration to provide integration between AVM Base classes and the Managed Object Interface.

4.2.1.2 Managed Object Interface

The Managed Object Interface (MOI) uses RMI to perform remote management of objects. A managed object is an abstraction of a resource (network element such as a hub or router, a service such as a DNS server, or anything else one may wish to manage) in the enterprise. Although the managed object is an abstraction of the resource, the actual resource can be managed from a JMAPI point of view only if it also implements a JMAPI appliance.

4.2.1.3 Java-enabled Browser

Any JDK 1.1 compliant browser, such as Netscape Navigator, HotJava, or Internet Explorer, can be used to run the AVM and MOI. If a Java application is used the browser will not be necessary.

4.2.2 ADMIN RUNTIME MODULE

The Admin Runtime Module (ARM) is the component that provides active instantiated managed objects for a number of appliances. The ARM is the component that will carry out the management operations requested by the BUI through the Managed Object Interface. The ARM and all of its associated functionality is referred to in JMAPI terminology as a long-running process called the JMAPI server or the managed object server. Please note that a BUI can communicate with more than one ARM, but ARMs do not communicate amongst them. Also note that one ARM can communicate with multiple appliances, and one appliance can communicate with multiple ARMs.

4.2.2.1 HTTP Server

The HTTP server is only used to load the applet, the AVM classes, and the network management application built on top of them into the browser. The benefits of this approach is that a new version of the network management application or JMAPI can automatically be acquired and software distribution and versioning are no longer an issue.

4.2.2.2 Managed Object Factory

The Managed Object Factory (MOF) is the JMAPI component responsible for creating instances of managed objects. Managed objects are RMI remote objects. The creation of a managed object is a two-step process. First, the new object is created by calling MOFactory.newObj(). Then the new object is made persistent by calling its addObject() method. This component is what actually constitutes the managed object server.

The MOF is also the component responsible for actually implementing the management operations requested by the BUI through interactions with Agent Objects Interfaces and Managed Data Interfaces.

4.2.2.2.1 Managed Data Interfaces

The Managed Data Interfaces (MDI) is responsible for keeping persistent information about managed objects using any commercial JDBC (Java Database Connectivity) compliant relational database management system.

4.2.2.2.2 Agent Object Interfaces

The Agent Object Interfaces (AOI) are the interface to agent objects residing on the appliances. These interfaces provide the MOF the ability to control and do the actual management of appliances or network resources.

4.2.2.2.3 Notification Dispatcher

The Notification Dispatcher (ND) is the JMAP component which filters and forwards events from the appliances managed by the ARM to the interested parties (BUI) that have registered for notification with the ARM.

4.2.3 APPLIANCES

4.2.3.1 Agent Object Factory

As you may recall from earlier, a managed object instance in the ARM has no remote management capabilities unless it can communicate with a remote agent object running on the appliance. In order to support this remote management capability the appliance must implement the Agent Object Factory which creates and maintains Agent Object Instances.

4.2.3.2 Agent Object Instance

Each Agent Object Instance (which corresponds to RMI objects) will call Java classes or Native Methods (using Java Native Interface, or JNI) to implement the management operation. If necessary, the Class Loader will download the Java classes and the Library Loader will download the Native Methods.

4.3   SIMPLE NETWORK MANAGEMENT PROTOCOL (SNMP) AND JMAPI

The importance of the SNMP protocol as the most prevalent management protocol was recognized and is integrated with JMAPI. Although it is possible for the BUI interface to directly interface with SNMP entities, the most useful application of the SNMP support provided by JMAPI is at the ARM or MOF level. In that context, SNMP is being used (instead of RMI) to communicate to SNMP agent entities (instead of JMAPI appliances).

4.4   JMAPI AND THE INDUSTRY

The potential for JMAPI lies in the fact that this technology can be used to manage a heterogeneous environment, which is more the norm than the exception in today's enterprises. Its proposal for a unified management API is a key contributor to being capable of managing systems of different operating systems, different hardware platforms, and different protocols. Last but not least, the portability inherent in Java imposes few restrictions on where the JMAPI management platform can run.

Before JMAPI can realize its full potential, vendors of hardware and software must embrace it and provide the necessary interfaces to implement appliances. In the case of network hardware vendors, this might not be a trivial task or feasible in the restricted hardware footprint where such implementation must reside. For example, a Java Virtual Machine would have to be implemented as well as management software interfaces to query or modify the operation of the device (to be made accessible to JMAPI via JNI). This is, of course, in addition to the necessary SNMP agent entity implementation.

The all-Java aspect of JMAPI is both a blessing and a curse. The fact that a single portable language is used throughout the environment means that development time and skills required are reduced. However, in the case of the ARM, which is meant to be a server type component, unacceptable performance is most likely to appear in medium-sized environments and almost guaranteed in large environments.

The JMAPI issue that has caused the most difficulties for people trying to evaluate or investigate the technology is its requirements for a commercial relational database management system (RDBMS, such as Sybase, Oracle, etc.). Some people have incorrectly interpreted this issue to mean that JMAPI requires a method of providing persistent storage for the states and attributes of managed objects. This is actually one of JMAPI's strong points. The real issue is that this requirement could not be achieved using a less costly and less administration-intensive solution than a commercial RDBMS. The change to support any JDBC-compliant RDBMS is one step in the right direction, but the fact that no recommendation using a non commercial JDBC-compliant solution exists from Sun makes this still a very resource-intensive approach.

4.5 WHAT'S AHEAD FOR JMAPI

Given that Sun has not modified JMAPI in the last six months it is difficult to assess where the technology will go in the future. According to Bruce Boardman[1] JMAPI "has a significant and growing list of supporters — including Bay, Cisco, IBM, Novell, PLATINUM Technologies, Sun, and 3Com." Paula Musich,[2] on the other hand, indicated in December of the same year that although Cisco and Bay had pledged their allegiance to JMAPI in May 1996, in December 1997 they had no products shipping based on it. To this date I am not aware of any such products either. The fact that the specification is not final probably has a lot to do with this. I think that a potential customer interested in using JMAPI is probably even more puzzled by the fact that Sun's own Solstice Enterprise Manager product comes with Java add-on technology called Java Dynamic Management Kit (JDMK) with no reference altogether in its product literature to JMAPI.

Bill Gates' (Microsoft CEO) statement[3] in April 1998 to the effect that he had not heard of the JMAPI acronym and Microsoft's direction to push for Web Based Enterprise Management (WBEM, a competing technology to JMAPI) are certainly not good news. The industry would not be served by two new management standards when it is looking for a unifying platform-independent solution.

REFERENCES

1. Bruce Boardman, "The Dawning Of The Age Of Java Management," *Network Computing*, May 15, 1997.
2. Paula Musich, "Sun Rises — Finally," *PC Week*, December 5, 1997.
3. Steven Burke, "Gates Takes Jab At Java Network Management," *Computer News Resellers*, April 27, 1998.
4. Luca Deri, "Rapid Network Management Application Development," IBM Zurich Research Laboratory, University of Berne.
5. Cameron Sturdevant, "Room for growth," *PC Week Labs*, July 14, 1997.
6. Paula Musich, "Web management specs inch forward," *PC Week*, July 11, 1997.
7. Paula Musich, "Next Java jolt: Management," *PC Week*, July 11, 1997.
8. Paula Musich, "CA announces Solaris version of Unicenter TNG," *PC Week Online*, July 16, 1997.
9. Paula Musich, "The Golden Age of Enterprise Systems," *PC Week Online*, October 6, 1997.
10. *Java Management API Architecture*, SunSoft, September 1996.
11. *Java Management Programmer's Guide*, SunSoft, September 1997.
12. *Java Management API, Help Developers' Guide*, SunSoft, May 1997.
13. *Java Management API User Interface Visual Design Style Guide*, SunSoft, September 1997.
14. *Java Management API User Interface Style Guide*, SunSoft, May 1997.

5   Cable Modem and HFC

Albert Azzam

CONTENTS

0-8493-9594-1/00/$0.00+$.50
© 2000 by CRC Press LLC

5.1 OVERVIEW

Developing a high performance National Information Infrastructure (NII) for the information age is a national goal. For the United States, one of the major objectives of the *Telecommunication Act of 1996* is to promote a competitive environment in which old and new communications providers build a network of interconnected networks. This new infrastructure should support new interactive multimedia services that are becoming popular with a phenomenal growth on a widespread basis.

In this chapter, the cable modem is described. Making the cable modem functional will require costly modernization efforts of the cable network so it can communicate bidirectionally. To fully describe the technology of the cable modem, its environment is briefly described so the reader has a better appreciation of the various assumptions made to develop this technology.

This chapter is organized in two parts:

1. the cable network environment in which the cable modem must operate
2. the two cable modems the industry is specifying mainly:

- ATM-centric cable modem
- IP-centric cable modem

5.2 MARKET PULL/TECHNOLOGY PUSH

Today's networks, worldwide, are service specific. Cable TV networks were optimized for video broadcasting (one way). Telephone networks, including wireless, were deployed specifically to handle voice traffic. Both platform and fabric were optimized in the design to switch voice traffic efficiently. The Internet was initially developed and optimized for data transport. None of these service specific networks can cope or was designed to provide these emerging interactive services.

In 1993, the Internet emerged not just as a way to send e-mail or download an occasional file but as a place to visit, full of people and ideas. It became *cyberspace*. Its impact quickly spread as everyone wanted to experience this virtual community. The Internet frenzy is a worldwide phenomenon. Europe, Asia, and developing countries are building the Internet infrastructure at a faster pace than the telephone companies and creating a new market for high speed Internet connectivity. The high demand market for digital data, voice, image, and video transmission necessitates the latest and greatest access technologies. Moreover, today's users are becoming knowledgable about the performance that network services must provide if their bandwidth hungry applications are to work adequately. Advanced Internet and broadband applications are being developed in various research centers, including government agencies through Next Generation Internet funding (NGI), Internet 2 Consortium, and universities all over the globe.

Service specific optimization is most prevalent in the local access networks where deployment and operating costs are directly associated with individual customers. The AT&T and TCI merger is a case in point. Hence, the value of a communications network increases with the number of locations it serves and the number of individual users. The two emerging high speed interface technologies are ADSL and cable modem.

The information age has penetrated our society at all levels: education, business, government, marketing, entertainment, and global competition. All these factors are clear to the long term planners for both the telephone and cable operators. The business reality, however, may take precedence over long term strategies, and as always the free market will determine the outcome.

To meet this business model, the cable multiple system operator (MSO) developed the high speed cable modem as the technology of choice. The telephone companies, on the other hand, opted to provide high speed access via ADSL over copper lines.

5.3   CABLE NETWORK AND EVOLUTION TO HFC

5.3.1   HISTORY OF THE CABLE NETWORK

The cable network was originally deployed to perform a very simple task. Reception of TV signals was very poor, especially in suburban areas where the middle class began moving in the sixties. This CATV network used coax shielded cable to deliver strong and equal in strength TV signals to the home. A good quality antenna tower received TV channels from the airwaves and mapped them in the cable spectrum. In North America, bandwidth 50 to 550 MHz is reserved for NTSC analog cable TV broadcasts as shown in Figure 5.1. The 50 to 550 MHz range of frequencies is

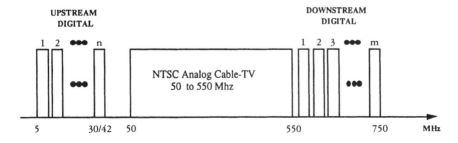

Figure 5.1 CATV cable spectrum

divided into 6-MHz channels (8-MHz for Europe). TV analog signals are modulated in each of the 6 MHz channels.

The TV signals in the cable coax are replicas of the ones broadcast through the airwaves, so no modifications were needed to the television set. CATV brought about yet another advantage. It could provide more channels, with signals of equal strength, to the end user than those delivered conventionally. TV signals lose power more readily in the air than in cable, and TV receivers and tuners cannot cope with the interference of more powerful adjacent TV signals.

Channel allocation in the spectrum is regulated by the Federal Communications Commission (FCC) which also regulates the frequency location and signal power used by TV broadcasters. These FCC rules guarantee that stations that use the same TV channel are far enough apart so as not to interfere with one another. Those rules coupled with the rapid attenuation of signal power in the air, enables cable operators to deliver more channels with equal signal power to homes.

Cable operators, by virtue of the increased market demand due to TV program variety and flexibility, and premium channel availability, were elevated from CATV operators to broadcasters, and later they became content providers. From the outset, the cable network evolved little if any in terms of two-way communication media. Lately, however, fiber optics installation, increased system reliability, and reduced operating and maintenance costs have accelerated. HFC modernization plans not only to set the stage for providing an infrastructure for bidirectional communication but also to increase channel capacity. With HFC modernization as a prerequisite, cable modem became practical to develop because the infrastructure was conducive to bidirectional communication and the high-speed interactive market was finally in demand and considered a normal part of society.

Regulation: Cable operators became more powerful as more and more homes subscribed to the service, and TV signals from metropolitan areas replaced local channels to make CATV more attractive to subscribers. At that time, the FCC began to regulate the industry and to dictate what an operator and which channels must be carried to serve the local community. The Cable Communications Policy Act of 1984 eased price control due to competition from broadcasters and encouraged growth. By that time, cable operators became content providers as well.

In 1992, the U.S. Congress passed the Cable Television Consumer Protection and Competition Act and reenacted price control regulation with some exceptions.

With the Telecommunication (reform) Act of 1996, Congress deregulated this industry with the proviso that the telephone industry could then compete in video services and the cable operators could also enter the local telephone service market.

5.3.2 LEGACY CABLE NETWORK

The technologies of the sixties and seventies were readily available to provide CATV broadcasting services (one way). The networks were built independently to serve particular communities, so the economic model, more or less, dictated a simple and somewhat organized topology of a branch and tree architecture. A point-to-point approach was economically prohibitive and did not offer an advantage over a shared

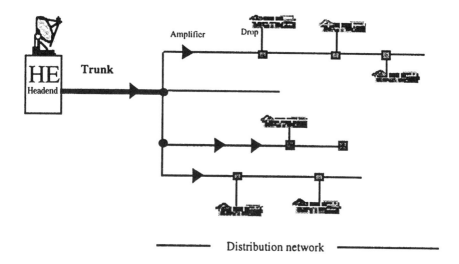

Figure 5.2 Cable system topology

medium access specially for broadcasting applications. Figure 5.2 illustrates a traditional cable network. The functional elements are

1. Cable TV headend
2. Long haul trunks
3. Amplifiers
4. Feeders
5. Drops

Headend: The CATV headend is mainly responsible for the reception of TV channels gathered from various sources, such as broadcast television, satellite, local community programming, and local signal insertion. These 6-MHz TV analog channels are modulated, using a frequency division multiplexing technique, and are placed into the cable spectrum as shown in Figure 5.1. This central control headend can serve thousands of customers using a simple distribution scheme. To achieve geographical coverage of the community, the cables emanating from the headend are split into multiple cables. When the cable is physically split, part of the signal power is split off and sent down the branch. The content of the signal, however, stays intact.

Trunks: High quality coax cables are used as trunks to deliver the signals to the distribution network and finally to its intended destination. The trunk can be as long as 15 miles. Lower quality coax is commonly used in the distribution and drop portions of the plant.

Amplifiers: TV signals attenuate as they travel several miles through the cable network to the subscribers' homes. Therefore, amplifiers have to be deployed throughout the plant to restore the signal power. The more times the cable is split and the longer the cable, the more amplifiers are needed in the plant. Excessive cascade of amplifiers in the network creates signal distortion. Amplifiers are also located in the distribution network (sometimes referred to as the *last mile*). These amplifiers, used in the traditional cable network, are one-way (amplifying signal from the headend to the subscriber). This scheme introduces several potential problems when a network needs to be upgraded to provide bidirectional communication. In such cases, these amplifiers need to be replaced with new two-way amplifiers.

Feeders: Feeders are sometimes referred to as the distribution network that serves the residential market. The term *home passed* usually refers to homes that are near the distribution network. The coax cables in the distribution network (branch/ tree) are usually short and are in a range of one to two miles.

Drops: Drops are usually located on telephone poles or more recently in a residential pedestrian area. A lower quality coax is used to connect from the drop to the home.

5.3.3   HFC Network

HFC (Hybrid Fiber Coax) was the next generation cable network (shown in Figure 5.3). HFC is the first step needed to provide bidirectional communications and it paved the way for serious cable modem deployments.

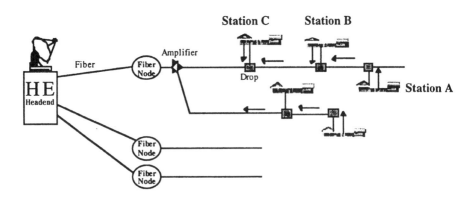

Figure 5.3 Hybrid Fiber Coax topology

HFC enhances a bidirectional shared-media system using fiber trunks between the headend and the fiber nodes, and coaxial distribution from the fiber nodes to the customer locations. The fiber extends from the access node to a neighborhood node. This fiber node interfaces with the fiber trunk and the coaxial distribution. It typically serves about 500 to 2000 (optimistic) subscribers via coaxial cable drops. These connected subscribers share the same cable and its available capacity and bandwidth. Because several subscribers share the same downstream and upstream bandwidth, special requirements such as privacy and security have to be taken into account.

Moreover, a special medium access control (MAC) scheme is required in the upstream direction, mainly acting as a traffic cop. MAC controls and mediates information flow, e.g., to prevent collision of information that is transmitted from users to the headend.

There are several other advantages of this HFC topology. The fiber trunk no longer needs amplifiers. Fiber is less immune to noise, and signal attenuation is practically nonexistent. These characteristics have the obvious advantage of increasing reliability, hence an amplifier failure affects only that particular residential area. Fiber deployment also means that far more bandwidth /channels will be at the disposal of the cable operator than would be available in the network for the subscriber.

5.3.4 UPSTREAM/DOWNSTREAM CABLE SPECTRUM

In Figure 5.1, Frequency Division Multiplexing (FDM) is the scheme employed. In the upstream direction, the 5-42 MHz range is dedicated to digital transmission. In this direction, the cable modem as transmitter uses this range to transmit digital information from the users to the headend. In the downstream direction the 450-750 MHz frequencies are restricted for downstream digital transmission.

Cable modems must tune their receivers between 450 and 750 MHz to receive data digital signals. The digital data is modulated and placed into the 6 MHz channel (traditional TV signal). A cable modem, therefore, functions as a tuner. The QAM modulation scheme was selected by the industry for the downstream direction. In the upstream direction, the cable modem transmits the signal between 5 and 42 MHz. The data is modulated and placed in the 6 MHz channel using the QPSK modulation technique. At this frequency range the environment is very polluted and noisy because of interference from CB and HAM radios and impulse and ingress noise from home appliances. For that reason QPSK was selected as the modulation scheme. QPSK is more robust in terms of its immunity to noise, but at the cost of delivering data at much lower speed than other modulation techniques.

5.3.5 DIGITAL CABLE NETWORK

All forms of communication today migrated or are migrating into digital format, e.g., CDs, cellular, voice, video. Most, if not all, future communication services are likely to be in digital format. The cable companies are under great competitive pressure to go completely digital. Digital transmission results in a noticeably better quality picture, at least noticable enough to be a differentiating factor for the consumer.

There is nothing inherent in the characteristics of cable or fiber pipes that prevents signals from being carried in digital format. Today's cable system can carry digital signals without modification as long as the modulated signal fits within the bandwidth and power constraints that the cable system carries. Digital communication can also co exist with analog TV signals as long as the digital signals are contained in their own 6 MHz band.

Using the cable network to transmit digital signals, including broadcast video signals, is of course possible. There is nothing in the cable network that specifically prevents such migration. Analog amplifiers in the system will be replaced with digital repeaters much like what the telephone network uses today to recondition the T1 digital signals. The advantage of going digital, in addition to improving signal quality due to noise, is the increased capacity of the cable system.

5.3.5.1 Potential Capacity

The cable system capacity will increase enormously if digital signals are transmitted instead of analog TV signals. The cable networks are built to support about 50 television channels. Cable channels should maintain 48 to 50 dB SNR in each 6 MHz channel. Modern modulation techniques such as Quadrature Amplitude Modulation (QAM) encoding, can achieve 43 Mbps capacity in a 6 MHz channel. The compression scheme used in the MPEG-2 standards for audio and video dramatically reduce the data rate required for transmission. A digitally compressed video signal of 3 to 6 Mbps can deliver an excellent quality broadcast video. So digital capacity of the cable system can potentially reach over 500 channels with existing cable bandwidth

5.3.6 CABLE NETWORK MODERNIZATION EFFORT

Modernizing the cable network to provide high speed interactive service is underway by most MSOs. HFC appears to have advantages over the other networks: its bandwidth capacity is enviable, and deploying a cable modem over a modernized HFC might be all it takes to be able to provide today's demanding, high speed interactive applications. A well-engineered cable modem can provide not only bidirectional data transmission but all the TV cable analog channels, high speed Internet access, voice, and high quality interactive video.

5.3.7 HFC ACCESS DRAWBACKS

HFC is evolutionary and can be accommodated in a stepwise approach. However, access over HFC can also introduce a host of technically challenging problems. Cable operators must first and foremost address the service affecting problems when introducing integrated digital services. The most crucial ones are reliability, security/privacy, and operation and maintenance.

The HFC architecture, although cost effective, is not ideal when it comes to service reliability. The critical drawbacks are

- component failure in an amplifier in the distribution network can render the entire neighborhood out-of-service.
- AC power failure (powering the amplifiers) is a more serious problem that must be resolved.
- AC power outage can render the entire area out-of-service. AC backup for powering the amplifiers must be provided so customers can still make voice calls during power outage.
- Because of the shared medium topology, the action of a malicious user can affect the operation and communication of all those connected users

in the branch or tree in both directions. A failed cable modem may have the same effect (of disrupting the shared bus), but it is expected that the cable modem will be designed to isolate such failures.

- The upstream transmission path is prone to noise of all kinds. The entire cable network must be well-maintained to ensure that ingress noise is not leaking into the system, causing failures to users who are on the bus.

5.3.8 Factors Influencing Cable Modem Operation

5.3.8.1 Amplifiers Bidirectional Issues

Modern cable systems (HFC) with bidirectional communication must use amplifiers that work in both directions. To accomplish this, back-to-back amplifiers with filters are arranged so that downstream and upstream signals are first filtered then amplified. The upstream path has an inherent disadvantage because of the branch-and-tree topology. During amplification of the upstream, the splitter outputs become its input; the splitter simply combines the incoming signals and noise, hence both are amplified. In the downstream direction, the signals passing through a splitter are attenuated on the splitter outputs, but the noise carried downstream is also attenuated.

5.3.8.2 Frequency-Agile

A modem that is frequency-agile capable can tune into any one of the downstream or upstream frequencies. The cable modem in the upstream is able to transmit on whatever frequency the cable system is equipped to handle. This gives cable operators the tools to change the upstream and downstream bandwidth allocation spectrum in their system due to changing traffic demand, without user intervention or worse, having to change the terminal equipment. Excessive noise due to ingress (temporarily or long term) of an upstream channel can be dynamically isolated by simply retuning the cable modem to other downstream and associated upstream channel(s). A wider range frequency-agile cable modem for a single carrier (beyond the 5 to 42 MHz range) is implementation dependent. The expense of providing more complex agility may not justify the development cost. It will, however, offer a very flexible and robust cable modem.

5.3.9 Noise

The upstream channel in HFC networks has been the source of great concern. The channel frequency in which it must operate positions it in a very hostile noise environment. Ingress noise in the upstream direction is the main cause of impairments in an HFC system. This noise comes in different flavors and severity.

The industry developed a channel model that mathematically defined the nature and physics of the cable network noise. This model was used to refine the specifications of the physical and MAC layers for the cable modem.

The noise phenemenon environment in the cable network is unique. The cable system acts as giant antennae for various noises and impairments that are additive, especially in the 5 to 42 MHz band of the RF spectrum. Each type of noise must be combatted from its source before it propagates further into the network and mutates. Just as challenging is that the noise phenomena in the cable network are time dependent. What is measured in the morning is quite different from measurements made in the peak TV viewing hours. Moreover, these measurements are different from one region to the next. The age of the cable plant and drops in particular, humidity of the region, number of subscribers in the drop, inside-home wiring, and past maintenance practices, all play a part in how the network behaves under different loads. To say that the system must be developed for a worse-case scenario is not the optimal solution. In most situations, a field technician can enhance video signal quality and reduce noise measurably by mechanically and electrically securing the cable plant.

This presents a unique problem for the industry: noise measurements, to great extent, are based on field measurements. Hence, a cost effective solution in one region may unduly penalize other solutions in a less-current region.

In general, network noise problems come from three areas: the subscriber's home, drop plant, and rigid coaxial plant. Seventy percent of the problem comes from inside the home, 25% is generated from the drop portion of the network, and 5% is from rigid coaxial plant. Troubleshooting intermittent problems is costly and time consuming, and finding the problem does not always mean it can be fixed.

5.3.9.1   Noise Characteristics in the Upstream Direction

In the upstream direction, there are several noise sources that can impair upstream communications. A channel model was developed by the industry identifying these sources:

- **Hum Modulation** — Hum modulation is amplitude modulation due to coupling of 60 Hz AC power through power supply equipment onto the envelope of the signal.
- **Microreflections** — Microreflections occur at discontinuities in the transmission medium which cause part of the signal energy to be reflected.
- **Ingress Noise** — Ingress noise is the unwanted narrowband noise component that is the result of external, narrowband RF signals entering or leaking into the cable distribution system. The weak point of entry is usually drops and faulty connectors, loose connections, broken shielding, poor equipment grounding, or poorly shielded RF oscillators in the subscriber's household. Since the upstream transmission is at the lowest frequency of the network's passband, the noise summates at the trunk. Ingress noise contribution includes most, if not all, FCC-conforming RF power levels, such as hair dryers, power line interference, electric neon signs interference, electric motors, vehicle ignitions, garbage disposals, washers, passing nearby airplanes, high voltage line, power system atmospheric noise, bad electrical contact, and any open-air RF transmission such as CB and HAM radio transmission, leaky TV sets, RF computers, civil defense, aircraft guidance broadcasts, international shortwave, and AM broadcasters.

- **Common-path Distortion** — Common mode rejection is due to nonlinearities in the passive devices of corroded connectors in the cable plant.
- **Thermal Noise** — White noise is generated by random thermal noise (electron motion in the cable and other network devices) of the 75-ohm terminating impedance.
- **Impulsive Noise/Burst Noise** — Burst noise is similar to the impulse noise, but with a longer duration. It is a major problem in the two-way cable systems and the most dominant peak source of noise (a short burst duration — less than 3 seconds). Impulse noise is mainly caused by 60 Hz high voltage lines and any electrical and large static discharges such as lightning strikes, AC motors starting, car ignition systems, televisions, radios, and home appliances such as washers. Loose connectors also contribute to impulse noise.

  There are two kinds of impulse noise: Corona noise and Gap noise. *Corona noise* is generated by the ionization of the air surrounding a high voltage line. Temperature and humidity play a major role in contribution of this event. *Gap noise* is generated when the insulation breaks down or via corroded connector contacts. Such failures pave the way to the entry of lines discharge of 100 Kv lines. This discharge or arc has a very short duration (in μ sec) with a sharp rise- and-fall time period. The sources are most likely to be automobile ignition and household appliances, such as electric motors.

- **Phase Noise and Frequency Offset** — Phase noise arises in frequency-stacking multiplexers, which occur in some return path systems.
- **Plant response** — The cable plant contains linear filtering elements that are dominated by the diplex filters that separate upstream frequencies from downstream frequencies.
- **Nonlinearities** — Nonlinearities include limiting effects in amplifiers, laser transmitters in the fiber node, and the laser receiver in the headend.

5.3.9.2 Noise Characteristics in the Downstream Direction

In the downstream direction, there are several noise sources that can impair downstream communications. The noise sources, described below, are additive.

- **Fiber cable** — The fiber affects the digital signal in two ways:
  1. Group delay is due to the high modulation frequency of the signal in the fiber.
  2. White Gaussian noise is added to the power.
- **Plant Response** — Impulse response is defined as *tilt* and *ripple*. The tilt is a linear change in amplitude with frequency and is an approximation to the frequency response of the components in the network. The ripple is a sum of a number of sinusoidal varying amplitude changes riding on top of the tilt and is a measure of the effect of microreflections in the network.

- **AM/FM Hum Modulation** — AM/FM hum modulation is amplitude/frequency modulation caused by coupling of 110 Hz AC power through power supply equipment onto the envelope of the signal or shift both up and down in frequency.
- **Thermal Noise and Intermod** — Thermal noise is modeled as white Gaussian noise with power defined relative to the power at the output of the plant response. Intermod is caused by nonlinearities in the system generating harmonics of other channels.
- **Burst Noise** — Burst noise is due to laser clipping which occurs when the sum total of all the downstream channels exceeds the signal capacity of the laser.
- **Channel Surfing** — Channel surfing causes microreflections to appear and disappear. Because the significant sources of channel surfing are close to the receiver, a large but slowly changing ripple in the frequency domain will appear and disappear.

5.3.10 Approaches to Suppress Noise

There are many approaches to suppress or avoid ingress in HFC networks. Since these approaches are not mutually exclusive, they could be combined to improve the performance of the network. The guidelines shown below should be followed on a plant-by-plant basis. Network performance is affected for both the upstream and downstream channels, although the upstream channel is more pronounced in the overall performance.

- Aligning the amplifiers properly in the reverse direction.
- An important aspect of cable plant installation or modernization is to ensure the system is both mechanically and electrically sealed. For it to be otherwise will invariably cause a significant contribution to ingress and impulse noise within the system.
- All powered devices and the cable plant must be electrically grounded appropriately. This may prove difficult in arid and/or rocky climates due to the inability to establish a good electrical ground.
- Almost 70% of the source of ingress noise is generated at the subscriber drops. Low quality coax are used for the subscriber drops, and radial cracks and cracks in the shield's foil are the main source of leaks, hence ingress noise. The do-it-yourselfers are also contributing to system leakage when installing their in-home wiring, using older, bad, or loose connections. One effective approach to improving network performance upgrading adequate coaxial residential wiring and adding good connector and good grounding practices. This, however, may prove costly.
- Reducing the channel bandwidth adds robustness to the system because it reduces the group delay distortion and enables the use of higher order modulation schemes. This approach may not be economically feasible in some regions.

- A frequency agile cable modem (in a multitone carrier) is one method used to reduce (skip) noise impairments. It selects only the carrier frequency in the return path where noise is not present, that portion of the noisy return path spectrum will be marked as not usable. Fine frequency agile systems function well by avoiding noise; however, industry experts are divided and argue that, if such systems are fully adopted and deployed, they may become a liability when interactive services demand increasingly more bandwidth from the upstream resources. The argument maintains that frequency agility at the subcarriers is useful if the noise sources were a narrowband and not a broadband noise component. Ingress noise is the only type of noise that meets these criteria. Frequency agility is not an effective strategy for dealing with impulse or amplifier noise because the noise is broadband. This issue is described further in Section 5.2.5.2

5.4 CABLE MODEM

The previous section provided an overall picture of the cable environment and outlined the groundrules for developing and preparing requirements for the cable modem. This section describes the details of the cable modem.

Modern technologies, such as ATM, promise high speed integrated services for heterogeneous applications. With a broadband as the service kernel, network convergence will accelerate and, theoretically, one would no longer be able to distinguish one network from another. The Internet, a late market entry into the equation of the telephone and cable modernization plan, has a central but somewhat confusing role.

The current state of digital technology is such that we cannot fully predict how systems will be built. The quality of service needed, cost, and especially the uncertainty of the market demand for future applications and services are not fully predictable. With all such uncertainties, IEEE 802.14's approach was to develop a standard for a cable modem based on ATM and took advantage of our present understanding of quality of service (QoS) as it applies to ATM.

Multimedia Cable Network System (MCNS) and the Society of Cable Telecommunications Engineers (SCTE) developed a cable modem that is IP-centric, known as DOCSIS™ (Data Over Cable Service Interface Specifications). DOCSIS™ based cable modem bypasses the ATM layer (at least in its initial specification) hence it is IP-centric. In that context, two types of cable modems are under development by the industry:

1. Cable Modem that is ATM-centric
2. Cable Modem that is IP-centric

5.4.1 STANDARDS PERSPECTIVE

In 1994, the IEEE Project 802 executive committee approved the formation of IEEE 802.14 to develop a cable modem standard that is ATM friendly. That cable modem specification is expected to be released in the middle of 1999. IEEE 802.14 is presently working on a high speed physical layer (HI_ PHY) for the upstream channel (see Section 5.2.5.2).

Multimedia Cable Network System (MCNS) in conjunction with Arthur D. Little, Inc., developed a series of interface specifications for early design, development and deployment of data-over-cable systems. DOCSIS™ was one of the results of this group. The goal of the DOCSIS™ project is to rapidly develop an IP version of a cable modem and a set of communications- and operations-support interface specifications for cable modems and associated equipment over HFC.

The Society of Cable Telecommunications Engineers (SCTE) worked closely with MCNS and standardized the DOCSIS™ version of the cable modem. It is also involved in the development of standards for digital video signal delivery through coordination of efforts with NCTA, the FCC, and others.

5.4.1.1 ATM-based vs. IP-based Cable Modems

Architecturally and philosophically, ATM-centric and IP-centric cable modems use similar technology. The physical layer of both cable modems is identical and based on ITU J.83 (with some exceptions for North America). The fundamental difference between DOCSIS™ and the IEEE 802 cable modem is in the development of the MAC layer and the layers above. Unlike DOCSIS™ the ATM-based cable modem MAC layer contains the segmentation and reassembles (SAR) necessary for ATM end-to-end operation. For IP services, IP over ATM uses AAL5's SAR to contain the IP packet. DOCSIS™ on the other hand, uses variable packet (IP-packet) as the transport mechanism per ISO8802-3.

Other differences between the ATM and IP-based cable modems lie in the functionalities of the upper layer interface services as well as security, maintenance, and management messages (e.g., registration and initialization).

Such differences may be considered major, but some vendors are considering the idea of providing glues to accommodate both ATM-based and IP-based cable hardware in silicon. Although this is possible, especially because the physical layers are similar, it is unlikely to have a cable modem with dual personality if only because of the stiff competition that will be ruling the market.

The differences between the IP- and ATM-based cable modems will be noted when warranted, as we describe the various functional components.

5.4.2 ABSTRACT CABLE MODEM OPERATION

The technical challenge of developing a cable modem that will deliver integrated services over cable is significant. Sharing access among multiple users creates security and privacy problems. One user connected to a cable network can possibly receive transmissions intended for another or maliciously make transmissions pretending to be another user. As such, all components of the cable modem are providing the hooks for the management and security for the system. The basic generic description of the cable modem below hides all that complexity but provides an initial operational understanding of a cable modem.

The primary function of the cable modem is to transport high speed digital data from the cable network to the users, and from users to the cable network. Figure 5.4 depicts the physical landscape and interface negotiation between a headend and the cable modem. Cable Modem # x denotes the number of cable modems that are attached to the subnetwork. Depending on the traffic, 500 to 2000 typical homes might be served in this topology. At the channel level, the cable modem in the downstream direction must tune its receiver between 450 to 750 MHz, at the 6 MHz band, to receive data digital signals. In this example m downstream channels are available in the cable subsystem. The QAM modulation scheme was selected by the industry as the modulation technique for the downstream direction.

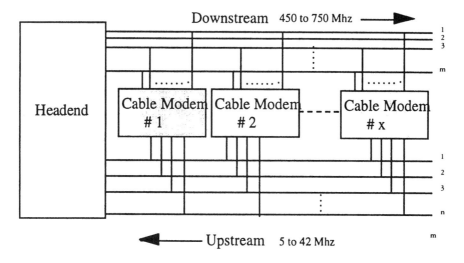

Figure 5.4 Cable Modem/ Headend Physical Topology

In the upstream direction, the cable modem performs the transmitter function, transferring information to the headend using the 6 MHz band between 5 and 42 MHz. In this example, n downstream channels are used to transmit upstream data. The data is modulated and placed in the 6 MHz channel using the burst QPSK modulation technique. At this frequency range, the environment is very noisy because of interference from CB and HAM radios and impulse and ingress noise from home appliances, as described in Section 5.3.10, which is why QPSK was selected as the modulation scheme. QPSK is more robust in terms of its immunity to noise, but at the cost of delivering data at a much lower rate.

Basic operation — A cable modem must be able to tune into any one of the downstream 6 MHz 1 to m bands to receive data from the headend. At the transmitting end, the cable modem must also be able to transmit at any of the downstream channels from 1 to n. A dialog between the cable modem and headend is triggered when a station requests registration to join the cable network (to be attached). There are certain associations between the downstream channels m and upstream channels *n*. Depending on the cable network span and traffic engineering, downstream channel 1 may be associated with one or more upstream channels.

During registration to join the network, a cable modem automatically starts listening to the downstream channels seeking entry to register its device. When the cable modem receives a strong signal, it reads the subsystem frequency allocation layout and sends a request to register to the headend using one of the assigned upstream channels. Once acknowledged by the headend, ranging, authentication, and initialization begins in order to legitimize the cable modem and provide the subscribed services. Plug-and-play is the philosophy used to get the modem operational.

Cable modem speed — Depending on the cable noise environment, typical bandwidths of 36 to 43 Mbps can be delivered to the cable modem in the downstream direction with the 64-QAM modulation technique. In the upstream direction, QPSK modulation can deliver up to 1.5 Mbps speed. In a shared-medium environment such bandwidth is shared by many users who will be competing for access to the same upstream channel or channels. The most critical path in the upstream direction (many-to-one) is the modem MAC. MAC arbitrates access in this shared-medium bus and allocates the requested bandwidth. The bandwidth allocated to a user in the upstream direction will invariably depend on the number of users sharing the bus. It will also depend, to a large extent, on the characteristics of the traffic being used by others who are sharing the bus. If congestion is encountered, a smart cable modem can retune its receiver and hop to a different upstream channel, when instructed to do so by the headend. This mechanism is referred to as a *frequency agile capable* cable modem.

5.4.3 CABLE MODEM LAYER ARCHITECTURE

The reference architecture is the building block and the blue print that is needed to construct any device. The reference layer architecture for a cable modem is as shown in Figure 5.5 and contains

- Physical (PHY) layer
- Mac Layer
- upper layers

The layers are described below, followed by details necessary for having an operational understanding of a functional cable modem.

5.4.3.1 Physical Layer

The physical interface for digital cable systems is the ordinary coax cable. This physical layer contains both the upstream and downstream channels. Upstream transmission is more difficult than downstream transmission because of the shared-medium access collision and the multitude of noise that pollutes that spectrum (below 42 MHz.) To some extent this noise problem can be compensated for by using complex encoding technologies at the cost of reducing the data rate. Hence,

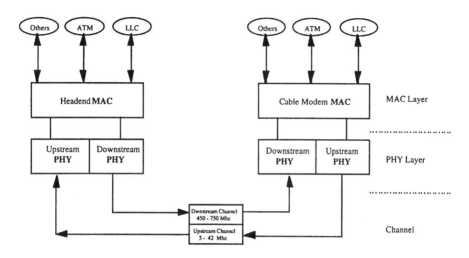

Figure 5.5 Cable Modem / headend Layered Architecture.

a digital cable architecture may use only one downstream transmitter at the header and has several associated upstream transmitters. Transmitters are more expensive than receivers.

5.4.3.2 MAC Layer

The MAC layer is the most challenging to ascertain and has been and continues to be the topic of discussion in the higher educational community. The complexity arose because of the shared medium coupled with the requirement to maintain QoS for each connection of the user application.

In the upstream direction, the communication path is shared by many active users, all transmitting data to the headend. A fundamental function of any MAC is to devise a mechanism that performs random access to the network, resolves contention, and arbitrates resources when more than one station wishes to transmit at the same time. The MAC is further burdened by the requirement to preserve the QoS of any and all specific applications. If real-time video or voice is being transmitted, the jitter must be minimized and a bandwidth of constant bit rate (CBR) must be allocated. Unlike data packets, if delayed only slightly a real-time voice packet becomes useless.

The MAC for the cable modem offers much more challenging opportunities because it must operate under a much more hostile environment than any previous MAC developed so far: a public environment where QoS and user expectations are of paramount importance. The cable modem MAC must deal with interactive and multimedia services with bandwidth-hungry applications requiring a multitude of service requirements.

Several cable modems are on the market today, and most if not all deal with service-specific applications. Market pressure also yielded MAC specifications that deal exclusively with service specific applications, particularly for the Internet. However, just speeding the access interface, such as for a cable modem, does not solve the multimedia service requirement. The concept of QoS must be embedded in the development of a MAC protocol so it can operate to the satisfaction of the end users. Market dynamics are very unpredictable and confuse short term planners.

5.4.3.3   Upper Layers

The cable modem should be designed to handle all management entities and service interfaces, be they IP, native ATM, or others.

IP interface — The cable modem can connect directly to a PC handling IP traffic. The physical layer, to the end-user, will most likely remain Ethernet 10BaseT, the predominant method. Although it probably would be less expensive to produce the cable modem as an internal card for the computer, doing such would require different modem cards for different computers. Moreover, it would further confuse the demarcation between cable network and the subscriber's computer.

ATM-native interface — The cable modem being standardized by IEEE 802.14 is designed to handle native ATM services. That means, ATM adaptation layers will be developed to handle ATM applications, including CBR, VBR, and ABR services, among others.

5.4.4   CABLE MODEM FUNDAMENTAL LAYERS

The critical components of any cable modem, as shown in Figure 5.6, can best be described mainly using the two fundamental layers: the physical layer and MAC layer. Both of these layers will be imbedded mostly in the hardware. Software in the cable modem will complement all other layers to give it its service personality. It is worth noting that the MAC/ATM convergence sublayer (SAR) will *not* be present for the IP-based modems. A bottom-up approach is used to describe details of the two layers. These layers and sublayers define the dialog needed between the headend and cable modem.

5.4.4.1   Physical Layer

The physical layer, shown in Figure 5.6, contains two sublayers: the physical medium-dependent (PMD) sublayer and the transmission convergence (TC) sublayer. These sublayers take on the personality of the attached transmission link to perform the needed bit translation, synchronization, orientation, and modulation functions. Figure 5.7 shows the landscape of the sublayers and associated functionalities of each.

5.4.4.1.1   PMD Sublayers

The main function of the PMD sublayer is to modulate/demodulate the RF carriers on the analog cable network into digital bit streams, perform synchronization,

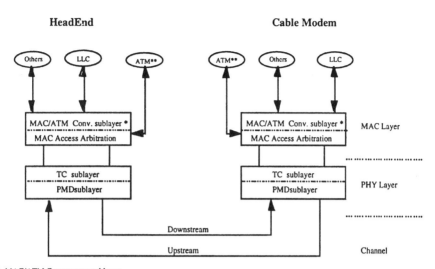

* MAC/ATM Convergence sublayer
  is present for ATM-centric modem only
** ATM native service for ATM-centric modem only

Figure 5.6 Fundamental layers of a cable modem

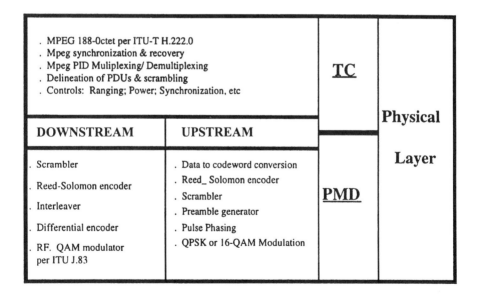

Figure 5.7 Sublayers of the physical layer (cable modem perspective)

coding, and error checks. The ITUJ.83 recommendation was adopted for the NA as the base. The PMD, as shown in Figure 5.7, is further subdivided into two sublayers:

- downstream PMD
- upstream PMD

Because of space limitations, a full discussion of the PMDs is not feasible. Readers are encouraged to read ITU J.83 for the full technical specifications and critical time requirements. However, a brief description of the downstream and upstream PDMs is provided below.

Downstream PMD — Downstream PMD modulates and demodulates the RF carrier using QAM modulation techniques, a means of coding digital information over radio. ITU-J.83 specifies three types of downstream interfaces: type A, B, and C. Type B of ITU J.83 is the downstream as shown for this example. The cable modem supports both 64 and 256-QAM. At 256-QAM the nominal symbol rate value is at 5.360537 Msym/sec (baud rate). At 64-QAM the nominal symbol rate value is at 5.056941 Msym/sec. Nominal channel spacing is 6 MHz and center frequency is specified at 91 to 857 MHz.

Upstream PMD — The upstream PMD sublayer supports two modulation formats: QPSK and 16-QAM. The modulation rates of the modulator provide QPSK at 160, 320, 640, 1,280, and 2,560 ksym/sec, and for 16-QAM at 160, 320, 640, 1,280, and 2,560 ksym/sec. The upstream PMD supports a frequency range of 5 to 42 at 6 MHz of the subsplit. The range for European cable plants is, 5 to 65 MHz, and the Japanese range is 5 to 55 MHz.

Other functions performed in the upstream and downstream PMD layers are designed to transform FDM to TDMA. It will also improve the efficiency and robustness of the transmission by mitigating the effect of burst noise, using encoders to make phase rotation insensitive to QAM constellation, randomizing the transmitted data payload, correcting symbol errors within the information block (codeword), synchronizing, and establishing a TDMA landscape. For the upstream, it performs pulse shaping and a variable-length modulated burst with precise timing beginning at boundaries spaced at integer multiples of 6.25 μsec apart.

5.4.4.1.2 Upstream Frame Structure

Immediately after the PMD sublayer's bit stream is processed, the frames begin to emerge and the information looks more comprehensive. The TC (Transmission Convergence) sublayer refines the information further, especially for the downstream, and processes it into formats readying it for further data processing. The frame format of the upstream is described below.

Characteristics of the Upstream frame — The headend generates a time reference identifying slot in the time-domain. The slot containers, in the upstream, enable a cable modem to transmit information to the headend. Based on that time reference measured in 6.25 μsec ticks, a minislot is then created whose size is of this time tick. The duration (in time) of one minislot will equal the time required to transmit 6 octets (programmable) of data plus the time required to transmit the physical layer overhead and the guard time.

A MAC layer Protocol Data Unit (PDU) occupies a single minislot known as a MiniPDU. The upstream landscape can be thought of as a stream of minislots. Minislots are identified and labeled using a free running counter assigned by the headend and incremented by the master clock tick. The headend determines the usage of each minislot in each of the upstream channels. This information is conveyed to the cable modems by broadcasting their usage using the downstream channel in a form of a map or image reflection.

Several minislots can be concatenated to form a packet, as shown in Figure 5.8. For an ATM-based cable modem, consecutive minislots are used to form and transport an ATM cell to the headend. For an IP-based modem, minislots are allocated consecutively to form a variable-length packet for transmission.

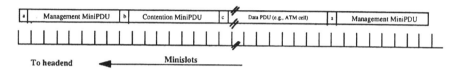

Figure 5.8 Minislots landscape for the upstream

There are various personalities defining each of the MiniPDUs. An information element in the MiniPDU (shown in Figure 5.8 with header a, b, or c) defines the various functions allocated. A MiniPDU can be used as management messages for ranging or RF power adjustment (between the headend and cable modem), the Mini PDU may be requested by the cable modem to vie for access to the shared medium, or the MiniPDU may be a portion of a payload of a data PDU.

The number of minislots required to carry an ATM cell depends on the length of PHY overhead and guard time required by the upstream PHY (per the MiniPDU burst profile). For the ATM-based cable modem, an integral number of minislots are allocated by the headend to transmit an ATM cell.

5.4.4.1.3   TC (Transmission Convergence) Sublayer
For the downstream, the TC sublayer refines the data further, and the bits are assembled and massaged to fit into a frame.

5.4.4.1.4   Downstream Frame Structure
For the downstream frame structure, the industry adopted ITU-T H.222.0. It is defined as the MPEG-2 (MPEG) packet format with 4-byte header followed by 184 bytes of payload (totaling 188 bytes). The header (PID field) identifies the payload as belonging to either DOCSIS™ or IEEE 802.14-based MAC. Figure 5.9 illustrates the cable modem MAC frame, interleaved with other digital information payload(s). The interleaving rate must take into account the jitter that may influence service profiles. A constant rate of interleaving (1: n) is suggested (i.e., one cable modem Mac payload for every *n* digital video payload).

The digital video information payload in the downstream, MPEG-based, was not an accident. It was designed so it can be provisioned for use when the cable

4 Bytes	184 Bytes
Header PID=0X1FFD= ATM-based cable modem PID=0X1FFE= IP-based cable modem	Cable Modem MAC payload
Header PID= Digital Video	Digital Video payload
Header PID= Digital Video	Digital Video payload
Header PID= Digital Video	Digital Video payload
Header PID=0X1FFD= ATM-based cable modem PID=0X1FFE= IP-based cable modem	Cable Modem MAC payload

Figure 5.9 Frame format for the upstream

network evolves into a digital format with common receiving hardware accommodating both video and data. This provides an opportunity for possible evolution to digital as described in Section 5.1.3.5.

The first order of business of the TC is to establish synchronization and identify the MPEG frame boundaries which is accomplished by the TC hardware (of the cable modem). When entering the hunt state, the hardware in the TC shifts, calculates, and seeks the correct CRC of the MPEG payload. Five consecutive correct parity checksums of the 188 bytes declares the MPEG packet *in frame*. *Out of frame* is declared when 9 consecutive incorrect parity checksums are received.

Once the MPEG frame boundary is established the TC extracts the cable modem packet data from the MPEG payload. The format is shown in Figure 5.10.

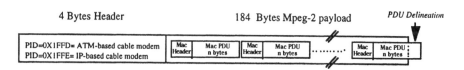

Figure 5.10 Downstream frame embedded in the MPEG frame

Beyond this point, the similarities between the ATM-based cable modem and IP-based cable modem end. Although philosophically and fundamentally they are similar, the data interpretation and manipulation differ in a number of ways. As Figure 5.10 shows, for an ATM-based cable modem each PDU within the MPEG frame is fixed in length and, in fact, is an ATM cell.

For the IP-based cable modem, the PDU payload is variable in length and conforms to ISO8802-3 type PDU.

PDU delineation (where a PDU traverses from one MPEG frame to the next) is supported on both cable modems. For the ATM-based cable modem, the PDU boundaries (ATM cell) are marked using the procedure described in ITU-T Recommendation I.432. In summary, HEC (Header Error Control) checksum will be responsible for identifying cell boundaries. Once in the hunt state, the reception of seven consecutive and correct HEC (on the five ATM header bytes), marks the boundary of the cells that are in a frame, in which case it declares the PDU *in frame*. *Out of frame* is declared when five consecutive HEC errors occur, in which case the TC will then go into hunt state, again, to establish the cell boundary.

MAC PDU boundary for the IP-based cable modem uses pointer_field. The pointer_field is the first byte of the MPEG payload (not shown in Figure 5.10) and may be present to point to the start of the next MAC PDU. The header indicates if the pointer_field must be used. With this approach MAC PDU may begin anywhere within an MPEG packet, a MAC PDU may span MPEG packets, and several MAC frames may exist within an MPEG packet.

Low level initialization — Once the PDU framing and formats are established the TC begins performing initial low level tasks such as synchronization, ranging, and power adjustments. To accomplish this, two fundamental pieces of information are needed by each cable modem: global timing reference to all modems and timing offset. Similarly, the attenuation from any cable modem is most likely different from another and from the headend. Therefore, each must properly adjust its power level for its transmitter such that all stations' transmissions reach the headend at approximately the same received signal level.

Synchronization — Once the cable modem successfully assembles the frames (by the TC as described above), it then must synchronize its clock with the headend clock. This is performed when the headend (periodically) sends a management message containing a global timing reference. The management message contains the timestamp identifying when the headend transmitted this reference clock. The cable modem compares it with its own time and adjusts its local time accordingly. The cable modem periodically adjusts its local clock.

Ranging and power adjustment — Once synchronization is established, the cable modem must then acquire the correct timing offset such that the cable modem's transmissions are aligned to the correct minislot boundary. In other words, it adjusts the cable modem timing offset such that it appears to be located right next to the headend (without delay).

First the cable modem must learn the map of the available upstream channels so it can send an initial management message to the headend to perform the ranging. The initial maintenance region slot demarcation (subsequent to ranging) is large enough to account for the variation in delays between any two cable modems. When the initial maintenance transmits opportunity occurs, the cable modem sends the ranging request message.

The headend responds with a ranging response message addressed to that particular cable modem. The response message contains the needed information on RF power level adjustment and offsets frequency as well as any timing offset corrections.

A dialog is then established again to fine-tune and correct both the power and timing offset of the cable modem.

5.4.5   HIGH SPEED PHYSICAL LAYER

IEEE 802.14 organization is working with SCTE/MCNS on a new HI_PHY for the upstream. Experts from both organizations are studying several proposals to specify a new high speed upstream physical layer that will accommodate, without changes to existing upper layers, DOCSIS™ cable modem, as well as the IEEE 802.14 cable modem. The HI_PHY specification is expected to be released in 1999.

Discussion revolves around whether a frequency band (e.g., the 6 MHz) should be divided further into a discrete set of frequencies and carry the data in the upstream direction. It is worth noting that 6 MHz is an arbitrary value the industry used; one can use any value.

Frequency agility, described in section 5.1.3.8, is in this context a coarse agile cable modem. For example, the cable modem, upon request from the headend, hops to a different upstream channel to a 6 MHz increment. The advance (fine) frequency agile cable modem can tune its carrier frequency to multiple subfrequencies with very low resolution. The intention is to evade and selectively ignore that portion of the band in which the noise becomes excessive or intolerable.

This DMT (Discrete Multi-tone) technology was adapted for ADSL but not necessarily because of the noise environment.

Other proposals argue for the use of a single carrier approach (e.g., 6 MHz) with a high density modulation scheme and a front-end filter. This scheme abates the ingress noise while deploying a higher coding efficiency of QAM that can be dynamically changed with the changing noise environment.

Preliminary analysis suggests that the varying solutions are not necessarily superior to the present single carrier approach (assuming higher order QAM) as described in Section 5.4.5.1.1.

Each of the approaches has benefits in terms of noise assumptions, but behave less so in the different noise characteristics, as described in Section 5.3.10.

The idea for the next generation cable modem, of using a TDMA system (single carrier) with adaptive and dynamic changing constellation size, is attractive. The headend continuously monitors the channel noise and adjusts the constellation size accordingly.

It is too early, however, to discount other solutions yet. Experts are studying the noise dynamics of the cable plant more closely so they can select the best suitable solution. Cost and complexity will no doubt play a role in the eventual outcome.

5.4.6   OVERVIEW OF MAC

One of the main functions of a MAC is a collection of upstream and downstream channels for which a single MAC allocation and management protocol operates. Its working environment includes the headend and all other connected modems.

The headend services all of the upstream and downstream channels.

One can think of a MAC protocol as a collection of components, each performing a certain number of functions. A cable modem MAC protocol can be broken into the following sets of critical components: acquisition process, message format, support for higher layer traffic classes, bandwidth request, bandwidth allocation, and contention resolution mechanism. The message format element of the MAC defines the upstream and downstream message timing and describes their contents.

The MAC layer in the cable modem may contain sublayers. For an ATM-based modem, the MAC sublayer contains the SAR which is used to assemble or disassemble ATM cells from the non-ATM service application (e.g., IP over ATM). Hence, it provides the interface for the upper service layers, be it IP-based or native ATM. This was illustrated in Figure 5.6.

The main feature of the MAC, however, is its ability to support the transfer of packets while maintaining the ability to provide QoS. The upstream channel is a precious resource, so collision and data flow must be managed very efficiently. The upstream channel is divided in time into basic units of minislots, of which there are several types. Their function is defined by the headend and conveyed to each cable modem by means of downstream control messages. Several minislots can be concatenated in order to form a single data PDU, or ATM cell. There is no fixed frame structure, and there is a variable number of minislots in any given time. Thus the upstream channel is viewed as a stream of minislots. The MAC layer also contains the controls and rules governing information processing and flow control. Management messages are defined to handle various tasks, primarily the interaction between the cable modem and headend for modem initialization, authentication, configuration, and authorization.

To the extent possible, plug-and-play is the philosophy adopted to perform these tasks.

Some critical examples of management routines and dialog between the headend and cable modem are provided below.

5.4.6.1   Initializations at the Upper Layers

Channel Acquisition — Channel acquisition is the process already described above. Once a cable modem accomplishes its synchronization and framing, and established communication with the headend it has completed channel acquisition. Depending on traffic distribution, the headend may request the cable modem to change its channel(s). The headend acts as the traffic cop and could instruct the cable modem to change either the upstream or downstream channel(s). The cable modem must respond and re-initialize at the PMD and TC layers. Once accomplished, channel acquisition is completed.

Registration — During registration several messages are exchanged between the cable modem MAC and headend to legitimize entry of the cable modem in the network so it can be declared operational.

If channel acquisition, ranging, and power leveling were performed, a cable modem must first register with the headend. This starts the MAC registration process. The cable modem is assigned a temporary service ID that has only local significance. This ID will be associated with the cable modem IEEE 802 - 48-bit MAC address which is assigned during the cable modem manufacturing process. It is used to identify the modem to the various provisioning security servers during registration.

5.4.6.2 Security and Privacy in the HFC Network

The security and privacy problems for HFC are different from the traditional point-to-point wireline networks. In the telephone environment, the copper wires are dedicated to the user and connected directly to a line card at the central office. Eavesdropping on a telephone line cannot be done as easily as in a shared-medium line. It certainly cannot be monitored by users in other homes. Registering a device illicitly (service theft) on a dedicated line is nearly impossible. The operator knows the identity of that line because it terminates physically at the site. In a cable network environment, the security problem is more difficult because many stations have physical access to the same wire.

Understandably some in the cable industry question the need for privacy and the added complexities to provide it. The cable telephony, for example, could fall under the same category as cellular, cordless phone, or PCS. Eavesdropping on these systems can be done just as easily, if not more easily, as in the cable modem. It can then be argued that if a cable phone is to be used, privacy and its associated complexity may no longer be considered necessary. After all, not long ago, even copper wires were shared among several telephone users. Distinctive ringing was used to identify the called party.

Security requirement at the MAC — Provisions were made on the cable modem to specify the access security mechanisms so as to make the security of shared-media access networks comparable to that of nonshared-media access networks. The process is basically to exchange a secret key during registration in which a cable modem sends it unique ID (IEEE 802 - 48-bit MAC address) to the headend, then proceeds with a secret key exchange to register. The certification ID is simply used as authenticity (e.g., initial password) prior to the secret key exchange that will follow. If the headend is not provisioned to accept this ID then registration fails. A hacker using a legitimate ID (an illicitly obtained secret key) would be able to register as long as the legitimate user had not registered first. During this registration process, the ID information is transmitted in the clear so a hacker would be able to listen to a successful registration transaction and record the ID information.

Secret key exchange — The secret key exchange uses the Diffie-Hellman exchange during the registration process to establish a common secret key. The authentication procedure incorporating the secret key is used to verify the identity of the station to the headend. A hacker who obtains a legitimate ID number of the device would also need to obtain the correct secret key. IEEE 802.14 adopted the Diffie-Hellman key exchange procedure.

Maintaining station keys — A cable modem is usually equipped with more than one separate encryption/decryption secret key. They are exchanged during registration by means of *cookies* which are exchanged between the cable modem and headend during registration/authentication routine when entering the network or at any time the network operator deems necessary. A 512-bit ephemeral Diffie-Hellman is used for main key exchange which produces a cookie.

This process, however, does not differentiate between a newly subscribed user with which it has not established a cookie yet. A hacker may very well be able to establish a network connection using a clone MAC address (or an illicitly obtained one) during registration. This, however, is a futile exercise for the hacker because the legitimate user will be denied access to the network when attempting to register. This denial of service to the legitimate user will prompt the operator to perform authentication by other means such as personal intervention, thus revealing the attacker and remedying the problem.

5.4.6.3 Fundamentals of Collision Resolution

MAC operation, in terms of flow control, describes the entry mechanism and steady state operation of the station's behavior. In the shared-medium environment communication is many-to-one. Subscribers compete on the shared-medium bus to get the attention of the headend so it can be granted permission to start sending the data. MAC controls the behavior of users who want to access the network as well as honor the service contract promised by the application and network. Hence, MAC arbitrates the communicating path and resolves any collision that occurs. The headend acts as a central office in that it controls and mediates all communications between, and/or from, all connected cable modems.

There are several MACs developed and specified in the public and private networks. The two contention/resolution mechanisms are the time division multiplexing access and collision resolution like contention and collision resolution mechanism (CDMA).

Time Division Multiplexing (TDMA) — In the TDMA approach, each connected device is allocated a timeslot in a specified timeframe. The frame contains a fixed number of time slots and each will be dedicated exclusively to one of the connected devices. When a device has data to send it uses its dedicated timeslot to send the information at its leisure. The obvious advantages of this mechanism are

- no collision is experienced in the shared medium so no contention resolution is needed,
- it is the ideal solution for constant bit rate traffic like voice or video telephony, and
- it gives fair access to all connected subscribers.

The disadvantages of TDMA are also obvious:

- An idle user's timeslot is unduly wasting network resources. In most multimedia applications the traffic is bursty and unpredictable, whether the subscriber is on the Internet, sending e-mail, or on the WWW clicking the icons. In this case the allocated timeslot is used occasionally. Providing full time access to stations in such a premium and limited upstream channel is a waste and may become prohibitively expensive.

Contention and collision resolution mechanism: In the contention and collision approach, the mechanism assumes that devices, when they need to, must vie for accessing the bus so it can send data. MAC responsibility is to arbitrate the access, resolve contention, and control the traffic flow. Consequently, the shared bandwidth is used only when needed, efficiently using the upstream resources. Collision is certain to occur, especially in a folded architecture. The device retries until it gets access. This mechanism is well known and serves the data transaction very well (it is not delay sensitive). QoS, fairness, and effective use of resources are not fully addressed in this mechanism.

5.4.6.4 Cable Modem MAC-Bandwidth Allocation

The above discussion suggests that a hybrid of both the TDMA and reservation/contention mechanism will best serve a multimedia application. Both the IP-based cable modem and ATM-based cable modem adopted this technique, but they differ in the collision resolution algorithm solution.

The headend algorithm is responsible for computing bandwidth allocation and granting requests. The flavor on the number of request slots/grant combination is vendor-implementation specific.

The MAC protocol description concentrates on the following issues:

- Upstream bandwidth-control formats define the request minislot types and structure.
- Upstream PDU format describes the protocol data unit format for ATM PDUs segmented into variable-length fragments for an efficient transport of LLC traffic types.
- Downstream format specifies the downstream data flow that can be seen as a stream of allocation units, seach 6-bytes long. As mentioned earlier, ATM cells can be sent by concatenating several basic minicells together. Each ATM cell can carry a number of information elements such as bandwidth information elements: grant information, allocation information (request minislot allocation), and feedback information (request minislot contention feedback).

5.4.6.5 Request for Upstream Bandwidth

The headend is responsible for allocating transmission resources in the upstream channel to cable modems that are queued for contention-based reservations. Each cable modem vies for access to obtain its share of transmission resources. One or more logical queues may contend for access from a single cable modem to serve the need of multimedia applications (several connections within a session with different QoS).

Each cable modem has various means of requesting bandwidth from the headend. Initially, the modem requests bandwidth through contention-based transmissions on the upstream channel. Once the headend grants the request, additional bandwidth may be requested by the cable modem by setting a bit (in the appropriate field) in the data PDU in transition. This method is known as *piggybacking*. It is most useful when contention access delay is high. A CBR permanent allocation can be requested by a cable modem for a logical queue, such that periodic grants at a desired frequency are allocated until the cable modem sends a CBR release message.

Request Minislots (RMSs) are allocated in the upstream channel by the headend to the cable modems for contention access. An RMS grant message identifies a number of RMSs divided into groups for different distinct sets of MAC users which are at various stages of contention resolution.

Typically, the headend allocates more than one RMS to each cable modem, and an RMS may be allocated to multiple MAC users. To reserve transmission resources in the upstream channel, a MAC user randomly selects an RMS from the group of RMSs available to it then attempts to send a request message in the selected RMS. The request message, referred to as a Request MiniPDU (RPDU) identifies the MAC user and the size of its requested allocation. Since the MAC user does not necessarily have exclusive access to the RMS, collisions can occur in which case a contention resolution algorithm is invoked to resolve it. The contention resolution algorithm for the ATM-based cable modem is based on a tree splitting algorithm and is specified with a flexible framework that permits a number of variational implementations. For IP-based cable modems, the contention resolution is based on a binary exponential backoff algorithm.

If the MAC needs to provide support for ATM, it also needs to differentiate between different classes of traffic supported by ATM, such as CBR, VBR and ABR. Bandwidth allocation represents an essential part of the MAC and controls the granting of requests at the headend. Finally, the contention resolution mechanism, which is the most important aspect of the MAC, consists of a backoff phase and retransmission phase.

5.4.6.6 Contention Resolution

A collision resolution algorithm needs to be implemented because request packets and possibly data packets are transmitted in a contention fashion. A wide variety of algorithms can be used. Because cable modems cannot monitor collisions, feedback information about contended requests is provided by the headend. The algorithm must also take fairness into account, such as compensating for the delay it receives through the feedback information. (See ranging procedure.)

Two mechanisms for contention resolution are described below:

1. Tree resolution and priority mechanism. Adopted by IEEE 802.14 for the ATM-centric cable modem
2. Binary exponential backoff. Adopted by MCNS/SCTE for the IP-based cable modem

Tree-based Contention Resolution Algorithm — The *n*-ary tree-based contention resolution algorithm was adopted for the ATM-centric modems. The principle behind it can best be described in terms of collision management, splitting the colliding entities into a smaller and more manageable subset, and building a hierarchy stack to throttle and control collision, classify the colliding entities, and compensate for the delay due to the HFC topology.

When a MAC entity (in cable modem) has data to send, it first must send the appropriate message requesting upstream bandwidth so it can be reserved by the headend. Since more than one entity is competing on this shared medium, collision will inevitably occur. A tree-based contention resolution algorithm is used to resolve this collision. The contention algorithm operates as follows: all the colliding entities will split into n subsets, and each of them randomly selects a number between 1 and n. This begins to form a hierarchy stack that will be managed by the headend in order to resolve the contention in an orderly and fair manner. The idea is to allow different subsets to retransmit first, while the subsets from 2 to n wait for their turn. In building this hierarchy, the QoS profile associated with the collided entity is classified further within that subset, forming yet another pecking order.

If there is another collision within a subset, then the first subset splits again, forming another subset, hence n-ary. The number of subgroups continues splitting further until the original collision is resolved. The subsets that are already waiting in the stack must have their positions shifted up in the stack accordingly so as to leave room for the new entities that collided. If no collision occurs, the entities with the lowest level in the hierarchy will get their opportunity to transmit, until all colliding entities are resolved.

The above algorithm works very well if one assumes that an entity received its collision feedback immediately. However this is not the case for HFC. Instead, the feedback is conveyed to the colliding entity via the next frame of the downstream frame. The tree-splitting algorithm is therefore modified further to accommodate this delay in feedback; that is, the level of the hierarchy stack is modified to accommodate the delay as well as the new collisions as the tree continues splitting.

There is a variety of mechanisms to limit entry of new packets to the system and thereby avoid congestion collapse. New requests (based on their admission priority) might be blocked during initial contention access to prevent excessive collisions. The admission control mechanism is based on pre-assigned priorities and is used to classify QoS in terms of contention access performance.

Or, new requests occupy different levels in the hierarchy based on their arrival time, or could send their requests on a first-come-first-served basis.

If the algorithm is blocking, the new requests are not allowed to use the slot reserved for collision resolution. Instead, they will be queued in the hierarchy of the stack and allowed to randomly select a slot among the remaining available slots.

IEEE 802.14 considered other allocation resolution algorithms such as the p-Persistence algorithm, an adaptation of a stabilized ALOHA protocol to frames with multiple contention slots. Newly active stations and stations resolving collisions have an equal probability of access p (p-persistence) to contention slot within a frame. p is determined by an estimate of the number of backlogged stations computed by the headend and sent to the station in the downstream frames.

After extensive simulation (performed by NIST and Georgia Institute of Technology), the tree-based blocking algorithm performance was found superior to the p-Persistence algorithm.

Another reason the contention resolution scheme was chosen for the ATM-centric modem is the probability density function of the access. Delay for a cable modem is computed, (the time a packet is generated until it is received at the headend). This measurement is considered especially important for issues related to Cell Delay Variation (CDV) in ATM environments.

Binary exponential back-off Contention Resolution — This contention resolution was adopted by MCNS/SCTE for the IP-based cable modem. The headend controls assignments on the upstream channel through feedback and determines which minislots are subject to collisions. This method of contention resolution is based on a truncated binary exponential back-off, with the headend controlling the initial and maximum back-off window.

When a cable modem enters the contention resolution process (due to collision), it sets its internal back-off window to the value conveyed that is in effect. The cable modem randomly selects a number within its back-off window. This random value indicates the number of contention transmit opportunities that the cable modem defers before transmitting.

After a contention transmission, if the request was not granted, the cable modem increases its back-off window by a factor of two, again randomly selects a number within its new back-off window, and repeats the deferring process. If the maximum number of retries is reached, then the PDU must be discarded.

5.4.7 CABLE MODEM OPERATION (SERVICE PERSPECTIVE)

We now provide a bird's eye view of what MSOs hope when they deploy cable modems.

5.4.7.1 Review of Cable Modem Operation

On power-up, the cable modem is in the unregistered mode. The PMD performs physical layer synchronization, followed by the TC, which performs synchronization as well as framing the information packet. The format for the upstream is MPEG, and a MiniPDU is created for it from timeslots controlled and programmed by the headend.

The cable modem then seeks and registers to join the network and exchanges security cookies and other parameters with the headend. Once ranged, a picture begins to emerge and operation commences. Mini PDUs can be concatenated to form an ATM cell for the ATM-centric modem, or IP packets for the IP-centric modem (as shown in Figure 5.11). Delineation of packets/ATM cells is performed with available HUNT state machine techniques.

When a cable modem needs to send PDUs and has no allocation pending, it requests an allocation by sending a request PDU in a MiniPDU. If the contention is successful, the headend will allocate the requested upstream in the grant information elements bandwidth and informs the cable modem, using the downstream. The cable modem may then send the PDUs in the allocated slots. If contention occurred, the cable modem invokes either tree resolution and priority mechanism (if ATM-based modem) or binary exponential back-off mechanism (if IP-based modem).

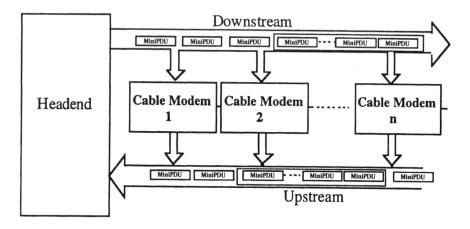

Figure 5.11 Cable modem operation

In the downstream direction, the cable modem receives a stream of ATM cells or PDU packets encapsulated over the MPEG packet. Management PDUs with an identifier address a particular cable modem that has only local significance. Each cable modem filters its incoming PDUs based on its identifier.

IP connectivity (to obtain an IP address) for the IP-based modem is invoked by the cable modem using the DHCP mechanism (RFC-1541).

5.4.7.2 Cable Modem Service Aspects

The MSOs have traditionally focused on quick returns to meet their short term business goal. MSOs attitude toward customer service is, admittedly, not good but they are becoming more receptive given the new dynamics of the market. The AT&T/TCI merger will set a new standard for the industry because AT&T has an excellent customer services reputation.

The key differentiated services cable operators are likely to provide are

- high-speed data services (Internet access)
- voice/video over IP
- broadcast, one-way entertainment
- telephony (wireline and wireless)
- digital NVOD
- work at home
- e-commerce/shopping

Convergence of these services across millions of U.S. households will change the way people live, work, and play. Byproducts of this convergence will be a major source of growth and opportunity for the U.S. economy.

The key differentiated market entries for MSOs are most likely to be

* time-to-market
* quality of service

Although time-to-market is important in the short term, it will not, however, be the ultimate deciding factor to win new customers, especially from the telephone companies. Time-to-market is important only if customer service and cost are satisfactorily met. Their competitors, the telephone operators, enjoy a very good customer service reputation, and that in itself will be a marketing tool they can use effectively to win cable customers. The telephone operators also have deep pockets to build a video infrastructure and to leverage the present infrastructure in billing, network management, and network reliability.

The battle for the MSOs to gain market share will not be easy, something of which the cable operators are very much aware.

GLOSSARY

Algorithm Well-defined rule or process for arriving at a solution to a problem. In networking, algorithms are commonly used to determine the best route for traffic from a particular source to a particular destination.

Amplifier A device that boosts the strength of an electronic signal. In a cable system, amplifiers are spaced at regular intervals throughout the system to keep signals picture-perfect regardless of the distance they must travel.

Asynchronous Transfer Mode (ATM) The transfer mode in which information is organized into cells; it is asynchronous in the sense that the recurrence of cells containing information from an individual user is not necessarily periodic.

ATM Cell A digital information block of fixed length (53 octets) identified by a label at the ATM layer.

Available Bit Rate (ABR) A service class that is an ATM layer service where the limiting ATM layer transfers characteristics provided by the network that may change subsequent to connection establishment.

Bandwidth A measurable characteristic defining the available resources of a device in a specific time period (typically in one second).

CATV (Community Antenna Television or Cable Television) A communication system that simultaneously distributes via a coaxial cable several different channels of broadcast programs and other information to customers.

Channel A communication path. Multiple channels can be multiplexed over a single cable in the cable television environment.

Coaxial cable Actual line of transmission for carrying television signals. Its principal conductor is either a pure copper or copper-coated wire, surrounded by insulation and then encased in aluminum.

Constant bit rate (CBR) A service class intended for real-time applications, i.e., those requiring tightly constrained delay and delay variation, as would be appropriate for voice and video applications. The consistent availability of a fixed quantity of bandwidth is considered appropriate for CBR service.

CRC Cyclic redundancy check. An error-checking technique in which the frame recipient calculates a remainder by dividing frame contents by a prime binary divisor and compares the calculated remainder to a value stored in the frame by the sending node.

Data link layer In Open System Interconnection (OSI) architecture, the layer that provides services to transfer data over the transmission link between open systems.

Delay The elapsed time between the instant when user information is submitted to the network and when it is received by the user at the other end.

Digital compression An engineering technique for converting a cable television signal to digital format in which it can easily be stored.

Downstream Flow of signals from the cable system headend through the distribution network to the customer.

End user A person, organization, or telecommunications system that accesses the network in order to communicate via the services it provides.

Feeder cable Coaxial cables that run along streets within the served area and connect between the individual taps which serve the customer drops.

Fiber node A point of interface between a fiber trunk and the coaxial distribution.

Fiber optics Very thin and pliable tubes of glass or plastic used to carry wide bands of frequencies.

Guardband Provides for slot timing uncertainty due to inaccuracy of the ranging.

Headend The central location, in an MSO environment, that has access to signals traveling in both the forward and reverse directions.

Header Protocol control information located at the beginning of a protocol data unit.

High split A frequency division scheme that allows bidirectional traffic on a single cable. Reverse path signals propagate to the headend from 5 to 174 MHz. Forward path signals go from the headend from 234 MHz to the upper frequency limit. A guardband is located in 174 to 234 MHz.

Home passed Total number of homes which have the potential for being connected to the cable system.

Internet Term used to refer to the largest global internetwork, connecting tens of thousands of networks worldwide and having a culture that focuses on research and standardization based on real-life use. Many leading edge network technologies come from the Internet community. The Internet evolved in part from ARPANET. At one time, it was called the DARPA Internet. Not to be confused with the general term *internet*.

IP Internet Protocol. Network layer protocol in the TCP/IP stack offering a connectionless internetwork service. IP provides features for addressing, type-of-service specification, fragmentation and reassemble, and security. Documented in RFC 791.

IP over ATM Specification for running IP over ATM in a manner that takes full advantage of the features of ATM. Defined in RFC 1577. Sometimes called CIA.

ISO International Organization for Standardization. International organization that is responsible for a wide range of standards, including those relevant to networking. ISO developed the OSI reference model, a popular networking reference model.

Layer A subdivision of the Open System Interconnection (OSI) architecture, constituted by subsystems of the same rank.

LLC Logical Link Control. Higher of the two data link layer sublayers defined by the IEEE. The LLC sublayer handles error control, flow control, framing, and MAC-sublayer addressing. The most prevalent LLC protocol is IEEE 802.2, which includes both connectionless and connection-oriented variants.

Medium access control (MAC) address An address that identifies a particular medium access control (MAC) sublayer service-access point.

Medium access control (MAC) procedure In a subnetwork, that part of the protocol that governs access to the transmission medium independent of the physical characteristics of the medium, but takes into account the topological aspects of the subnetworks, in order to enable the exchange of data between nodes. MAC procedures include framing and error protection.

Medium access control (MAC) sublayer Part of the data link layer that supports topology-dependent functions and uses the services of the physical layer to provide services to the logical link control (LLC) sublayer.

Mid split A frequency division scheme that allows bidirectional traffic on a single cable. Reverse channel signals propagate to the headend from 5 to 108 MHz. Forward path signals go from the headend from 162 MHz to the upper frequency limit. The guardband is located in 108 to 162 MHz.

Mini-slot An integer multiple of 6.25 μsec increments. It represents the byte-time needed for transmission off a fixed number of bytes.

MCNS Multimedia cable network system. A consortium of Comcast Cable Communications, Inc., Cox Communications, Tele-Communications, Inc., and Time Warner Cable, interested in deploying high-speed data communications systems on cable television systems.

MSO Multiple System Operators. Company that owns and operates more than one cable television system.

NTSC National Television Systems Committee. Committee that defined the analog color television broadcast standard used today in North America.

Network Collection of computers, printers, routers, switches, and other devices that are able to communicate with each other over some transmission medium.

NIC Network interface card. Board that provides network communication capabilities to and from a computer system.

Payload Portion of a frame that contains upper-layer information (data).

Physical layer In Open System Interconnections (OSI) architecture, the layer that provides services to transmit bits or groups of bits over a transmission link between open systems.

Protocol A set of rules and formats that determines the communication behavior of layer entities in the performance of the layer functions.

QAM Quadrature amplitude modulation. A method of modulating digital signals onto a radio-frequency carrier signal involving both amplitude and phase coding.

QPSK Quadrature phase-shift keying. A method of modulating digital signals onto a radio-frequency carrier signal using four phase states to code two digital bits.

QoS Quality of service. The accumulation of the cell loss, delay, and delay variation incurred by those cells belonging to a particular ATM connection.

SAR Segmentation and reassemble. One of the two sublayers of the AAL CPCS, responsible for dividing (at the source) and reassembling (at the destination) the PDUs passed from the CS. The SAR sublayer takes the PDUs processed by the CS and, after dividing them into 48-byte pieces of payload data, passes them to the ATM layer for further processing.

Scrambling A signal security technique for rendering a TV picture unviewable, while permitting full restoration with a properly authorized decoder or descrambler.

Sublayer A subdivision of a layer in the Open System Interconnection (OSI) reference model.

Subnetwork Subnetworks are physically formed by connecting adjacent nodes with transmission links.

Subsplit A frequency division scheme that allows bidirectional traffic on a single cable. Reverse path signals come to the headend from 5 to 30 (up to 42 on newer systems) MHz. Forward path signals go from the headend from 54 MHz to the upper frequency limit.

Synchronization Establishment of common timing between sender and receiver.

TCP Transmission Control Protocol. Connection-oriented transport layer protocol that provides reliable full-duplex data transmission. TCP is part of the TCP/IP protocol stack. See also TCP/IP.

TCP/IP Transmission Control Protocol/Internet Protocol. Common name for the suite of protocols developed by the U.S. DOD in the seventies to support the construction of worldwide internetworks. TCP and IP are the two best known protocols in the suite. See also IP and TCP.

TDM Time-division multiplexing. Technique in which information from multiple channels can be allocated bandwidth on a single wire, based on preassigned time slots. Bandwidth is allocated to each channel regardless of whether the station has data to transmit. Compare with ATDM, FDM, and statistical multiplexing.

Topology Physical arrangement of network nodes and media within an enterprise networking structure.

Transmission Medium The material on which information signals may be carried; e.g., optical fiber, coaxial cable, and twisted-wire pairs.

Upstream Flow of any information from the customer, through the cable system, to the headend.

WWW World Wide Web. Large network of Internet servers providing hypertext and other services to terminals running client applications such as a WWW browser.

WWW browser Client application, such as Mosaic, used to access hypertext documents and other services located on innumerable remote servers throughout the WWW and Internet. See also Internet and WWW.

10BaseT 10-Mbps baseband Ethernet specification using two pairs of twisted-pair cabling (Category 3, 4, or 5): one pair for transmitting data and the other for receiving data. 10BaseT, which is part of the IEEE 802.3 specification, has a distance limit of approximately 100 meters per segment.

ACRONYMS

AAL	ATM adaptation layer
ABR	Available bit rate
ADSL	Asymmetric digital subscriber line
ANM	Answer message
ANSI	American National Standards Institute
ATM	Asynchronous transfer mode
BER	Bit error rate
BW	Bandwidth
CATV	Community antenna television
CBR	Constant bit rate
CDMA	Code division multiple access
CPE	Customer premises equipment
CRC	Cyclic redundancy check
CSMA/CD	Carrier sense multiple access with collision detection
HE	Headend
HEC	Header error control
HFC	Hybrid fiber coax
IEEE	Institute of Electrical and Electronics Engineers
ITU	International Telecommunication Union

IP	Internet Protocol
LLC	Logical link control
MAC	Medium access control
MPEGx	Motion Picture Editor's Group compression algorithm x
OSI	Open system interconnect
PMD	Physical medium dependent
QAM	Quadrature amplitude modulation
QPSK	Quaternary phase shift keying
QOS	Quality of service
RF	Radio frequency
SAAL	Signaling ATM adaptation layer
TDMA	Time division multiple access
VBR	Variable bit rate

REFERENCES

1. ANSI/IEEE. "Carrier Sense Multiple Access with Collision Detection (CSMA/CD) Access Method and Physical Layer Specifications." ANSI/IEEE Std. 802.3-1985, 1985.
2. Azzam, A. and Brandt, M., Dajer, M., Eng, J., Lin, D., Mollenauer, J., Siller, C., Grobicki, C., Sriram, K., Ulm, J. "IEEE P 802.14 Cable-TV Functional Requirements and Evaluation Criteria." November 1993.
3. Azzam, A. et al. "ATM over ADSL" ATM Forum Magazine (53 Byte), 1997.
4. Bingham, J. and Jacobson, K. "A proposal for an MAC protocol to support both QPSK and SDMT." IEEE 802.14-95/137.
5. Currivan, B. "CATV Upstream Channel Model, Rev 1.0." IEEE 802.14-95/133, March 1997.
6. "Data-Over-Cable Service Interface Specifications Radio Frequency Interface Specification." SP-RFI-I02-971008 (Interim Specification). 1997.
7. De Prycker, M. "Asynchronous Transfer Mode Solution for B-ISDN." 1993.
8. Gingold, D. "Integrated Digital Services for Cable: Economics, Architecture, and the Role of Standards — MIT Research Program on Communications Policy." IEEE 802.14-96/230. September 1996.
9. Golmie, N. "Performance Evaluation of Contention Resolution Algorithms: Ternary-tree vs. p-Persistence." IEEE 802.14/96-241. September 1996.
10. Golmie N. "Performance Evaluation of MAC Protocol Components for HFC Networks." Broadband Access Systems, Proc. SPIE 2917, pp.120-130. Boston, Massachusetts. November 1996.
11. Grossman, D. "Security overview." IEEE 802.14-97/030. March 1997.
12. Hilton, R. and Prodan, R. "Evolving Cable Network Architecture Cable Televisions Laboratory." ATM Forum/95-0075. February 1995.
13. Karaoguz, J. Gottfried Ungerboeck. "Formal Proposal: Frequency Agile Multi-Mode (FAMM) Single-Carrier Modems for Upstream Transmission in HFC Systems." IEEE 802.14-95/131. November 7, 1995.
14. "Cable-TV access method and physical layer specification." IEEE Project 802.14/a Draft 3 Revision 1. March 1998.
15. "Information technology — generic coding of moving pictures and associated audio information systems." ITU-T Recommendation H.222.0 (1995) I ISO/IEC 13818-1:1996.

16. "Digital multi-programme systems for television sound and data services for cable distribution." ITU-T Recommendation J.83 October 1995.

17. "Information technology — Open Systems Interconnection — Local area networks — Medium Access Control (MAC) service definition." ISO/IEC10039 ISO/IEC 10039:1991.

18. Kwok, T. "Communications Requirements of Multimedia Applications: A preliminary study." International Conference on Selected Topics in Wireless Communications, Vancouver, Canada. 1992.

19. Laubach, M. "ATM HFC Overview." IEEE 802.14-95/022. March 1995.

20. _____. "MAC <> Phy-TC Sublayer interface." IEEE 802.14-97/025. January 1997.

21. Limb, J. "Performance Evaluation Process for MAC Protocols." IEEE 802.14-96/83R2. May 1996.

22. Prodan, R. "Letter From CableLabs - regarding Annex A/B." IEEE 802.14-96/205. July 1996.

23. Quinn, S. "Requirements on HFC Access Networks — A Public Network Operator's Perspective." IEEE 802.14-95/099. September 1995.

24. Sriram, P. "ADAPt MAC PDU and MAC-PHY Services in Support of ATM." IEEE 802.14-95/168. November 1995.

25. Van der Plas, G. "APON: An ATM-based FITL System." EFOC&N'93, 1993.

26. Vandenameele, J. "How to upgrade CATV networks to provide interactive ATM-based services." GLOBECOM 95. Singapore, November 1995.

27. Verbiest, W. "Integrated Broadband Access." Proc. Fourth IEEE conference on Telecommunication, Manchester 1993.

6   DVD Technology

Bruce C. Klopfenstein

CONTENTS

6.1   INTRODUCTION

This chapter reviews DVD, which is perhaps the most significant new communication technology since the videocassette recorder (VCR). DVD originally meant *digital video disc,* but advocates found this label too limiting and changed it to digital *versatile* disc. Some prefer restricting the name to DVD (just as we no longer refer to IBM as International Business Machines). DVD is a digital storage device that physically resembles a CD. DVD discs look like CDs and DVD players look like high end CD players. DVD-ROM devices are compatible with existing CD discs and are already well on their way to replacing CD-ROM drives on high-end personal computer systems. This technology substitution (Klopfenstein, 1989b) has begun despite the lack of DVD-ROM specific software. Many believe the CD's days are numbered, with DVD its likely, if not inevitable, successor. A DVD's visual similarity

0-8493-9594-1/00/$0.00+$.50
© 2000 by CRC Press LLC

to CDs bodes very well for its rapid adoption by users (Rogers, 1995), and because DVD players are backward compatible with CDs of varied formats, the stage is set for rapid DVD technology substitution.

As is the case with broadcasting technologies, video recording and transmission technologies are in a transitional stage from the past's analog technologies to the very near future's digital technologies. DVD is actually not the last stop in the transition to a completely digital video world. All-digital television systems are still a few years away. Cable television and telephone companies eventually may encroach into the video delivery business, but telephone companies in particular have been slow to do that (O'Shea, 1998). In the short-term, recorded video retains many advantages over digital transmission systems. Video storage technology has an established history of market success (VCR), market failure (RCA CED videodisc and Sony Betamax), and limited adoption (laserdisc) (Klopfenstein, 1989a). For a complete review of all current home video technologies, see Klopfenstein (1998).

6.2 DVD BACKGROUND

Those unfamiliar with history are doomed to repeat it. There are a number of historical parallels between DVD and its predecessors that allow for easier understanding of where DVD technologies are headed today. Home video recording dates back to the earliest days of magnetic recording technology, (see Schoenherr, 1996), while VCRs can trace their heritage to the professional Sony U-Matic, the first videotape cassette recorder (Klopfenstein, 1985). The first consumer Betamax was introduced in 1975 at a list price of $2295.

While DVD proponents today will note how DVD diffusion is much faster than that of the VCR, the DVD has been introduced at prices dramatically lower than that of the first VCRs. A continuing theme in the history of home video has been format standards battles. The most famous is the Sony Beta versus JVC VHS battle of the 1970s that was won rather handedly by VHS. Since then, there has been a camcorder battle where the tide was turned: Sony's 8mm camcorders have been displacing VHS and VHS-C camcorders. In 1998, the new digital disc format, DVD, was facing incompatible recording and limited play (Divx) versions of the standard read-only system as well as some competing standards in rewritable DVD technologies.

The growth of the VCR in the late 1980s was impressive. Indeed, it equaled and then exceeded comparable sales for color television sets in the 1960s. Many believe for good reason that DVD will diffuse more rapidly than either VCR or color television. Table 6.1 shows this intriguing growth.

In the early 1980s, the VCR was an expensive item that seemed destined for elite households (Klopfenstein, 1989b). A less expensive alternative that could exceed the video quality (but could not record) was the videodisc player (VDP). One VDP was developed by RCA, which needed a new product to follow the success of color TV in the market (Graham, 1986). Another was developed by MCA (in partnership with Philips) as an outlet for its movie inventory. Their Laservision videodisc could hold up to 54,000 separate video images, or up to one hour of

TABLE 6.1
VCR Versus Color Television Set Sales

	Total VCR Sales		Color TV Sales
1975	30,000	1959	90,000
1976	55,000	1960	120,000
1977	160,000	1961	147,000
1978	402,000	1962	438,000
1979	475,000	1963	747,000
1980	805,000	1964	1,404,000
1981	1,361,000	1965	2,694,000
1982	2,035,000	1966	5,012,000
1983	4,091,000	1967	5,563,000
1984	7,616,000	1968	6,215,000
1985	11,853,000	1969	6,191,000
1986	12,500,000	1970	5,320,000

Source: Klopfenstein (1998)

Note: VCR was introduced in 1975. Color TV was introduced in 1954.

moving pictures, on a side. When Philips' Magnavox $695 *Magnavision* was marketed in late 1978, only two hundred disc titles were available, mostly old movies. As it turned out, the limited number of titles available early in the history of the laserdisc was critical to its lack of adoption (Klopfenstein, 1985). There are direct parallels between the situation of videodisc players and today's DVD; literally hundreds of times more titles are available on VHS cassettes than on DVD.

Never having reached mass market status, the laserdisc now clearly appears to be well on its way to the same fate as LP audio records. RCA's heavily promoted, non-optical Selectavision VDP lost the company $580 million in its three years on the market in the early 1980s (Klopfenstein, 1985). Today's Divx version of DVD is backed by a $200 million promotional program as well as partnerships with many consumer electronics manufacturers. Students of consumer electronics history will see that such strong promotion does not guarantee market success.

Many lament the lack of standards in video recording and storage technologies because of the belief that lack of standards slows market adoption. Indeed, lack of a technical standard prevented the adoption and diffusion of AM stereo (Klopfenstein and Sedman, 1990). On the other hand, the standards battle between Beta and VHS led both sides to technological innovations, such as longer recording times, high fidelity sound, and lower prices, that probably accelerated VCR adoptions (Klopfenstein, 1985). The same drive to innovation is evidenced in the battle between the quasi-compatible Netscape and Microsoft WWW browsers. The introduction of Divx by a segment of the consumer electronics industry may have contributed and detracted from DVD, diffusion as will be discussed later.

Another issue associated with video recording technologies is copyright (see the pro-electronics manufacturers' "Home Recording Rights Coalition" web site for a history of events in recording rights: http://www.hrrc.org). The relevance of this concern is dramatized by the plight of the digital audio tape (DAT) format. Challenges by recorded music copyright holders slowed the diffusion of this audio technology and may have effectively killed it as a consumer audio format (Cohen, 1991). The Motion Picture Association of America (MPAA) and the Consumer Electronics Manufacturers Association (CEMA) have worked together in an attempt to avoid this happening again with new digital video recorders (Pietrucha, 1996). Look to these organizations to see their latest stances on recording and antipiracy technologies, topics of direct relevance to DVD.

6.2.1 RECENT HOME VIDEO DEVELOPMENTS

DVD is expected to eventually challenge the VCR as the movie playback device of choice in many homes. Rather than fight the new DVD technology, video stores are likely to embrace it. The stores can rent and sell DVD discs just as they do now with VHS tapes. Several chains are experimenting with renting both discs and DVD players. This move is significant because it addresses one of the variables trialability research has shown to be directly correlated with adoption of new technology (Rogers, 1995). Trialability involves removing barriers from potential adopters so they may experience a new technology without fully committing to it. This disc-with-player rental does precisely that.

Early studies showed that the primary use of VCRs was to record broadcast programs for later viewing, a practice known as *time shifting*. However, the VCR really made its presence felt in the U.S. film industry. Video rentals in 1995 reached about $8 billion with cassette sales climbing over $7 billion ("Rental stores," 1995). Home video revenues appear to have reached a high plateau (King, 1996), and the enormous number of tapes now available would seem to assure the VCR some life despite the encroachment of the DVD.

Ironically, the home video industry is in position to embrace the new DVD format. The home video market did not grow much in 1997. A widely documented downturn in the rental business sent video retail stocks plummeting and played a role in retailers' aggressive lobbying for longer windows of video exclusivity before titles are released to pay-per-view (retailers want 60 days rather than the current average of 38 days). Despite differing views among researchers, a wide cross section of Hollywood studio executives and retailers believe that consumer video rental activity declined significantly in 1997 (Klopfenstein, 1998). The most often cited explanation for the decline was a shortage of box-office blockbusters. Prominent Wall Street investment analyst Tom Wolzien believes as many as 5 million rentals a month evaporated from the market because more households are watching movies on cable pay-per-view and digital satellite systems. Hollywood studios and video retailers reacted by working together to offer consumers more in-store copies of new releases ("A year long battle," 1998).

Two other studies confirm significant sales declines for 1997 as well as the first quarter of 1998. Alexander & Associates' Video Flash, a weekly phone survey of

1000 consumers, reports consumer video spending at $9.3 billion in 1997, down about 10% from $10.38 billion in 1996. It estimates that for the first 12 weeks of 1998, consumers spent $2.2 billion buying videos, down 3.1% from 1997. Similarly, VideoScan, which tracks point-of-sale data from 16,000 nationwide retail stores that account for about 70% of all video sales, says purchases for the first quarter of 1998 were down 5% from 1997. Annual sales for 1997 fell by a similar percentage from 1996 figures (Arnold, 1998). According to VideoScan, the top seller in 1997 was *Bambi* with *Ransom* taking the top spot in 1997 rentals ("A Year Long Battle," 1998).

According to a number of sources, video stores supply approximately 54% of movie studios' revenues by paying more than $60 per video. The store sees a profit after renting the tape 25 times which usually happens because demand remains high for up to six weeks after a hit is released. About 8% of store revenue comes from fees charged for late tapes. In 1995, Media Group Research estimated total video revenue would grow about 8% a year the next few years with rentals rising about 3% a year and cassette sales leaping 15% a year ("Rental stores," 1995). That prediction looked foolish in 1997.

While viewing movies at home has become a way of life, the movie theater still has some advantages over home tape viewing: 1) going to the movies is a social occasion, 2) movies appear in theaters before they appear on cassette, and 3) theaters offer the large, wide screen. As noted at the conclusion of this chapter, home theater technology threatens to erode the nonsocial reasons for going to the movie theater.

6.2.2 DIGITAL VCRs

A digital VCR standards group representing about 50 companies announced it had agreed on technical specifications for recording transmission signals from the U.S. high-definition TV (HDTV) system. The U. S. HDTV system will operate at 19.4 megabits/s. JVC jumped ahead of digital VCR competitors by rolling out its own digital VHS (D-VHS) format in 1994. D-VHS players will initially work with Thomson Consumer Electronics' Digital Satellite System (DSS) set-top boxes, which are deployed as part of Hughes' DirecTV digital broadcasting satellite (DBS) system. VCR manufacturers believe the VCR is on track to survive its 25th anniversary and life beyond millennium, if only due to the lack of consensus on a rewritable DVD format as a replacement for analog tape. The prospects for advanced digital VCR formats such as D-VHS or W-VHS are negligible in 1998, but they may be introduced along with a new M-DVD magnetoresistive format ("VCRs proliferate," 1998). JVC demonstrated a prototype D-VHS deck operating in HD (high definition) mode needed for terrestrial DTV (digital television) in January 1998. Digital VHS recorders were available for under $500 by spring 1999 (Hara, 1999). At the same exhibition, Hitachi touted the merits of its current D-VHS model, codeveloped with Thomson, for recording HD-DSS signal ("Novel VCRs," 1998).

Smart VCRs were available by 1999 (Hall, 1999). These VCRs will automate many aspects of installation including setting the clock, locating the available channels, and will even accept voice commands (an interface likely to begin showing up almost as an afterthought on many home appliances in the next few years). Not only are there VCRs that skip commercials in recording and/or playing back, Thomson markets one that also fast forwards over the prerecorded promotional material at the beginning of most feature films on cassette (Cole, 1998). Sanyo's VCR has a feature called Speed Watch that allows users to watch a tape at twice the normal tape speed while the audio remains at normal levels. It seems likely that only the huge installed base of videotapes will assure some future for the VHS VCR, but its fate is tied to the success of recordable disc technology ("VCR near end," 1998).

6.2.3 PERSONAL COMPUTING AND DVD

Continued advances in processing speed and personal mass storage bode well for computer applications in media including video. At the start of 1998, about 45% of U.S. households had computers ("PCs in Over 45 Percent," 1998; Lanctot, 1998), and various manufacturers offer combination TV/PCs intended for the home's current television viewing room. In terms of new technology, the trend of the Internet and WWW is clearly toward multimedia applications (audio, video, and animation). The day is rapidly approaching when we may be able to access hundreds then thousands of video titles through video servers made available from Hollywood studios, telephone companies, and cable television companies (Klopfenstein, 1997; 1998).

CD-ROM drives have become standard equipment on personal computers, but the substitution of DVD-ROM drives for CD-ROM drives is well under way. The sales push is coming from high-end personal computer manufacturers offering DVD-ROM drives for a small price increase over CD-ROMs. While CD-ROM drives cost PC manufacturers $25-$35, DVD-ROM drives cost about $100. This cost is more of a problem at the low-end computers. Software titles available only on DVD-ROM are appearing slowly, but DVD movies are being produced more rapidly. Unfortunately, incompatibility among DVD drives from different manufacturers is inhibiting the growth of DVD-ROM. A title might work on one drive, but not another. PC OEMs and drive manufacturers must resolve compatibility issues before consumers rush to embrace the products. Moreover, computer and video game manufacturers are taking a wait-and-see attitude ("DVD-ROM Watershed," 1998).

Despite a lack of DVD-ROM disc titles, sales for DVD-ROM devices in 1997 and 1998 were surprisingly strong. According to industry observers, worldwide market demand for DVD-ROM was expected to quadruple in 1999 to more than 20 million units, and increase to over 40 million in the year 2000 (Digital Video Systems, 1998). CEMA reported that 34, 000 DVD players were sold in January 1998, and more than double that (81,000 units) were sold in August. Pre-Christmas sales of DVD players for 1998 were about 500,000 units, according to CEMA (Koenig, 1998). Various research reports predicted that by 1999 DVD-ROM drives would outsell CD-ROM drives in the U.S. (Kovar, 1998b).

6.2.4 DVD

In 1995, consumer electronics manufacturers, including old VCR rivals Sony and Matsushita, bombarded the media with descriptions of their next-generation recording medium, DVD. They touted DVD and DVD players as digital replacements for VCRs and VHS tapes, laser discs, video game cartridges, and compact discs (CDs) — both audio CDs and computer CD-ROMs — are to be subsumed as well. As vendors did with videocassette tapes, DVD vendors proposed different standards: Sony's Multimedia Compact Disc (MMCD) versus Matsushita's DVD, called Super Density DVD (SD-DVD) (D'Amico, 1995). A format war was avoided when the industry players agreed to support a format that combined a Toshiba design with a Sony/Philips (the original CD partners) encoding scheme (Braham, 1996).

The DVD is a variation on the now-ubiquitous CD. DVD, introduced in March 1997, can store complete feature films, music, video games, and multimedia computer applications. This new disc is identical in shape and size to CDs and CD-ROMs, but it has a much more storage capacity. The 1998 DVD discs hold 133 minutes or 4.7 gigabytes of video per side; double-layered DVDs capable of holding 241 minutes or 8.5GB of video became available soon thereafter. These higher-capacity discs are most likely to be used in computers. For example, if stereo music is the stored information, a single DVD can hold the contents of more than a dozen CDs. The movie studios consider the DVD disc a major opportunity because it will allow them to resell many of their existing inventory of films, just as music companies have done with the audio CD.

DVD specifications compared with the standard CD, are shown in Table 6.2

TABLE 6.2
DVD and CD Specifications

	CD	DVD
Disc Diameter	120 mm	120 mm
Disc Thickness	1.2 mm	1.2 mm
Disc Structure	Single Substrate	Two bonded 0.6 mm substrates
Laser Wavelength	780 nm	650 and 635 nm
Numerical Aperture	0.45	0.60
Track Pitch	1.6 um	0.74 um
Shortest pit/land length	0.83 um	0.4 um
Reference Speed	1.2 m/sec, CLV	4.0 m/sec, CLV
Data Layers	1	1 or 2
Data Capacity	680 megabytes	Single layer: 4.7 gigabytes
		Double layer: 8.5 gigabytes

Source: Sony (n.d., 1996).

Basic DVDs have one or two layers per side and are one- or two-sided. Each variation is given a code matching the rough capacity: DVD-5, 4.7GB (1 side, 1 layer); DVD-9, 8.5GB (1 side, 2 layers); DVD-10, 9.4GB (2 sides, 1 layer); and DVD-18, 17GB (2 sides, 2 layers). The first iteration of a write-once disc is called DVD-R, with 3.9GB per side (one layer only), and the first iteration of read-write DVD is called DVD-RAM, with 2.6GB per side, one layer only (Dvorak, 1998).

6.3   DVD FORMATS

Various DVD formats are defined by function rather than storage capacity as shown in Table 6.3 (based on "State of the DVD Union" (1998) and other sources.).

6.3.1   DVD-VIDEO

The DVD Forum's DVD-Video standard, the basis for today's DVD movie discs, is being challenged by the controversial new Divx format, supported by retailer Circuit City and several leading movie studios. Promoted as a new way to rent movies, and as a possible future mechanism for music and software distribution, Divx DVDs will cost about $5 and allow unlimited playback for 48 hours. Viewers would then discard the disc or use a modem-equipped Divx player to buy more time. Today's DVD-ROM drives and set-top players won't play Divx media.

6.3.2   DVD-ROM

Second-generation DVD-ROM drives have little trouble reading the various CD formats, as well as DVD-Video discs and interactive DVD-ROM titles created exclusively for playback on a PC. Format compatibility, however, has been an issue.

6.3.3   DVD-RECORDABLE (DVD-R)

The DVD Forum's first DVD-R specification defines a write-once format storing 3.95GB of data per side, but two rival proposals define 4.7GB capacities. Recently accepted for consideration by the DVD Forum's Working Group 6, DVD-RW is the rewritable version of DVD-R, the "other" rewritable format endorsed by the DVD Forum. The next generation of 4.7GB DVD-R drives was expected to reach the market in 1999 at a price between $3000 and $5000 with the 4.7GB DVD-RW coming along as well. Philips, Sony, and Hewlett-Packard have all revealed plans to deliver drives by mid-1999. The expectation was that the drives and media will be compatible with DVD-RAM. (Parker 1999). Original DVD-R drives cost well over $10,000, making them of interest only to DVD content creators.

6.3.4   DVD-RAM

Rewritable DVD suffers from a format war between the DVD Forum's announced 2.6GB-per-side DVD-RAM spec and at least three competing proposed formats, including a 3GB-per-side version, DVD+RW, backed by Sony and Philips. With no clear resolution, the forum was working to upgrade to its existing spec. Initial DVD-ROM drives probably will not be able to read at least a few of the proposed formats. A number of DVD-RAM units are shipping, but shifting standards might give these early models short lifespans.

6.3.5   DVD-AUDIO

The DVD Forum recently announced a draft DVD-Audio specification defining discs that can hold up to 30 hours of six-channel sound. A final draft is expected later this year, but it could be derailed by a competing Sony/Philips proposal called Direct Stream Digital. Today's DVD-ROM equipment might lack the copy-protection circuitry needed to play DVD-audio discs, but the demand for DVD-audio may be limited by marketing rather than technological concerns. Marketers commonly sell 2-disc sets of CD-audio when one CD is capable of holding all the audio. It's not clear where the demand for 30 hours of audio will come from.

TABLE 6.3
DVD Versions as of 1998

Type	Sides or layers used	Capacity
DVD-ROM	Read-only 1 side, 1 layer	4.7GB
DVD-ROM	Read-only 1 side, 2 layers	8.5GB
DVD-ROM	Read-only 2 sides, 1 layer	9.4GB
DVD-ROM	Read-only 2 sides, 2 layers	17.0GB
DVD-R	Write-once 1 side	3.9GB
DVD-R	Write-once 2 sides	7.8GB
DVD-RAM	Rewritable 1 side	2.6GB
DVD-RAM	Rewritable 2 sides	5.2GB
DVD+RW	Rewritable 1 side	3.0GB
DVD+RW	Rewritable 2 sides	6.0GB
DVD-R/W	Rewritable 1 side	4.0GB
MMVF	Rewritable 1 side	5.2GB
MMVF	Rewritable 2 sides	10.4GB

Source: DVD (1998).

Sonic Solutions announced in 1998 that its technology was being used to produce the first individual DVD disc that combines DVD video, DVD audio, and DVD-ROM. Fans of Travis Tritt, a country music singer, will be able to use the same disc to view a concert from multiple camera angles on their home DVD-video system, interact with him on their home computer's DVD-ROM drive, or listen to the full, uncompressed high-density audio program in surround-sound on their DVD-audio player. The company's unique DVD-video and new DVD-audio production technology make it possible for single, hybrid discs containing both formats, plus computer data, to be formatted on the same DVD disc (Sonic DVD Technology, 1998). This product may be a gimmick, but it does show the multimedia possibilities for DVD.

6.4 FORECASTING DVD ADOPTION

As of August 1998, an estimated 800,000 DVD home video players had been sold (Patrizio, 1998). DVD-ROM drives are expected to outship CD-ROM drives by 2001 (Kovar, 1998b). DVD player sales through the first quarter of 1999 were 300% higher than in the same period in 1998 (DVD Player, 1999). (Online: http://www.twice.com/html/statistics.html. [May 6, 1999]. The total installed base of DVD players was at about the one million unit mark at the beginning of 1999 (Scally, 1999).

InfoTech predicted that worldwide DVD-ROM title revenues would increase from $3.5 million in 1997 to $567 million in 1998, largely because of 7 million PCs equipped with DVD-ROM by year-end. DVD-ROM is expected to become a mainstream PC component. By 2003, worldwide DVD-ROM title revenue across both PC desktop and TV set-top platforms is forecast to exceed $70 billion, as DVD-ROM replaces CD-ROM as the principal format for packaged interactive media, including applications software, business information, games, education and computer-based training.

A key to the early success of DVD will be the perceived difference in the sharpness of the user's television picture. The CD was clearly an improvement over easily scratched LP records. DVD carries images with smaller and more varied pixels (720 pixels per horizontal line versus the standard 240) than a CD, which means greater clarity and detail than today's best VCRs can produce. If viewers can easily see the difference, it will bode well for DVD (Vizard, 1997). Given the steady increase in television picture resolution, these differences will become more visible. An easily overlooked problem, however, could be the durability of DVD discs and how well they can stand up to the rigors of video and DVD-ROM rentals. On the other hand, DVD discs may allow longer archiving than is possible with videotape.

The following summary of DVD's key advantages is based on Johnson (1995) but remains valid as of this writing:

- better video and audio quality
- backwardly compatible with existing CDs
- movie discs more convenient and durable, and potentially less expensive than cassettes
- able to cut directly to particular scenes, with no need to rewind
- more storage capacity (the same is true for DVD audio)
- negligible cost differential between a DVD player and high-end VCR

DVD offers other advantages including new flexibility in home video use. During playback of movies, DVD discs can provide a choice of viewing options. For example, there's standard 4:3 pan-and-scan viewing, the way most movies are displayed from broadcast and tape sources. By pressing a button on the player's remote control, users can switch to letterboxed viewing or high resolution pictures on advanced widescreen (16:9) sets. DVD movie discs also have the capability to present soundtracks in eight different languages and up to 32 distinct subtitles. A feature announced by Toshiba is a built-in parental control system that allows selection of the rating level to be viewed: PG, PG-13, R or NC-17; the player automatically shows a version of the movie edited to that rating level by the producers of the film.

DVD liabilities include the following:

- limited software available, and it will take years to approach the number of titles available on VHS
- disc players do not record; this fact combined with the previous makes software availability a limiting factor in the adoption of DVD players as was the case with the laserdisc around 1980
- durability: CDs and CD-ROMs are only marginally durable enough to withstand the abuses of public library and commercial rentals
- current CD-ROM drives will not play DVD discs
- DVD systems will eventually be challenged by delivery of movies-on-demand over cable or phone lines

The potential of DVD includes a broad range of multimedia and computer applications. Because DVD consists of a suite of disc types, each with increasingly higher storage capacities, the format holds tremendous growth potential for data-intensive home and business applications. DVD players also come in a variety of models. The first in the DVD family of drives were read-only, a one-time recordable DVD-R, and a rewritable DVD-RAM. Philips correctly believed the computer DVD-ROM would be more important to early sales than the consumer video DVD player (Oosterveld, 1996), a position supported by evidence form the marketplace to date.

Introduced in 1997 at prices around $1500, by mid-1998 DVD player prices were under $400. Both the rate of decline and the price in real dollars are dramatic in historical terms. DVD is simply far less expensive than were its VCR, laserdisc, and CD predecessors. This relative coast strongly suggests the potential for accelerated diffusion patterns. Retail discount chains began selling DVD players in 1998 (Scally, 1998) with all Wal-Mart stores expected to have them by the end of that year (Koenig, 1998). This is a very significant development because it would seem to already portend the DVD's movment from videophile status to that of a mass market product.

In 1998, all major Hollywood film studios announced support for the DVD format. Initially, some (especially Dreamworks SKG) were reticent to support this digital version of their product. One problem that remained was the deliberate pace at which most studios were releasing titles as well as a tendency to release older films rather than new releases on DVD.

Somerfield (1996) predicted a number of obstacles for DVD to overcome. The short-term success or failure of the DVD rests on the availability of movie discs, until the time comes when inexpensive recordable discs are available. The movie studios that control the production and sale of movies on video have their own agendas. First, movies released on video in this country may not yet have been distributed to theaters on other continents. The studios are concerned that a DVD released in the U.S. could be sold abroad and hurt foreign theatrical sales. To prevent this, the studios want the discs encoded to prevent playback in certain regions of the world. This regional coding system has become the norm.

The studios also want copyright legislation in place similar to the Audio Home Recording Act that delayed the launch of the DAT format several years ago. That legislation took 18 months, from proposal to passage, to get through Congress. No new DVD legislation has apparently been introduced. To compound the issue, the MPEG-2 encoding that must be implemented to compress the movies to DVD is quite complex and can be done by only three facilities: Warner, Sony, and MCA. The number of titles that can be made ready for disc in the few months before late 1996 fall launch was quite limited. Paucity of discs was one major factor in the quick demise of RCA's videodisc player in the early 1980s (Klopfenstein, 1985; 1989).

6.5   DIGITAL TRANSMISSION CONTENT PROTECTION (DTCP)

According to DVD Frequently Asked Questions (Taylor, 1998), in order to provide for digital connections between DVD components without allowing perfect digital copies, a digital copy protection system has been developed. The 5CP draft proposal (for "five-company proposal") was made by Intel, Sony, Hitachi, Matsushita, and Toshiba in February 1998. Content is marked with standard CGMS flags of *copy never* or *copy once*. Devices that are digitally connected, such as a DVD player and a digital TV, will exchange keys and authentication certificates to establish a channel. The DVD player will encrypt the encoded video signal as it sends it to the receiving device, which must decrypt it.

Digital display devices will be able to receive and display all data. Digital recording devices will be able to receive only data that is not marked *copy never*, and they must change the CGMS flags to zero copies if the source is marked for one copy. Digital CPS is designed for the next generation of digital TVs and digital video recorders. It will require new DVD players with digital connectors (such as those provided on DV cameras and decks). These new products probably won't appear before the middle of 1999. Since the encryption is done by the player, no changes are needed to the existing disc format. Movie studios and consumer electronics companies want to make it illegal to defeat DVD copy protection, and are pursuing legislation in the U.S. and other countries.

CSS is allowed for DVD-video content only. Because a DVD-ROM can hold any form of computer data, any desired encryption scheme can be implemented.

Watermarking, which will be added to DVD at some point, permanently marks each digital video frame with noise that is supposedly visually undetectable. Watermark signatures can be recognized by video playback and recording equipment to prevent copying. New players and other equipment will be required to support watermarking. It is possible to make new watermarked discs compatible with existing players, but movie studios will probably not allow it. There are reports that the watermarking technique used by Divx causes visible raindrop or gunshot patterns (Taylor, 1998).

6.6   RECENT DEVELOPMENTS IN DVD-ROM

Two factors will pull DVD in different directions. First, if Disney and other studios continue to limit their releases on DVD, its value as a home video technology will be greatly diminished. On the other hand, as computer manufacturers begin to substitute DVD-ROM drives for CD-ROM drives, the cost of DVD player manufacturing will go down. It is not unreasonable to believe that in the next 5 years, although

VCRs will remain the home video technology of choice, a significant number of homes will add a DVD player to their home entertainment system.

DVD movie players and titles have been available in the U.S. since March 1997, and PCs equipped with DVD-ROM drives began shipping during the third quarter. The DVD-ROM market will enhance economies-of-scale in manufacturing and should lead to lower DVD player prices. The initial signs for the DVD player in 1997 were mixed. While DVD player sales in its first year compared very well to those for CD players, the meaningfulness of that comparison is limited severely by the lack of affordability of CD players in that technology's first year.

A survey from the Yankee Group research firm found consumer awareness of the new format (the first stage of adoption) surprisingly low despite significant publicity surrounding the launch of DVD hardware and software. Only 28% of the more than 1900 U.S. households surveyed by the Yankee Group were familiar with DVD. Among consumers who had heard about DVD, only 13% said they were very or somewhat likely to purchase a DVD player within the next 12 months.

TABLE 6.4
Consumer Awareness and Purchase Intentions for DVD

	% of Total Households	% of Total Households Aware of DVD
Have you ever heard of DVD?	28.3	
How likely are you to buy a DVD player?		
Very likely	0.5	1.8
Somewhat likely	3.1	11.1
Total	3.6	12.9

Source: Yankee Group Study (1997).

Perhaps more interesting from this survey of potential adopters is that approximately equal numbers of respondents indicated an interest in purchasing a DVD video or a DVD-ROM device ("Study indicates," 1998). One might easily argue that 50% awareness of the DVD product within a year of its launch is not all that low.

According to *GameWeek* (McGowan, 1998), DVD finished 1997 with an estimated U.S. household population estimated to be somewhere between 100,000 and 200,000. There were 350,000 DVD players shipped (not consumer purchases) to retailers in 1997, according to the Consumer Electronics Manufacturers Association. In terms of software, there were 1.5-2 million DVDs sold in the U.S. in 1997 at stores tracked by VideoScan. Warner Home Video claimed that Warner alone shipped more than 3 million DVDs (worth $50.6 million at wholesale) to retailers in 1997 including 92,000 copies of *Batman and Robin*. More than 500 DVD-Video releases were available at the end of last year and the DVD Video Group predicted that total would rise to 1500 at the close of 1998. The DVD system price point to reach mass consumer market acceptance is said to be $299 (Bismuth, 1998).

Cable and DBS are becoming far more serious competitors to home video than ever before. Much to the chagrin of the Video Software Dealers Association (VSDA), Hollywood has been allowing shorter windows between the time a movie comes out on video and the time it's available on pay television. Hollywood stands to benefit if the CD audio library rebuilding phenomenon is repeated in DVD. Just as CD adopters bought new copies of music they already owned in 33 1/3 rpm record albums, so might VHS tape owners choose to buy better quality DVD copies of movies they already have on tape.

6.7   FACTORS TO WATCH

Although sales of DVD players in 1998 will pale compared to those for VCRs, we are likely to see the beginning of a product substitution of DVD for VCRs. The random access capability of a disc cannot be matched by a tape cassette, and most home video movie buffs will not miss having to rewind a tape. If the CD market is any indication, software manufacturers will keep DVD prices higher than those for prerecorded videocassettes. This may well be based on psychological rather than economic considerations.

On the other hand, observers should not be fooled by inappropriate comparisons between 1997-1998 sales of DVD players and those of VCRs in the mid-70s or CD players in the early 80s ("DVD Posts Impressive Gains," 1998). The DVD player costs in real terms significantly less than 10% of what VCRs and CD players cost in their first year of introduction, so the perceived risk in adoption is dramatically less than what it was for its two predecessors. After videophiles purchase their player, the next set of adopters will look at how many disc titles are available. With only a few hundred select titles available at any one store in 1998 and many blockbuster movies still available only on VHS, growth of the DVD player will remain restrained.

As noted earlier in this chapter, Divx technology was created to allow limited viewing of digital movies on disc. This was attractive to studios who feared the notion of selling their movies in a format that allows for ready duplication. This author expected the marketplace to reject Divx. Indeed, web sites were produced that decried the introduction of Divx. Divx probably served one of two purposes: 1) it may have confused the marketplace, or 2) it may have simply increased overall awareness of the DVD format.

In spring 1998, Circuit City delayed the rollout of Divx until September 1998. Divx players were priced about $100 more than DVD players for what many certainly would view as a less functional system (e.g., a Divx disc cannot be played on a regular DVD player). The company expected to lose money on the format for two years (the consumer electronics retailing business operates on very thin profit margins, so this sounds very risky). History has shown that announcements of participation by other market players (Hollywood studios, in this case) do not portend market success (Klopfenstein, 1985), and Divx's prospects were further limited because Circuit City did not find other retail competitors to carry its product. Finally, in June 1999, Circuit City announced that Divx would be discontinued.

DVD players will continue to add features, perfect the technology, and lower the player price. Recordable DVD machines are in development although exact predictions of market introduction are all but impossible. A good rule of thumb is to ignore bold predictions about recordable DVD introduction dates and instead watch for the date when they are actually available. Creative Labs introduced the only DVD-RAM available in fall 1998 for a list price of $499.

In addition, there may be technical obstacles to reliable DVD recording if recordable CDs are any indication. One study found that only 41% of the CD-R discs evaluated from 13 major manufacturers passed a quality test, while almost 32% were marginal and 27% failed (Williams, 1998). Until one standard DVD recording technology is established, DVD recorders will not be an important factor. Indeed, if incompatible standards are sought, the life of the VHS VCR will be extended. The day is coming, however, when we will be transferring our home movies from analog tape to digital disc. DVD-RAM camcorders were available by late 1998.

The future of DVD was complicated by the discovery that certain DVD drives in computers could not read certain CD-recordable discs. More recently, the market has been confused by the approach of digital HDTV. DVD will not support the digital HDTV format, which has minimally five times the resolution of DVD. Promoters of DVD say HDTV will not make DVD obsolete, but one could argue that DVD will survive during the transition to HDTV and eventually die a slow death (Dvorak, 1998). Perhaps a more likely scenario is that future DVD technology will be adapted to meet the needs of HDTV.

6.8 DVD-RAM

DVD-RAM is the latest generation of high-capacity rewritable 5.2GB double-sided DVD-RAM discs that provide much greater storage capacity than existing CD-R systems. One of the advantages that DVD-RAM offered when introduced was its functionality as a 2X DVD-ROM drive capable of reading DVD-Video, DVD-ROM, DVD-R, CD-ROM, CD-R/RW, Video CD, and CD Audio. DVD-RAM also provides PD read/write and playback capability. Future developments of this type of storage technology will include the ability to read a removable 2.6GB DVD-RAM disc using DVD-ROM, thereby making DVD-RAM an even more flexible storage solution (Kwong, 1998).

Four camps were promoting different rewritable DVD standards in late 1997: DVD-RAM, which is championed by Matsushita's Panasonic and approved by the industry's DVD Forum; DVD+RW, which is endorsed by Sony and Philips; the DVD-R/W format proposed by Pioneer; and the DVD MultiMedia Video File Format (MMVFF) proposed by NEC. A major concern of three of the four competing camps was maintaining compatibility with some existing specific CD or DVD format, at least in the first generation of rewritable DVD systems. The DVD-RAM format by Panasonic will play CD-PD and most other CD format discs. Sony and Philips said they see DVD+RW as a natural extension of the CD-RW format, and in Pioneer's DVD-RW system, compatibility with DVD-R is considered critical to its niche of potential customers. The first DVD+RW drives were scheduled for sample shipments in mid-

1998, with retail shipments planned for the third quarter; the specifications call for a 3GB per side disc that does not require a caddy. It will play most CD formats, including DVD-Movie, DVD-ROM, DVD-R, CD-ROM, CD-R, CD-RW, and CD Audio.

One of the first DVD-RAM drives was available in fall 1998. The Creative Labs' drive, manufactured by Matsushita, reads DVD-ROM and DVD movie discs, reads all CD formats, and records over 5 GB of information on a $30 cartridge. The drive works exactly like a hard disc. Windows users could use the Universal Disc Format (UDF) to format the entire 2.6GB per side of the double-sided media as a single partition. According to *PC Magazine*, the drive was fast by optical disc standards when reading discs, but had fairly slow access times, about one-third the observed sustained read rate. To use DVD movie discs, an extra card is needed (Poor, 1998).

The DVD-RAM format that is supported by Panasonic, Hitachi, and Toshiba and also has the backing of eight of the 10 original DVD Forum members, features single and dual-sided discs which require a caddy and 2.6GB per side. The format's strength is said to be its random access characteristics and backward compatibility with discs using most of today's CD and DVD formats. Panasonic, Toshiba, and Hitachi planned to deliver drives in 1998 at a street price of about $799. Blank DVD-RAM media is sold in single-sided ($24.95) and dual-sided ($39.95) configurations. Both will require a caddy, although caddies for the single-sided discs are removable, allowing the bare disc to be played in future versions of DVD-ROM and DVD Video players. DVD-RAM promoters see their format as a data storage device first, but Panasonic has also made it clear that the specifications will serve as a platform for audio and video recording systems coming in the next five years, when greater storage capacities can be realized from new technologies such as blue laser, and when MPEG-2 encoded chips are more economically manufactured (Tarr, 1997).

Pioneer released DVD-R drives in October 1997 for $17,000; this price could drop within a few years to less than $5000. The initial price for blank DVD-Rs is $50. DVD-RAM drives will be introduced for less than $1000, with blank discs at about $30 for single-sided and $45 for double-sided. Disc prices for both DVD-R and DVD-RAM will drop quickly, but DVD-R discs will probably be cheaper in the long run. Toshiba, Pioneer, and Hitachi expected DVD-RAM to be available in 1998 (see "DVD FAQ," 1998).

While DVD-R may compete eventually with speedy and convenient removable magnetic discs like Iomega's Jaz drive, benefits must be weighed against the comparatively low cost of CD-R media and the huge installed base of CD-ROM drives. Optical formats such as CD-R and CD-RW are readable in the great installed base of CD-ROM drives worldwide. Indicating a widespread demand for a standard rewritable data-exchange format, CD-RW drives were outselling CD-R by two to three times in 1998. DVD-RAM already may be a viable alternative to CD-RW. Based on Matsushita's phase-change technology, first used in the 650 MB Panasonic Phase Change Dual (PD) rewritable drive, a device that debuted in 1995 and could read and write PD media as well as read CD-ROM, DVD-RAM is a two-sided, 5.2 GB disc. DVD-RAM drives can read DVD-RAMs, DVD-ROMs, DVD-Videos, DVD-Rs and PDs. They also can write to DVD-RAM and PD media. But DVD-RAMs can not be read in current, second-generation DVD-ROM drives.

Manufacturers developing these DVD-ROM drives say that they are fixing the problem (Gustavson, 1998).

Like DVD-ROM, DVD-RAM is based on the Universal Disc Format (UDF) and supports the MultiRead specification, which makes it backward-compatible with CD, CD-ROM, CD-R and CD-RW. Just as CD-RW drives now outsell CD-R drives, it is likely that DVD-RAM could become a preferred SCSI replacement for generic IDE DVD-ROM drives in high-end PCs. Creative Labs' $500 PC-DVD-RAM drive had a write speed nine times faster than that of CD-R (Gustavson, 1998).

6.9   DVD-AUDIO

Finally, another standards battle to watch is that which is going on with DVD audio. This format would seem to elicit less interest than DVD video because music copyright holders have little to gain with DVD. Copy protection is a key concern, and few audio applications require the hours of recording capacity that DVD audio will allow.

6.10   SUMMARY AND CONCLUSIONS

The outlook for DVD technology in the next 5 years appears to be very bright if not guaranteed. The diffusion of innovations is one way to understand how new communication technologies spread through or are rejected by the marketplace. The five attributes of innovations are relative advantage, compatibility, trialability, complexity, and observability (Rogers, 1995). The only way to know how an innovation rates on each of these attributes is to conduct research about the innovation among potential adopters. Speculation about DVD is possible based upon these attributes.

RELATIVE ADVANTAGE (IS IT BETTER THAN PREVIOUS IDEAS?)

DVD has clear advantages over CD. It has dramatically superior storage capabilities, faster read/write times, and the ability to display feature films in various formats on one disc. However, until titles become available, CD-ROM and VHS tapes will continue a substantial relative advantage to users.

COMPATIBILITY (DOES IT FIT WITH HOW ADOPTERS USE THE TECHNOLOGY?)

Once again, because there already is so much familiarity with CD technology, DVD would seem to score highly on the compatibility variable.

COMPLEXITY (CAN I UNDERSTAND IT?)

To the extent that an innovation is perceived to be complex, the less likely it is to be adopted. DVD may come down on both sides of this attribute. Once again, the widespread existence of CD technology should make DVD seem less complex. On the other hand, the availability of various DVD formats adds to complexity. Squabbles over standards within the DVD variations (DVD-audio, DVD-RAM, etc.) will increase.

REFERENCES

1. A year-long battle to re-energize (1998, January 5). *Video Business*, 18(1), 1, 7.
2. Arnold, Thomas K. (1998, April 16). Industry Alarmed Over Slowing Video Sales. *Los Angeles Times* [Online http://www.hollywood.com:80/news/topstories/04-16-98/html/1-2.html].
3. Arnold, Thomas K, (1997, December 14). Analyst Says Consumer Video Spending Is Up 6 Percent This Year, *Video Store Magazine,*) 8.
4. Bismuth, Alain (1998, March 30). PCs, players seek one DVD solution. *EETimes* [Online http://www.techweb.com/search/search.html].
5. Braham, Robert (1996, January). Consumer electronics. *IEEE Spectrum*, 33(1), 46-50.
6. Changing of the Guard: CD to DVD (1998, March 10). *PC Magazine* [Online http://www.zdnet.com/products/stories/reviews/0,4161,286447,00.html as of 10 October 1998].
7. Churchill, Sam (1996). Interactive TV trials [Online http://www.teleport.com/~samc/cable4.html].
8. C'City Earnings Up; Divx Delayed (1998, April 13). *TWICE (This Week in Consumer Electronics)* [Online http://www.twice.com/domains/cahners/twice/archives/webpage_2220.htm].
9. Cohen, J. (1991, December). Making sense of new electronics products. 74, 34
10. Cole, George. (1998, February 5). VCRs get smart (about time). *Electronic Telegraph (London Daily Telegraph)* [Online http://www.telegraph.co.uk:80/et?ac=000647321007942&rtmo=flfsvDVs&atmo=flfsvDVs&pg=/et/98/2/5/ecvid05.html].
11. Connolly, Daniel W. and Berners-Lee, Tim (1996, January 3). Names and Addresses, URIs, URLs, URNs, URCs [Online http://www.w3.org/pub/WWW/Addressing/Addressing.html].
12. Consumer Poll Gives Thumbs Down To Divx (1997, September 26). *TWICE (This Week in Consumer Electronics)*[Online http://www.twice.com/domains/cahners/twice/archives/webpage_1084.htm].
13. D'Amico, Marie (1995, June 5). Digital video discs. *Digital Media*, 5, 13.
14. David, Ted (1999, April 9). Power Lunch WSJ Personal Technology Columnist Mossberg: VCR Challenged? April 8, 1999 Interview. CNBC/Dow Jones Business Video.
15. Digital Video Systems Introduces Super Fast 5.2X Speed DVD-ROM Drive to Meet Escalating Demand (1998, September 28). *Business Wire*. [On-line http://www.businesswire.com, as of 6 October 1998].
16. Doyle, Bob (1997, July). DV Camcorders Take Stills Too [sic] [Online http://newmedia.com/NewMedia/97/07/td/stills.html].
17. DVD (1998). TechEncyclopedia. [Online http://www.techweb.com/encyclopedia/defineterm?term=DVD, as of 6 October 1998].
18. DVD Frequently Asked Questions (1998, April 7). [Online http://www.videodiscovery.com/vdyweb/dvd/dvdfaq.html].
19. DVD player sales through the first quarter of 1999 were 300% higher than in the same period in 1998 (1999). [Online http://www.twice.com/html/statistics.html, as of May 6, 1999].
20. DVD Posts Impressive Gains During First Year On Market, Far Outpacing Early Sales Of VCRs And CD Players (1998, April 3). *Business Wire*. [Online http://www.digitaltheater.com/news/apr3.html].

21. DVD-ROM Watershed: No Time Soon Software Publishers Have Yet to Embrace Format. (1998, October 5). *Phillips Publishing International.* [On-line via Northern Lights search engine http://www.nlsearch.com, as of 6 October 1998].

22. Dvorak, J. C. (1998, September 14). DVD Debacle. *PC Computing.* [OnLine http://www.zdnet.com/zdnn/stories/zdnn_display/0,3440,347530,00.html, as of 8 October 1998].

23. Everything About Television Is More (1998, March 6). *Research Alert,* 16(5), 1.

24. Gateway 2000 (1996, March). Gateway 2000 Launches Destination Big Screen PC Featuring 31-inch Monitor [Online http://www.gw2k.com/corpinfo/press/1996/destnew4.htm].

25. Gustavson, R. (1998, September). Special Report: Big Feats. *AV Video & Multimedia Producer,* 20(9). [Online http://www.kipinet.com/av_mmp/avmmp_sep98/special.htm, as of 8 October 1998].

26. Halpin, J. (1998, June). Pioneer First to Market. *Computer Shopper.* [Online http://www.zdnet.com/products/stories/reviews/0,4161,309321,00.html, as of 10 October 1998].

27. Helm, Leslie (1996, January 5). DVD Is Poised to Steal the Consumer Electronics Show. *Los Angeles Times,* Business 1.

28. Henrik Herranen, Henrik (n.d.). The Laserdisc FAQ [Online http://www.cs.tut.fi/~leopold/Ld/FAQ/Introduction.html].

29. Ideal Hardware (1996). The Video Age. Change magazine [Online http://www.worldserver.pipex.com/ideal/story3.htm].

30. Hall, R. (1999, February 12). Smart VCR Links to Net. *The Ottawa Sun,* 59. [See also online www.replaytv.com].

31. Hara, Y. (1999, April 5). International: Pair will codevelop interface spec based on 1394 - JVC, Sony push D-VHS as home digital recorder. *Electronic Engineering Times,* 26.

32. Home Recording Rights Coalition (n.d.). Selected Chronology of the Home Taping Controversy [Online http://www.access.digex.net/~hrrc/history.html].

33. Intercast Industry Group (n.d./1996). Internet and Television Technology Timeline. [Online http://www.intercast.org/timeline.html].

34. Johnson, Bradley. (1996, February 5). Counting eyeballs on the Net. *Advertising* Age. [Online http://www.adage.com/bin/viewdataitem.cgi?opinions&opinions148.html].

35. Johnson, Greg (1995, September 15). Agreement reached on a new format for video; Technology: Toshiba and Time Warner end battle with Sony and Philips on disc for use in devices to replace VCRs. *Los Angeles Times,* D-1.

36. King, Susan (1996, April 19). Home video rentals drop but revenues are still up. *Los Angeles Times,* 26.

37. Klopfenstein, B.C. (1985). *Forecasting the Market for Home Video Players: A Retrospective Analysis.* Unpublished doctoral dissertation. Columbus: The Ohio State University.

38. Klopfenstein, B.C. (1987). New technology and the future of the media. In A. Wells (Ed.), *Mass Media and Society* (19-37). Lexington, Mass.: Lexington Books.

39. Klopfenstein, B.C. (1989a). The diffusion of the VCR in the United States. In Mark Levy (Ed.), *The VCR Age.* Newbury Park: Sage Publications.

40. Klopfenstein, B.C. (1989b). Forecasting consumer adoption of information technology and services — lessons from home video forecasting. *J. Am. Soc. Inf. Sci,* 40(1), 17-26.

41. Klopfenstein, B.C. (1997). The future of new media technologies. In Alan Wells and Earnest Haakanen (Eds.). *Mass Media and Society,* 19-50. Greenwich, Conn.: Ablex.

42. Klopfenstein, Bruce C. (1998). Digital revolution in home video technology. In August Grant (Ed.), *Communication Technology Update* (6th Edition), 174-189. Boston: Focal Press.

43. Klopfenstein, B.C., and Sedman, D. (1990). Technical standards and the marketplace: The case of AM stereo. *J. Broadcasting & Electron. Media*, 34 (2), 171-194.

44. Koenig, S. (1998, October 5). Price Cuts, New Merchants Boost Sales — DVD Reaches A Critical Mass. *Computer Retail Week*, 1.

45. Kovar, Joseph F. (1998a, March 23). ATI Gives VARs The Option Of DVD-ROM Software. *Computer Reseller News* [Online http://www.tech-web.com/se/directlink.cgi?CRN19980323S0139].

46. Kovar, J.F. (1998b, August 24). Sourcing: Rapid Growth Seen For DVD Market In Next Five Years, *Computer Reseller News*, 141.

47. Krol, Ed and Bruce Klopfenstein (1996). *The Whole Internet, Academic Edition: User's Guide & Catalog*. Sebastopol, CA : O'Reilly & Associates.

48. Kwong, R. (1998, August 17). DVD-RW makes digital video editing faster and more flexible. *Computer Dealer News*, 14, 37.

49. Labriola, Don (1998, May). Drive DVD to Your Desktop. *Computer Shopper* [Online http://www.zdnet.com:80/cshopper/content/9805/297262.html].

50. Lanctot, Roger C. (1998, March 23). Household PC Penetration Disputed at 45 Percent. *Computer Retail Week* [Online http://www.techweb.com/se/directlink.cgi ?CRW19980323S0026].

51. Patrizio, A. (1998, September 10). DVD Dominates Digital-Video Market. *TechWeb News*. [On-line http://www.techweb.com/wire/story/TWB19980910S0011, as of 5 October 1998].

52. Pool, A. (1998, September 20). DVD-RAM Drives Arrive. *PC Magazine*. [Online http://www.zdnet.com/products/stories/reviews/0,4161,348616,00.html, as of 10 October 1998].

53. PCs in Over 45 Percent of US Homes in 1997. (1998, March 10). Reuters news service. [Online http://my.excite.com/news/r/980310/06/business-pc].

54. McGowan, Chris (1998). 350,000 DVD-Video Players Shipped in 1997. [Online http://www.gameweek.com/news/1_21/news1.htm].

55. Mmwire buy the number: 35% of U.S. Have computers, Says survey (1996, March 17). *Multimedia Wire* [Online http://www.mmwire.com:80/archive/03_17_96.html].

56. Myslewski, Rik (1996, March). Future Tech: We Have Seen the Future, and It's Huge. *MacUser* [Online http://www.zdnet.com/macuser/mu_0396/news/news02.html].

57. New products. (1995, April 1). *Popular Electronics*, 12, 85.

58. Novel VCRs and Camcorders Make CES Debut (1998, January 19). *Consumer Electronics*, 16(36).

59. Oosterveld, Jan (1996, March 13).CeBIT '96 Opening Speech. Hanover, Germany. [Online http://www-eu.philips.com/pkm/laseroptics/news/cebit.htm].

60. O'Shea, D. (1998, April 6). Further adventures in video. *Telephony*, 234(14),72.

61. Patrizio, A. (1998a, October 1). DVD Adoption Outpaces VHS, CDs. *TechWeb* [Online http://www.techweb.com/wire/story/TWB19981001S0022, as of 6 October 1998].

62. Pandemonium Productions (1996). Technofile: How to buy a camcorder. [Online http://www.technofile.com/guides/camguide.html; see also http://www.techno-file.com/articles/viewcam.html].

63. Parker, Dana J. (1999, January). Writable DVD: A guide for the perplexed. *E Media Professional*, 12(1), 30–39.

64. Pietrucha, Bill (1996, April 1). Video industries seek copyright, digital video legislation. *Newsbytes News Network*.

65. Pioneer New Media Technologies (1998). An Introduction to DVD Recordable (DVD-R). [Online http://www.pioneerusa.com/dvdwhite.html, as of 6 October 1998].

66. Prange, Stephanie (1997, December 14). Universal Readies Program to Revitalize Rental Market. *Video Store Magazine*, 1.

67. Rental stores keep packing in the crowds (1995, October 23). *USA Today.*

68. Rogers, E.M. (1995). *Diffusion of Innovations*, 4th Ed. New York: The Free Press.

69. Scally, R. (1998, September 7). Large-scale hardware rollouts confirm DVD's arrival. *Discount Store News.* 37(17), 73-76.

70. Scally, R. (1999, April 5). DVD music video sales upbeat as format recognition grows. *Discount Store News*, 38(7), 37-39

71. Schoenherr, S. (1996). Recording technology history: A chronology with pictures and links. [Online http://ac.acusd.edu/History/recording/notes.html].

72. Schwartz, Evan I. (1994, September). Fran-On-Demand [Online http://www.hotwired.com/wired/2.09/departments/electrosphere/cable.labs].

73. Snow, Shauna. (1996, March 7). Morning report; TV & video. *Los Angeles Times*, F-2.

74. Somerfield, H. (1996, March 25). Trouble in DVD paradise: Problems may delay Digital Video Disc launch [Online http://e-town.myriadagency.com //html/news_html/articles/9613hsc.html].

75. Sonic DVD Technology Used to Create World's First Multi-Format DVD Release (1998, September 30). *Business Wire.* [Online http://www.businesswire.com].

76. Sony (1998). About DVD Features [Online http://www.sel.sony.com/SEL/consumer/dvd/about_feat.html].

77. Spiwak, Marc (1995, December). Gizmo's holiday gift guide. *Popular Electronics*, 12, 6.

78. State of the DVD Union (1998, May). *Computer Shopper* [Online http://www.zdnet.com:80/cshopper/content/9805/297878.html].

79. Statistical Research, Inc. (1997). 1997 TV Ownership Survey [Online URL http://www.sriresearch.com/pr970616.htm.]

80. Study indicates DVD adoption to be slower than expected (1998, February 19). [Online http://www.cybersurvey.com/2-19.htm, as of 17 October 1998].

81. Tarr, Gerg (1997, December 8). Recordable DVD Erupts: What If They Gave a Format War and Nobody Came? [Online http://www.e-town.com/news/articles/dvd120897 gtt.html].

82. Taylor, J. (1998, October 7). DVD Frequently Asked Questions. [Online http://www.videodiscovery.com/vdyweb/dvd/dvdfaq.html, as of 17 October 1998].

83. VCRs in 88% of Homes (1996, January). *Television Digest with Consumer Electronics*, 36:2, 17.

84. VCR Near End of Long and Rewinding Road (1998, January 10). *Atlanta Journal-Constitution.* [Online http://www.accessatlanta.com/business/news/1998/01/10/ces 2.html].

85. VCRs Proliferate, As Do Their Brands (1998, January 5). *Video Week*, 16(33).

86. Williams, L. M. (1998, August). All CD-Rs Are Not Equal. *Tape/Disc Business*, 12(78). [Online http://www.kipinet.com/tdb/tdb_aug98/aug_index.html, as of 8 October 1998].

87. Williams, Stephen (1996a, January 7). The hard drive / new video disc calls the tune / DVD is star of electronics expo. *Newsday*, 07.

88. Williams, Stephen (1996b, January 28). Plugged in/Cyberscene/DVD is the talk of the CES. *Newsday*, 24.

89. Yankee Group Study Finds Low Awareness for DVD (1997, December 15). [Online http://www.yankeegroup.com/press_releases/lowAwareDVD.html].

7 Switched Network Carrying Capacities

George Scheets and Mark Allen

CONTENTS

7.1   INTRODUCTION

Effectively delivering bandwidth today is more important than ever before. Networks that deliver bandwidth to users and their applications in the most efficient manner with the appropriate Quality of Service (QoS) will likely prevail over less effective bandwidth delivery schemes. Until recently, voice made up the bulk of a network's traffic. Today, data dominates. Tomorrow's traffic will likely be multimedia with varying levels of tolerance for loss, delay, and variation in bandwidth. Systems of the future must allow for this mixture. Clearly, the network that can most effectively satisfy the user's requirements will ultimately provide the service at the lowest price, a key factor being the ability not to over-engineer the system when a less expensive solution will satisfy the customer. For example, the success of the frame relay service from public carriers is due to the user's willingness to share a bandwidth pool with others and to tolerate occasional loss and delay in exchange for significantly lower data transport costs compared to DS1 (1.544 Mbps) and fractional DS1 private line service.

The increasing emphasis on mixed traffic requiring QoS guarantees and ever-expanding amounts of bandwidth means that protocols that rely on inefficient sharing of the physical media are less likely to be acceptable for end-to-end delivery. One consequence of these demands is that today's telecommunication networks are increasingly switched in nature. WANs have relied on switches to consolidate and move traffic to their end destinations for years, but only recently have classical shared Ethernet and Token Ring LANs given way to switched Ethernet and Token Ring, and only recently have shared LAN backbones, such as FDDI, given way to switched Fast Ethernet, Gigabit Ethernet, and ATM systems.

This chapter examines the ability of switched networks to carry end users' application traffic, given a multiplexing choice and an offered load mix ranging from 100% voice or video to 100% bursty data traffic. The focus is on the backbone where fiber tends to dominate, but the concepts discussed are applicable in *any* environment where the effective use of bandwidth is required, including down to the desktop.

Today a network provider essentially has four trunking choices regarding what combination of switching and multiplexing schemes to use for delivering bandwidth to the customer: circut-switched TDM trunking, hybrid TDM trunking, packet-switched StatMux trunking, and ATM StatMux trunking.

Circuit-Switched TDM Trunking — In this configuration, the network delivers bandwidth to the customer at a constant rate with no buffering or bursting capability and with a minimal fixed delay. Figure 7.1 shows a simplified block diagram of an edge switch in this type of network. The input arrows represent *multiple* time-sensitive and data traffic sources. Traffic from these sources is multiplexed onto a trunk connection for transport to the next switch. Input voice or video, which will frequently be referred to in this chapter as *time sensitive traffic* (TST), could be either fixed, or variable rate in nature. Input from any data source is assumed to be bursty in nature. Figure 7.2 shows what the traffic from a typical data source might look like. The input is either active, in which case data traffic is entering the switch from this particular source at the line speed, or inactive, in which case no data is entering the switch from this source. To successfully move all of the offered traffic, a TDM circuit switch, which does not include buffering and is unable to handle traffic bursts, must assign dedicated trunk bandwidth to each source based on the peak (line) rates of each input. A 64 Kbps fixed rate voice conversation must receive 64 Kbps of trunk bandwidth. Data traffic offered at an average rate of 154 Kbps, and a peak rate of 1.54 Mbps, must be assigned 1.54 Mbps of trunk bandwidth.

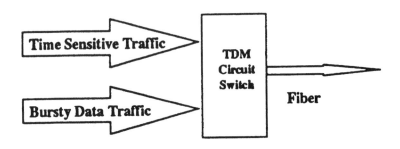

Figure 7.1 Circut-Switched TDM Trunking

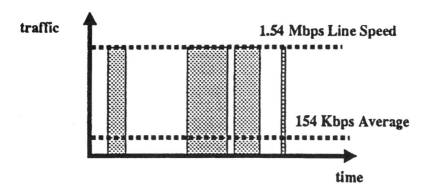

Figure 7.2 Bursty Data Traffic with a 10 to 1 Peak to Average Ratio

Hybrid TDM Trunking — Figure 7.3 shows a simplified block diagram of a hybrid network edge switching node. It is similar to the circuit-switched TDM configuration of Figure 7.1 except that *two separate networks are maintained.* All bursty data traffic is ideally groomed onto a packet-switched, statistically multiplexed (StatMux) network such as frame relay or the Internet, providing better utilization of network backbone resources. Fixed-rate time sensitive traffic remains on the circuit-switched TDM network. Variable rate voice and video could go either way, depending upon whether timely delivery or bandwidth efficiency is more important. Fiber bandwidth is assigned to the resulting packet-switched and circuit-switched bit streams on a dedicated circuit-switched TDM basis based on the peak rates of the resulting traffic.

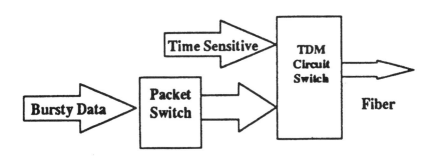

Figure 7.3 Hybrid TDM Trunking

Packet-Switched StatMux Trunking — All traffic, including that originating from fixed or variable rate time sensitive sources, is packetized and StatMuxed onto high speed trunks prior to insertion into the fiber, as illustrated in Figure 7.4. Packet Switching and StatMux are the foundations upon which the Internet Protocol (IP) and the Internet, as well as frame relay, are based. The Internet model is claimed

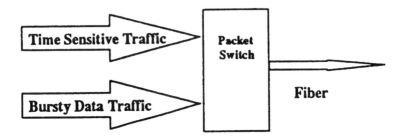

Figure 7.4 Packet Switch StatMux Trunking

by many to be the networking choice for the 21st century from which all telecommunications services will be delivered.

ATM StatMux Trunking — This is the technique perceived by others to be the network model for the 21st century. Traffic from all sources is segmented into fixed-size cells and StatMuxed onto high speed trunks prior to transmission over the fiber. Fixed-rate traffic is assigned constant bit rate (CBR) virtual circuits which are capable of providing TDM-like QoS. Bursty traffic is assigned variable bit rate (VBR), available bit rate (ABR), or unspecified bit rate (UBR) virtual circuits and is StatMuxed onto trunk bandwidth not reserved for CBR traffic. Figure 7.5 shows a simplified block diagram of this configuration.

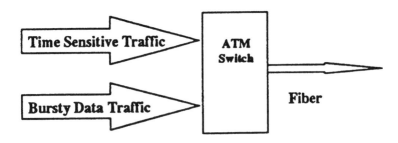

Figure 7.5 ATM StatMux Trunking

7.2   MEASURING A TRUNK'S ABILITY TO CARRY TRAFFIC

What parameter best measures a network trunk's carrying capacity? The trunk *Efficiency*, which is often defined as

$$\text{Efficiency} = \frac{\text{bits per second carried by trunk under heavy load conditions}}{\text{trunk line speed}} \quad (7.1)$$

is a parameter frequently touted. Figure 7.6 shows what a plot of trunk efficiency might look like. If the offered load consists of 100% time sensitive traffic, all of the previously mentioned configurations are able to completely load the trunk output lines, although it should be noted that the circuit switch TDM configuration can do so only if the voice and video are fixed rate. If any bursty traffic is offered, packet and ATM networks are more efficient as they are able to completely load the output trunk under heavy load conditions, but a circuit switch TDM backbone will have gaps in the traffic, as noted in Figure 7.2, and hence will have an efficiency less than 100%.

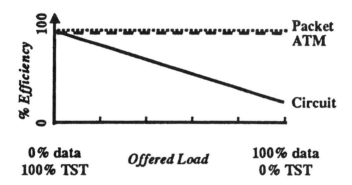

Figure 7.6 Switched Network Efficiency

However, the trunk efficiency does not tell the whole story. It does not account for the fact that a real-world StatMuxed trunk line carrying a 100% load is unusable as it either would have high queuing delays or would be dropping excessive amounts of offered traffic due to buffer overflows. As defined above, the efficiency also does not account for packet or cell overhead, although it should be noted that some definitions of efficiency do account for this overhead.

A more accurate measure would be the *carrying capacity* or *utilization,* which is defined here as

$$\text{Carrying Capacity} = \frac{\textit{carriable end user application traffic} \text{ in bits/second}}{\text{trunk line speed}} \quad (7.2)$$

The carrying capacity accounts for packet and cell overhead, and it accounts for the inability of StatMux switches to fully load output lines and have a usable system. Figure 7.7 shows what a plot of trunk utilization might be expected to look like. Note the differences between the packet switch and ATM utilization, and the packet switch and ATM efficiency.

The following sections provide details as to how the carrying capacity for each of the four different trunking options can be computed. They examine the issues that affect the amount of overhead consumed and how fully a trunk circuit can be loaded as the traffic mix changes between TST and data traffic. The overhead and

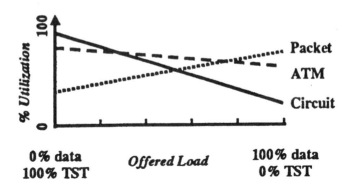

Figure 7.7 Switched Network Carrying Capacity

the StatMux queuing delays impose some severe penalties on a packet switch network's ability to carry time sensitive traffic, lowering the carrying capacity. ATM, which was originally designed to carry mixed traffic, not surprisingly shows high utilization when the offered traffic load consists of a combination of time sensitive and bursty data sources. ATM's ability to give CBR traffic TDM-like QoS gives it a high utilization when the offered load is all fixed-rate TST, and its ability to StatMux bursty traffic gives it high utilization when the offered load is all data.

The following discussion and examples focus somewhat on WANs, but the results can easily be extended to the MAN or LAN by appropriately adjusting the overhead and line speeds.

7.3   CIRCUIT-SWITCHED TDM TRUNKS

Traffic sources, be they fixed-rate voice or video, variable rate voice or video, or bursty data traffic, are all assigned trunk capacity based on the peak rates of each input circuit in a circuit switch TDM backbone network (see again Figure 7.1). The overall carrying capacity can be calculated based on knowledge of the *average* peak-to-average ratios of injected data traffic, the *average* peak-to-average ratios of the injected time sensitive traffic, traffic overhead, and knowledge of the ratio of data to TST being moved over the trunk, via the equation

$$\text{CapCSTDM} = \frac{(\% \text{ traffic to overhead})(\% \text{ usable line speed})}{(\text{peak-to-average ratio})} \tag{7.3}$$

An example of the calculations required is shown in Figure 7.8, which itemizes sources of bandwidth loss when the offered load is 100% bursty data traffic being carried over a SONET-based fiber system. On a typical 810 byte SONET frame, 36 bytes are set aside for operations, administration, and maintenance (OA&M) overhead purposes. Assuming the average packet size of data traffic is 300 bytes, as has recently been measured on the MCI Internet backbone,[1] data traffic originating from routers would require 6 bytes of Level 2 overhead for High Level Data Link Control (HDLC), 20 bytes of Level 3 overhead for the Internet Protocol version 4 (IPv4), and 20 bytes of Level 4 & 5 overhead for Transmission Control Protocol. Hence, 46 out of 300 bytes (15%) are lost for overhead for each packet, on average. Assuming that a weighted average of all input circuits carrying packet traffic indicated that, on average, 83% of the time the input packet circuits have idle bandwidth, and 17% of the time traffic is actually moving, then a 6-1 peak-to-average ratio is indicated. The overall result would be a trunk utilization of

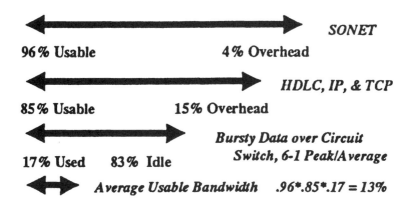

Figure 7.8 Usable Bandwidth: 100% Data over Circuit Switch TDM SONET

$$CapCSTDM = \frac{(254/300)(774/810)}{6} = 0.1348$$

in this situation. In other words, if the offered load to the switch is 100% bursty data being injected at an average rate of 100 million bits *of end user application traffic* each second with a 6-1 peak-to-average ratio, 100 Mbps/.1348 = 742 Mbps of trunk bandwidth would be required to carry this load. This is not a very effective way to haul data!

At the other extreme, if powerful add-drop multiplexers are available to multiplex individual 64 Kbps fixed-rate voice conversations onto SONET, the primary overhead would be the SONET OA&M traffic, allowing a carrying capacity near 96% to be achieved for TST.

Figure 7.9 shows several plots of circuit-switched TDM utilization as the switch offered load varies from 100% time sensitive to 100% bursty data traffic, for different data peak-to-average ratios. The TST is fixed rate for these graphs, as that is what this type of network most effectively transports.

7.4 HYBRID TRUNKING

In this configuration, the goal is to operate two distinct networks: a TDM-based network for transporting TST and a packet-based network for carrying bursty data

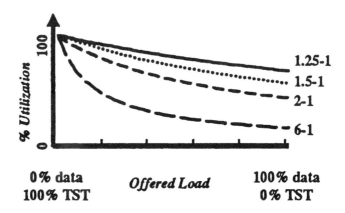

Figure 7.9 Circuit Switch TDM Trunk Utilization for various data peak-to-average ratios

traffic that lends itself to traffic shaping and StatMux. The key difference between this technique and the previous is that ideally *all* bursty data traffic is aggregated onto a packet-switched network (see Figure 7.3). StatMuxing many high peak-to-average ratio circuits together will generate fewer, more heavily utilized packet-switched output trunks, with lower peak-to-average ratios. Backbone capacity is again assigned on the basis of the peak traffic rates of the resulting circuits. As before, the overall carrying capacity can be calculated based on knowledge of the *average* peak-to-average ratios of injected data traffic, the *average* peak-to-average ratios of the injected time sensitive traffic (which ought to be 1-1 if *all* bursty traffic is shipped to the packet switch), traffic overhead, and knowledge of the ratio of data to TST being moved over the trunk.

Figure 7.9 may also be used to estimate the utilization for a hybrid network, as a key function of the hybrid system is to consolidate and shape the packet traffic, thereby reducing the peak-to-average ratio of bursty traffic injected onto the fiber. The consolidated traffic still utilizes dedicated circuit-switched TDM trunk connectivity to adjacent switches, so using the peak-to-average ratios as in Section 7.3 is appropriate for this discussion. It should be noted, however, that the techniques discussed for calculating the ATM carrying capacity in Section 7.6 could be modified to calculate the carrying capacity for hybrid networks, yielding slightly more accurate results.

As an example, if a circuit-switched TDM system with a mixture of fixed-rate voice and bursty data traffic with an *average* input peak-to-average ratio of 6-1 is replaced with a hybrid system capable of consolidating the data traffic onto a smaller number of high speed channels with an 80% load (a peak-to-average ratio of 1.25 to 1), the lowest line of Figure 7.9 would apply to the circuit-switched TDM system and the highest plotted line would apply to the hybrid system. A network that does not fully off-load all the data traffic onto the hybrid network packet switch would lie somewhere between these two extremes.

Examine this graph for an offered load mix of 70% data and 30% voice. The circuit switch system has a utilization of 18% and the hybrid system has a utilization of 72%. This means that for this example, a circuit-switched TDM backbone would require .72/.18 = 4 times the trunk bandwidth and higher speed switches, than a hybrid system hauling the *same* offered load. Depending on the exact equipment costs associated with each network, the hybrid system is likely to offer considerable installation cost savings. The key problem faced here would be properly segregating the traffic so that the highest possible utilization is actually achieved.

Many of the established public carriers originally deployed circuit switch TDM networks in the seventies and eighties, as that was the most economical choice for the voice-dominated systems of the time. Increases in computing power accompanied by simultaneous decreases in the cost of that power resulted in a rise in data traffic and the realization that circuit-switched TDM backbones were not a good choice in an increasingly data intensive environment. Eventually carriers began deploying hybrid systems and made a concerted effort to move as much data traffic as possible onto packet networks, such as frame relay, in order to better utilize their trunk bandwidth and offer lower cost connectivity to their customers. Today, the older carriers commonly deploy some sort of hybrid network to satisfy the continually growing demand for voice and data transport, with varying degrees of success in moving bursty traffic onto the packet side of the house.

7.5 PACKET-SWITCHED STATISTICAL MULTIPLEXED TRUNKS

As shown in Figure 7.4, in this technique traffic from *all* sources is packetized and StatMuxed onto trunks. Carrying capacity can be calculated based on knowledge of the average packet size of the injected data traffic, average packet size of the injected time sensitive traffic, tolerable delays through a typical network switch, ability of the network to prioritize traffic, knowledge of queuing theory and the recent discoveries of self-similarity in network traffic, and some knowledge of the processing limits associated with each switch or router.

In a manner analogous to what is shown in Section 7.3 and Figure 7.8, the carrying capacity of a packet-switched StatMux network can be calculated via

$$CapPSSM = \frac{(\text{Average } application \text{ traffic per package}) \times (\% \text{ Usable Line BW}) \times (\text{Trunk Load})}{(\text{Average Packet Size})} \quad (7.4)$$

Everything in this equation is relatively straightforward except for the trunk loading parameter, which is the inverse of the peak-to-average ratio. Determining the tolerable trunk loading requires a knowledge of queuing theory, a field which is currently somewhat unsettled due to discoveries in the last few years that data traffic has self-similar characteristics, meaning that many of the 'old reliable' (and inaccurate) queuing results have gone out the window. Some of the key results are briefly summarized here. The interested reader is referred to Stallings[2] for a very readable overview.

Queuing theory predicts that if the size of input packets is exponentially distributed and independent of the size of previous packets, and if the time between packet arrivals is also exponentially distributed and independent of the previous inter-arrival times, then the average queuing length in a switch is

$$\text{Average Queue Length (in packets)} = \frac{\text{Trunk Load}}{1 - \text{Trunk Load}} \qquad (7.5)$$

Experience has shown that these assumptions are not quite true for real-world traffic, with the result that this equation tends to predict overly optimistic small queue sizes. More recent work indicates that under certain circumstances, the following equation provides a more accurate estimate of the average queue length

$$\text{Average Queue Length (in packets)} = \frac{(\text{Trunk Load})^{\,0.5/(1-H)}}{(1 - \text{Trunk Load})^{\,H/(1-H)}} \qquad (7.6)$$

where H is the Hurst parameter, a value which lies between .5 and 1.0. A Hurst parameter of .5 implies that no self-similarity exists, and Equation 7.6 then simplifies to Equation 7.5. A Hurst parameter value of 1.0 implies that the traffic is completely self-similar, which essentially means that a traffic trace viewed on any time scale (any zoom factor) would look somewhat similar. Figure 7.10 shows a plot of Equation 1.6, for Hurst parameter values of .5 and .75. The key point to note here is that self-similar traffic (such as with H=.75), which has burstiness that is more 'clumped' than the 'smooth' burstiness associated with the exponentially distributed model (H=.5), has queues that tend to build more rapidly under smaller loads. This translates directly into higher queuing delays at a switch for packets that are not dropped, as the

$$\text{Average Queue Delay (in seconds)} = \frac{(\text{Average Queue Length}) \times (\text{Average Packet Length})}{\text{Trunk Line Speed}} \qquad (7.7)$$

While the jury is not yet completely in, initial studies indicate that the Hurst parameter for typical packet and cell traffic is probably somewhere between .7 and .9.[2-3]

A StatMux network switch can be considered to be operating in one of two modes:

(1) *low load*, where delay and not loss is a problem, or
(2) *heavy load*, where loss and not delay is a problem.

The Hurst parameter of the offered traffic will impact both modes. Using Equations 7.6 and 7.7 the Hurst parameter can be used to estimate the average queuing delay for the low load instance. Of equal importance is the heavy load case. Here the Hurst parameter will impact the probability that a buffer overflows.

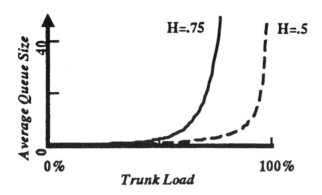

Figure 7.10 Queue Length vs. Trunk Load for H = .75 and H = .5

Figure 7.10 shows plots of the average queue lengths for switches with infinite length buffers. At any specific instant in time, the actual queue length is likely to be greater than or less than this average. To determine the probability that a switch with a finite length buffer overflows, which will impact the QoS hence the allowable load, what is needed is the distribution of the queue lengths as a function of the offered load traffic mix and the H parameter of that mix. Real-world distributions are generally *extremely* difficult, if not impossible, to find because they are directly impacted by the queue handling schemes of particular manufacturers and protocols, which are often quite complicated. Until research yields a simple and reasonably accurate solution, we suggest setting the maximum trunk load such that the average queue size predicted by Equation 7.6 is significantly less than the trunk queue size available in the switch. For comparison purposes, this chapter has standardized on an 80% maximum trunk load for all systems.

Considering the above information, estimates of a packet-switched network's carrying capacity can be obtained in the following manner:

1. Choose the target system-wide average end-to-end delays for *both* your time sensitive and data traffic, and estimate the average queuing delay allowable through a typical switch.
2. Estimate the average packet size and overhead associated with bursty data traffic and time sensitive traffic.
3. Estimate the Hurst parameters associated with your traffic. Doing this accurately may be somewhat difficult as determining the Hurst parameter from finite amounts of data is notoriously inaccurate.[4] What is known is that a Hurst parameter of .5 (meaning no self-similarity) *is known to be inaccurate for data.* A Hurst parameter of 1.0 must also be inaccurate, because it would imply that traffic plots would look similar if plotted on any scale. This is clearly incorrect for real-world traffic, as different 'zooms' will yield nonsimilar plots. Consider Figure 7.2 if you've 'zoomed' down to a single bit. A value of .75 is tentatively suggested

for use as a compromise in the event that additional information is lacking, as this value lies in the middle of the extreme Hurst parameter values and is also near the middle of the ranges noted for actual traffic from preliminary studies.

4. Estimate the maximum load your switches can reliably place on the output trunk lines. Trunk loads exceeding this value are assumed to result in intolerable amounts of packets being dropped due to finite buffer sizes. This parameter will impact the carrying capacity under heavy load conditions, where the queuing delay is easily met but the fear of overflowing the switch buffer limits the trunk loading.

5. Use weighted averages of steps 1–3, above, to account for the appropriate traffic mix.

6. Then use Equation 7.7 to solve for the average queue lengths.

7. Use Equation 7.6 to solve for the trunk loads.

8. Bound the Trunk Load by the value in step 4 if necessary.

9. Use Equation 7.4 to compute the carrying capacity.

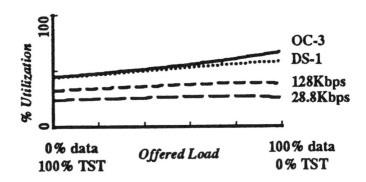

Figure 7.11 Packet Switch StatMux Trunk Utilization

Figure 7.11 shows some plots of packet-switched StatMux utilization as the switch offered load varies from 100% time sensitive to 100% bursty data traffic, for different trunk line speeds. These plots are based on the following assumptions:

- Average queuing delay through a network packet switch for time sensitive traffic is 20 msec, 40 msec for data. IPv4 is being used with no QoS provisions enabled, meaning all traffic must be moved through a switch with an average queuing delay of 20 msec in order to meet the tighter TST requirements.

- The Hurst parameter associated with both the data and time sensitive traffic is .75, a value believed to be a reasonable compromise based on some preliminary studies.

- Maximum reliable load that a packet switch can place on its output trunk is 80%.

- Average packet size of the data traffic is 300 bytes.[1] As mentioned earlier, the overhead would consist of 46 bytes, 6 bytes of Level 2 overhead for HDLC, 20 bytes of Level 3 overhead for IPv4, and 20 bytes of Level 4 & 5 overhead for TCP, leaving 254 bytes for the application.
- Time sensitive traffic is assumed to be mainly 8 Kbps compressed voice being moved at an average rate of 20 packets/second (50 bytes of voice + 8 bytes of user datagram protocol (UDP) overhead + 20 bytes of IPv4 overhead + 6 bytes of HDLC overhead).

Note that of the parameters listed above, the values that most affect the carrying capacity at broadband rates are the packet sizes (smaller packets have a larger percentage of overhead), and the maximum reliable load that the switches can support. With high speed trunks the carrying capacity will often not be limited by the allowable average switch queuing delays, but instead will be limited by switch buffer sizes, i.e., the switch will often be operating under heavy load conditions.

Figure 7.11, shows that with high speed trunks the small packet sizes required for timely delivery of digitized voice adversely impact the network's carrying capacity. Larger voice packets would improve the utilization, but at the same time they would drive down the quality perceived by the end user by increasing the end-to-end delivery delay. Broadband packet-switched StatMux networks offer the highest carrying capacities if they carry the type of traffic they were originally designed for, bursty data traffic.

Not evident from this plot is that increasing the trunk line speed to greater than OC-3 rates will not yield any additional utilization benefits, if the heaviest load that a switch can reliably place on the trunk line is 80%. Under this condition, a plot of OC-12 carrying capacity is virtually identical to that of OC-3. If a switch could handle a trunk load greater than 80%, which, depending upon the switch configuration, may very well be possible due to increased buffer sizes or the increased StatMux gains available using larger trunk sizes, these systems would show slight utilization increases per Equation 7.4.

At lower line speeds, the packet sizes, coupled with the choice of average switch queuing delay for this example, require that the trunks be lightly loaded, limiting the overall utilization.

7.6 ATM STATISTICAL MULTIPLEXED TRUNKS

As is noted in Figure 7.5, in this technique all traffic is inserted into fixed-size 53-byte cells and multiplexed onto a high speed trunk prior to insertion into fiber for transmission.

Fixed-rate traffic is best treated as a native ATM application hauled via CBR using ATM Adaptation Layer One (AAL1), which adds one byte of overhead per cell for sequencing purposes. As a result, 47 of the 53 bytes are available to carry traffic. ATM switches can offer TDM-like services to CBR traffic, reserving an appropriate number of cells at regular time intervals for this class of service.

Bursty traffic is normally carried via either VBR, ABR, or UBR classes of service, which are StatMuxed onto the remaining trunk bandwidth not reserved for

CBR traffic. In this chapter, bursty traffic is assumed to be passed down to AAL5 in the form of IP packets. AAL5 adds 16 bytes of overhead to each packet prior to segmentation.

Similar to what we saw in Section 7.5, the carrying capacity of ATM trunks can be calculated via

$$\text{CapATM} = \frac{(\text{Average } application \text{ traffic per cell}) \times (\% \text{ Usable Line BW}) \times (\text{Trunk Loading})}{(53 \text{ bytes})} \quad (7.8)$$

The key difference between Equations 7.8 and 7.4 is how the trunk loading is treated. In ATM, since fixed rate sources can be given TDM-like service by reserving specific cells for CBR traffic, the trunk loading for CBR under heavy load conditions is 100%. Bursty traffic would be StatMuxed onto the remaining trunk bandwidth not reserved for CBR service. Note that for a trunk with a fixed-amount bandwidth, as the offered load is varied from 100% bursty traffic to 100% fixed-rate traffic, the bandwidth available for StatMux use will decrease as more and more will be reserved for the fixed-rate traffic. Otherwise, the same technique used in Section 7.5 is used to estimate the carrying capacities here.

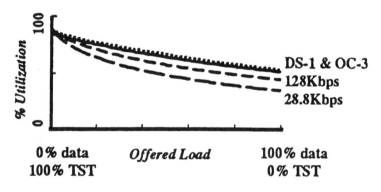

Figure 7.12 ATM Switch StatMux Trunk Utilization

Figure 7.12 shows a plot of ATM utilization as the switch offered load varies from 100% time sensitive to 100% bursty data traffic, for different trunk line speeds. This plot is based on the following choices:

- Average tolerable queuing delay through a network StatMuxed cell switch is 40 msec for data traffic, the same as in the previous section. These delays would be the average delay of all moved VBR, ABR, and UBR cells.
- The Hurst parameter associated with the bursty traffic is .75.
- The maximum reliable load that a cell switch can StatMux onto its output trunk is 80% of the line speed not reserved for CBR traffic.

- Average packet size of the data traffic offered to AAL5 is 300 bytes.[1] An ATM switch would first drop the overhead associated with HDLC and, as mentioned earlier, would then add 16 bytes of AAL5 overhead to each packet. The result would then be segmented into 48-byte chunks for insertion into ATM cells.
- Voice and video traffic is a fixed-rate native ATM application.

As with the packet-switched StatMux case, of the parameters listed above the values that most affect the carrying capacity at broadband rates are the packet sizes (smaller data packets offered for segmentation have a larger percentage of overhead) and the maximum reliable load that the switches can support. Note the ability of ATM to offer reasonably high utilization at low speeds. The smaller fixed-sized cells allow a higher load to be placed on the outgoing trunk while still meeting switch average delay specifications.

7.7 HEAD-TO-HEAD COMPARISON

It is illuminating to plot the carrying capacities of the four types of networks on a single graph for comparison purposes, similar to Figure 7.7. Figure 7.13 does so for OC-3 trunks. Note the following:

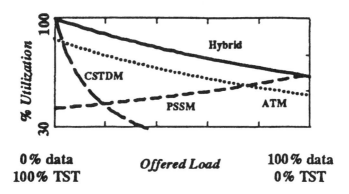

Figure 7.13 OC-3 IPv4 Head-to-Head Comparison

- The circuit-switched TDM backbone offers its highest carrying capacities if the offered load is almost 100% fixed rate. It *rapidly* falls off as bursty data becomes a larger percentage of the load, due to the well-known inability of this technique to efficiently carry bursty traffic. It is capable of hauling fixed-rate voice and video with a minimum amount of overhead.
- Packet switching and StatMuxing, which were originally designed to haul bursty data, not surprisingly haul this type of traffic best. However, when time sensitive traffic such as voice is offered, the overhead associated with packetizing this traffic seriously impacts the utilization. Given voice traffic with either fixed or variable bit rates, a packet-switched StatMuxed

network *cannot* match the utilization that a circuit-switched TDM network can achieve with fixed-rate voice traffic, provided that the average bit rates of the voice sources are the same.

• The hybrid backbone uses the best of both worlds, circuit switching and TDM for fixed-rate time sensitive traffic, and packet switching with Stat-Muxing for bursty data traffic. Provided that the load is segmented properly, this technique can *potentially* offer the highest possible overall utilization. Note, however, that if the traffic is not properly segmented, if some of the bursty data traffic is transmitted over the circuit-switched TDM network, then the *average* peak-to-average ratio will go up, reflecting the fact that more of the data traffic will not have been consolidated onto a small number of heavily used StatMuxed trunks. Depending on the degree of segmentation, the utilization of a hybrid backbone could lie between the plotted values (for 100% segmentation) and the circuit-switched TDM backbone (for 0% segmentation).

• ATM hauls no specific type of traffic best, but instead is clearly well suited for the mixed traffic environment for which it was designed. Its ability to offer different classes of service to different traffic sources allows it to follow in the shadow of the hybrid network in terms of carrying capacity. It cannot quite match the hybrid network's utilization due to the additional AAL and cell overhead, as well as the fact that it is a compromise, not tailored to a specific type of traffic, as are the circuit-switched TDM and packet-switched StatMux techniques. While ATM suffers a common problem with the hybrid network in that improperly segmented traffic will reduce the system carrying capacity, it is potentially far easier to properly classify the traffic because each flow can be assigned an appropriate class of service by the end user based on cost and desired quality, without constant carrier oversight.

Clearly, in terms of the network carrying capacity, different traffic mixes are best served by different trunking technologies.

Figure 7.14 shows essentially the same plot except that IPv4 has been replaced with IPv6. Shortcomings associated with IPv4 have resulted in the development of IPv6 which adds additional features at the cost of additional overhead — primarily larger source and destination address fields. IPv6 is expected by many to see significant deployment around the turn of the century. This change balloons the IP header from 20 to at least 40 bytes. Additionally, in the plots shown, IPv6's priorities are assumed to be enabled such that packet switches are able to meet a 20 msec average delay for the time sensitive traffic, and a 40 msec average delay for the bursty data. The overall result is that the utilization crossover point of ATM and packet switching moves from about 75% data to 85% data. In terms of utilization, even though the processing requirements at packet switches are relaxed to 40 msec for bursty data traffic, the additional IP overhead clearly makes the case worse for a 100% Internet backbone. Not evident from the plot, however, is that the use of these priorities would improve the quality of TST.

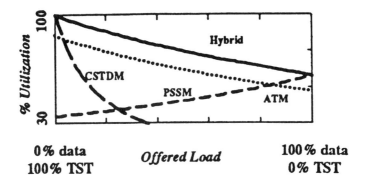

Figure 7.14 OC-3 IPv6 Head-to-Head Comparison

It was mentioned earlier that, at least on the WAN, carriers that have been around for awhile and have much capital sunk into older technology have tended to deploy hybrid-type networks. Recent buildouts of newer carriers, who don't have to worry about backward compatibility, have tended towards consolidated techniques whereby *all* traffic is carried on the backbone over a single, core technology. The two techniques most heavily touted have been 100% packet-switched StatMuxed backbones, specifically the IP-based Internet, and ATM. The carrying capacity provides useful insight into which technique has the potential to be the lowest cost solution, provided one can nail down the current and future offered traffic mix. Today, data traffic is clearly growing at a faster rate than that of time sensitive traffic, but will that remain the case in the immediate future? As the cost of bandwidth declines, how will real-time video traffic grow? Plenty of science fiction movies show high fidelity, interactive video conferencing as commonplace as the current telephone system. Can they all be wrong? Future communications will certainly require interactive video that is both time sensitive and possibly bursty in nature. Clearly, TDM-based networks will be left behind. The only remaining question is what technologies (IP, ATM, or both) are best suited to implement the necessary flow-based queuing and bandwidth reservation schemes.

Figure 7.15 shows a plot of the carrying capacities of circuit-switched, ATM, and packet-switched trunks running at DS-1, a speed commonly used for corporate enterprise WAN connectivity. Shown is the case when IPv6 and priorities are enabled. A plot for IPv4 looks almost identical; the inability of IPv4 to load trunk lines more heavily by prioritizing traffic (meaning that in order to meet time sensitive traffic delay criteria it must also whisk data through switches at TST rates) is compensated by the smaller percentage of packet overhead. Interestingly, at slow speeds ATM's utilization tends to dominate for almost any traffic mix. Given a target average queuing delay through a switch, slower speed connections are more likely to be delay-constrained than buffer-constrained. Equation 7.7 indicates that for identical target queuing delays, an ATM switch will have on average about six times the 53-byte cells queued up than the number of queued packets in a packet switch moving 300-byte packets. From Equation 7.6 and Figure 7.10 it can be seen

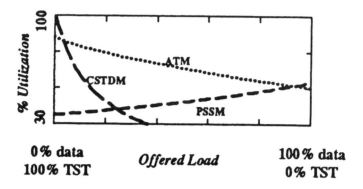

Figure 7.15 T-1 IPv6 Head-to-Head Comparison

that this allows an ATM trunk line to be more heavily loaded than the packet switch trunk line. The extra loading goes a long way towards canceling the extra overhead associated with chopping up the packet into smaller cells, resulting in ATM's carrying capacity almost matching that of packet switching for an offered load of 100% data. ATM warrants more consideration as the switching and multiplexing technique of choice for low speed connectivity than it is currently receiving.

7.8   CONCLUSIONS

The price a network technology pays to haul application traffic is clearly influenced by the value of the network's utilization, or carrying capacity. A network that requires a lot of overhead or is unable to load its trunk lines heavily is a network with a lowered carrying capacity. While this parameter, as discussed in this chapter, does not account for signaling overhead, the granularity associated with various trunking protocols such as SONET, or the impact of higher level protocols such as TCP, it nevertheless provides important insight into which techniques are best suited for hauling different traffic mixes. A technology with a higher utilization for a given traffic mix requires less trunk bandwidth and lower switch speeds to move a given amount of application traffic than a protocol with a lower utilization. This directly impacts the bottom line. Depending on the relative equipment costs, the network with the higher utilization potentially costs less to deploy.

Of the switching and multiplexing mixes discussed, the hybrid network clearly offers the highest *potential* utilization, but it suffers two key drawbacks. It requires twice the hardware of all-ATM, all-packet-switched, or all-circuit-switched networks, and it requires careful grooming to achieve these high utilization values.

Of the consolidated networks, a circuit-switched TDM backbone works best for fixed-rate time sensitive traffic, which makes it a *horrible* choice for today's traffic mix which is increasingly dominated by bursty data. The packet-switched StatMux network has the highest utilization for bursty traffic, and ATM is best for a mixture. Not surprisingly, each technology has the highest carrying capacity when used to haul the traffic mix for which it was originally designed.

The overall choice for deployment or upgrading of a network depends on at least three key issues: installation cost, maintenance costs, and reliability.

Many factors impact the installation cost of a network, including power requirements, bay size, backward compatibility needed, switch speeds and configuration, trunk sizes, and last-but-not-least, carrying capacity. Knowledge of the carrying capacity is vital here, as it enables an analyst to get a better idea of the trunk bandwidth and hardware speed requirements of networks hauling equivalent' loads, factors that directly impact the price of installing the network.

Maintenance costs reflect the organizational, administrative, and day-to-day operating costs associated with running the network after it is installed. A backbone having the fewest *types* of equipment has significant advantages here, as less effort will be required to integrate disparate hardware with the network control center, and fewer engineers and technicians with expertise on specific pieces of equipment will be required. A network technology ready-for-prime-time is also likely to have reduced maintenance costs compared to a network still in the 'Bleeding Edge' stages.

Reliability refers to the network's ability to maintain maximum uptime and its susceptibility to a catastrophic failure. Barring that catastrophic failure, all four backbone techniques should be engineerable to equivalent 99.99% uptimes. Catastrophic failures which bring down the *entire* network are extremely rare but can have disastrous effects. One has only to look at recent Internet or frame relay events to wonder whether putting all one's eggs in a single basket is a good idea. For example, in July of 1997 a large chunk of the Internet was isolated when Network Solutions botched a top level domain name server update. In April of 1998, AT&T's entire frame relay network went down for over a day during a switch upgrade of improperly tested software. There *is* something to be said about maintaining a hybrid network which would be more difficult to bring down totally.

Despite some pundits' claims to the contrary, the choice of the 'best' network technology is not clear-cut, as that choice depends on intangibles, that are frequently hard to quantize, and the cost that one is willing to pay. That cost can be severely impacted by the parameter which is the focus of this chapter, trunk carrying capacity. Can 155 mbps of Internet bandwidth haul the same amount of customer application traffic as 155 mbps of ATM? The often overlooked carrying capacity will tell you.

ACKNOWLEDGMENTS

This work was supported in part by a contract from the Williams Communications Group, Tulsa, Oklahoma.

REFERENCES

1. Thompson, K., et al., 'Wide Area Internet Traffic Patterns and Characteristics,' *IEEE NETWORK*, November/December 1997, p. 10-22.
2. Stallings, W., *HIGH SPEED NETWORKS: TCP/IP AND ATM DESIGN PRINCIPLES*, Prentice-Hall, 1998.

3. Crovella, M. and Bestavros, A., 'Self Similarity in World Wide Web Traffic: Evidence and Possible Causes,' *IEEE/ACM TRANSACTIONS ON NETWORKING*, December 1997, p. 835-846.
4. Michiel, H. and Laevens, K., 'Teletraffic Engineering in a Broad-Band Era,' *IEEE Proceedings*, December 1997, p. 2007-2033.

WEB ASSISTANCE

Interested in evaluating the carrying capacity for your current or proposed network? Don't agree with the choice of parameters used here and want to examine the results with different selections? A downloadable MathCad® executable file can be obtained at *http://www.mstm.okstate.edu/files/www/faculty/scheets/pub/wcg_cap4.html*. A non-executable Word document is available at *http://www.mstm.okstate.edu/faculty/scheets/pub/wcg_cap4.doc*

8 The Development Of Multiprotocol Label Switching

To Integrate IP With ATM For The Internet Backbone

Andrew G. Malis

CONTENTS

8.1 INTRODUCTION

This chapter describes the many methods developed to integrate IP routing with ATM switching, culminating in the development of MPLS, or Multiprotocol Label Switching. MPLS is a set of standards being developed by the Internet Engineering Task Force (IETF) to allow Layer 3 IP forwarding (traditionally performed by routers) to be combined with Layer 2 switching (such as ATM and Frame Relay) in a single combined system, to increase the scalability, speed, and traffic engineering capabilities of the Internet. It also adds additional functionality in routers such as traffic engineering (also known as service provisioning and bandwidth management) and virtual private networks that has traditionally been found only in switches.

This chapter provides an introduction to MPLS technology by describing the numerous methods of integrating IP and ATM that led to MPLS development. Sections 8.2 and 8.3 describe the overlay model of running IP on top of ATM which is used by classical IP over ATM, NHRP, MPOA, and simple interconnection of routers using ATM permanent virtual circuits. Section 8.4 describes the integrated model of combining IP and ATM, of which MPLS is one example, and Section 8.4.5 discusses MPLS in particular.

8.2   EVOLUTION OF THE INTERNET BACKBONE

The need for MPLS evolved from the rapid growth of the Internet and the inability of IP routers to keep up with the raw bandwidth requirements of the Internet backbone. Traditional router architectures included both routing (using a distributed algorithm to determine the path to a particular destination IP address) and forwarding (using the output of the routing algorithm to choose the output interface for a particular IP packet) as software tasks in a shared processor. Often forwarding would be a high-priority foreground task and routing a lower priority background task. More recent router architectures have improved their performance by moving the forwarding tasks to dedicated processors; often the number of forwarding processors increases linearly with the number of interfaces on the router. Most recently, gigabit routers have begun to move the forwarding task from general purpose processors to specialized application-specific integrated circuits (ASICs). Note that the routing protocols continue to run as a software background task in a general purpose processor; in almost all cases, this is the same processor also used for other background tasks, such as network management activities.

However, router forwarding speeds have still not been able to keep up with the raw bandwidth requirements of the Internet. In addition, the variable-sized packet nature of IP and the datagram hop-by-hop nature of IP routing protocols make it extremely difficult to support advanced features such as Quality of Service (QoS) and traffic engineering.

Consequently, a majority of the Internet backbone carriers have chosen to use an ATM-based network backbone. Using ATM to interconnect the routers at the edge of the carrier network has a number of advantages:

- Each ATM cell's short, fixed-length label, known as the virtual path identifier/virtual circuit identifier (VPI/VCI), is very simple to switch in hardware. IP's longer and hierarchically-structured addresses are much harder to switch. ATM switches generally switch cells at the line rate of the interfaces; routers often cannot keep up with the full rate of their input interfaces.

- The uniform cell length, 53 octets (including the five-octet cell header), simplifies buffering and queue management algorithms in the switch when compared to a router.
- The uniform cell length also makes it possible to support sophisticated QoS services in the network, especially if the same backbone is used for other services in addition to IP (such as providing native ATM services, Frame Relay, private line emulation, and so on).
- Because ATM uses virtual circuits rather than datagram forwarding, the routes that cells take through the ATM network are independent of the IP routing protocol, allowing network administrators complete control over traffic routing and the loads on network trunks.

However, because IP and ATM are so different, adapting IP to ATM networks has been easier said than done, and many approaches have developed.

8.3   INTEGRATING IP AND ATM: THE OVERLAY MODEL

8.3.1   MULTIPROTOCOL ENCAPSULATION

Before IP was able to be run within ATM circuits, a number of fundamental issues needed to be settled because of the basic differences between IP's connectionless datagram packet nature and ATM's connection-oriented cell nature. The first issue was basic encapsulation – how should variable-length IP packets be carried in ATM cells, and how should the higher-layer (with respect to ATM) protocol be identified, in order to be able to multiplex multiple higher-layer protocols over a single ATM virtual circuit (VC)?

The IETF, which produces all IP-related protocol standards, defined a method to encapsulate IP in ATM and to multiplex multiple protocols on the same ATM VC in RFC (Request for Comments) 1483.[1] For encapsulation, the IETF chose to use *ATM Adaptation Layer 5* (AAL5).[2] The decision of which protocol identification method to use was not as straightforward as it might seem because there were several among which to choose. WAN networking protocols, such as X.25 and Frame Relay, already used single-octet Network Layer Protocol IDs (NLPIDs) defined in ISO/IEC Technical Report 9577.[3] Some of the more popular NLPIDs are shown in Table 8.1.

TABLE 8.1
Selected NLPIDs

0xCC:	IP
0x81:	CLNP
0x08:	Q.933
0x80:	SNAP (see below)

Meanwhile, LANs used logical link control (LLC) and subnetwork attachment point (SNAP), as defined by the IEEE for protocol identification on 802 LANs such as 802.3 Ethernet and 802.5 Token Ring.[4] LLC uses three octets to define the following protocol in the LAN frame. These three octets are most often 0xAA-AA-03, which signify SNAP, which occupies another five octets. SNAP has two fields: a three-octet organizationally unique identifier (OUI) and a two-octet protocol identifier (PID). The meaning of the PID field is dependent on the value of the OUI field; this allows different organizations to define their own sets of protocols to be carried over LANs. LLC and SNAP are used together so often that they are usually referred to as LLC/SNAP. Interestingly, SNAP's use is not restricted to LLC; there is a NLPID, 0x80, which signifies that it is immediately followed by a SNAP header. This extends NLPIDs to be able to identify any protocol identified by SNAP.

The list of SNAP OUIs is administered by the IEEE, and any organization, company, or individual can be granted an OUI by paying a registration fee to the IEEE. However, the most complete list of OUIs is not published by the IEEE, but by the IETF's Internet Assigned Numbers Authority (IANA) in its document called "Assigned Numbers," which is periodically reissued as a new RFC. As of this writing, the most recent version of "Assigned Numbers" is RFC 1700,[5] but its most up-to-date contents can be found on the Internet at ftp://ftp.isi.edu/in-notes/iana/assignments/.

Some example OUIs are provided in Table 8.2.

TABLE 8.2
Selected OUIs

0x00-00-00	Xerox (see below)
0x00-00-5E	IANA
0x00-80-C2	IEEE 802.1 Committee
0x00-A0-3E	ATM Forum
0x02-60-8C	3Com (an example corporate assignment)

One of the OUIs is special: 0x00-00-00 is assigned to and administered by Xerox (one of the inventors of Ethernet), and it is used to record the list of protocol IDs used on the original Ethernet, which uses a simple two-octet protocol identifier rather than the eight-octet LLC/SNAP. As a result, most of the major multivendor internetworking protocols can be found in this OUI, including IP, IPX, Appletalk, etc. The list of protocol IDs in OUI 00-00-00 are also known as *Ethertypes* (after the original name of the Ethernet Type field), and new Ethertypes can be obtained from Xerox for a fee. Again, "Assigned Numbers" contains an extensive list of assigned Ethertypes.

Some of the more popular Ethertypes are listed in Table 8.3

After a considerable amount of debate, the IETF chose in RFC 1483 to use LLC/SNAP for multiprotocol identification. Their primary reason for this decision was that LLC/SNAP's fixed size of eight octets allowed more efficient processing

TABLE 8.3
Selected Ethertypes

0x08-00	IP
0x08-06	ARP
0x80-9B	Appletalk
0x81-37	IPX

than variable length (but more compact) NLPIDs. The IETF also had a greater degree of familiarity with LLC/SNAP. Of course, using NLPIDs would have eased ATM/Frame Relay interoperation, but this consideration was not given much weight. The one exception when NLPIDs are used for protocol identification over ATM is when Frame Relay is being layered over ATM via the frame relay service specific convergence sublayer (FR-SSCS), as specified in Appendix A of RFC 1483.

8.3.2 IP OVER ATM USING PERMANENT VIRTUAL CONNECTIONS (PVCs)

By far, the most popular method of carrying IP over ATM in the WANs that make up the Internet backbone is by interconnecting the routers at the edge of the WAN with a full mesh of ATM PVCs interconnecting the routers. Each ATM PVC emulates a point-to-point interconnection between the routers. Over each of the PVCs, RFC 1483 is used to encapsulate the IP packets in ATM cells and identify IP as the protocol being carried. In particular, the OUI 0x00-00-00 and the PID 0x08-00 are used to identify IP. Because a full mesh of PVCs is used, the physical topology of the network at the ATM layer and the logical topology as seen by the routers are quite different. Figure 8.1 shows an example physical topology of routers and ATM switches.

Note that the routers and ATM switches are numbered differently, to emphasize their independence from each other, and that the ATM switches are typically interconnected with a partial mesh of switch-switch trunks. The ATM network has a complete set of ATM PVCs configured between the routers, which produces the logical topology shown in Figure 8.2.

Because the logical topology is overlaid onto the physical ATM network, with separate addressing plans and topologies, this sort of arrangement is known as the overlay model. This model requires N*(N-1) PVCs to interconnect N routers, which does not scale well to large networks. It also produces a large number of routing adjacencies at the IP routing layer, which increases the overhead required to run the IP routing protocols. However, it does have one particular advantage, which is evident when the PVCs are shown overlaying the ATM network as in Figure 8.3.

Because the topologies are independent, the network operator can choose how to route the PVCs through the physical network. For example, the PVC between routers 3 and 5 can be routed at the ATM layer between S5, S6, and S8, or between S5, S7, and S8, whichever is more useful for producing the desired traffic loading on the ATM trunks. Network operators have found this ability to be extremely useful.

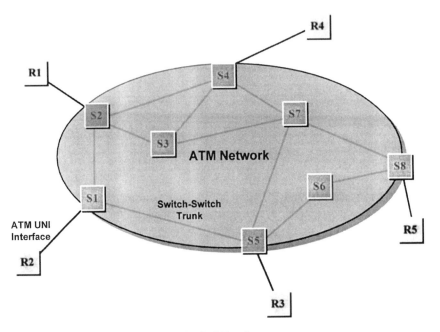

Figure 8.1 ATM-Connected Routers: Physical Topology

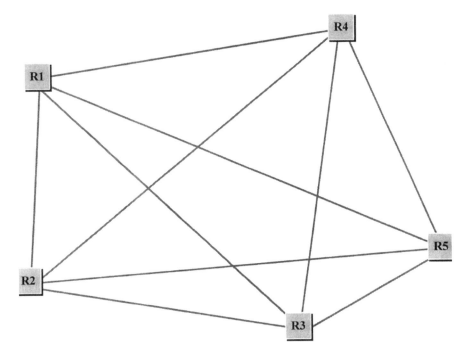

Figure 8.2 ATM-Connected Routers: Logical Topology

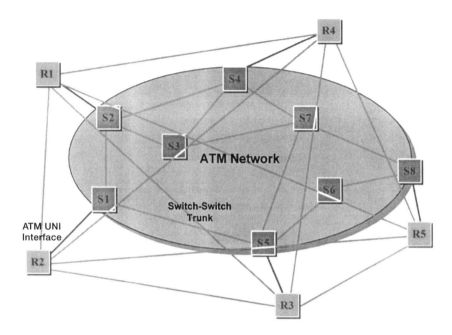

Figure 8.3 ATM-Connected Routers: Both Topologies

8.3.3 IP OVER ATM USING SWITCHED VIRTUAL CIRCUITS (SVCs)

8.3.3.1 Classical IP Over ATM and ATMARP

When there are too many ATM-attached IP end systems (whether routers or ATM-attached hosts, such as workstations) for a full PVC mesh to be practical, as might be found on an ATM LAN or in an enterprise ATM network, the overlay model can still be used, but with ATM SVCs rather than PVCs. The SVCs can be created when they are required to interconnect two end systems then destroyed when they are no longer needed. The IETF's *"Classical IP over ATM"* set of standards (RFCs 1755,[6] 2225,[7] and 2331[8]) form the IETF-approved method of carrying IP over ATM SVCs.

Classical IP over ATM is so named because it is the closest possible adaptation of the any-to-any connectivity provided by "classical" IP (such as used on a LAN) such that it would work over ATM's connection-oriented and nonbroadcast nature. As with PVCs, classical IP over ATM overlays IP over ATM; there is no direct relationship between the IP addresses in the IP packets and the ATM addresses used to open VCs to carry the IP packets.

There are three RFCs that together describe classical IP over ATM:

RFC 2225: How IP addresses are resolved to ATM addresses, and subsequently transported over ATM connections.

RFC 1755: How to use ATM Forum UNI 3.0^9 and 3.1^{10} SVC signaling when opening RFC 1577 connections.

RFC 2331: Updates to RFC 1755 to support ATM Forum UNI 4.0^{11} SVC signaling.

Because ATM networks can support many more stations (hosts and routers) than are found on typical LANs, a single ATM network can have more than one IP subnetwork layered on top of it, rather than using the typical strategy of having a one-to-one correspondence between the IP subnetwork and the physical network over which it is layered. Each of these subnetworks is called a logical IP subnetwork (LIS).

Figure 8.4 shows an ATM network with two IP LIS layered above it. LIS are purely logical entities that have been layered above the physical ATM network, and LIS membership is also purely logical (any host on the ATM network can belong to any LIS layered above it). Two IP stations in the same LIS communicate directly via ATM SVCs or PVCs, while IP stations in different LIS must intercommunicate via a router (such as R1). Multiple LIS are used for administrative requirements, such as security filtering, convenience, or to localize dependencies on (and reduce the load of) ATMARP and other network servers.

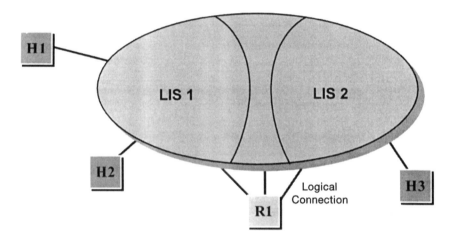

Figure 8.4 Two IP LIS on One ATM Network

Router R1, while having only one physical attachment to the ATM network, has been defined to logically belong to both LIS so that it may route between them. Because hosts H1 and H2 are in the same LIS, they intercommunicate directly by opening an ATM SVC from one to the other when there is data to send (see Figure 8.5).

Host H3, however, must indirectly communicate with hosts H1 and H2 by opening an SVC to router R1, which will in turn open an SVC to host H1 and/or H2 as required, as shown in Figure 8.6.

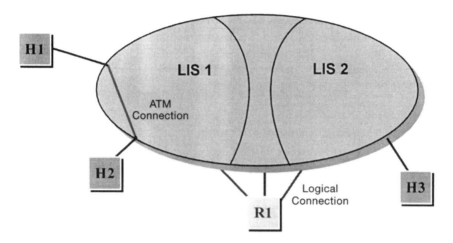

Figure 8.5 Connection between Hosts H1 and H2

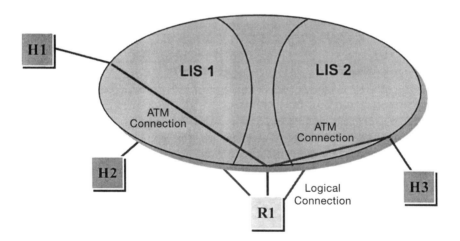

Figure 8.6 Connection between Hosts H1 and H3

Because there is no direct relationship between IP and ATM addresses, hosts need some method to resolve IP addresses to ATM addresses, in order to open SVCs to each other so they can communicate at the ATM layer when they have IP packets to send. RFC 2225 defines an ATMARP server and protocol to be used for such address resolution. This is necessary because IP over Ethernet (and other LANs) is able to broadcast ARP (Address Resolution Protocol) requests to all of the other stations on the LAN; however, because ATM does not have a broadcast mechanism, a server must be used instead. Each LIS requires an ATMARP server to translate IP to ATM addresses.

ATMARP operates as follows: When hosts H1 and H2 come up, they register themselves with their LIS' ATMARP server ("AS" in Figure 8.7) by opening connections to a configured ATMARP server, and issuing an ATMARP registration. Note that while the figure shows a stand-alone ATMARP server, they are typically implemented as a logical function in an ATM switch or an ATM-attached router, rather than as a separate piece of equipment.

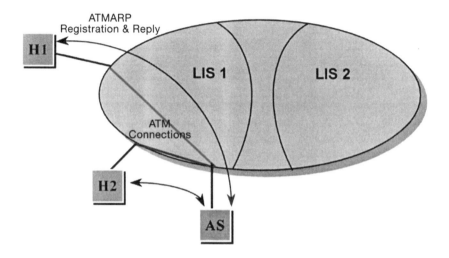

Figure 8.7 Connections to the ATMARP Server

The ATMARP server now contains the mapping between host H1 and H2's IP and ATM addresses. As shown in Figure 8.8, when host H1 needs to send IP packets to host H2, it sends an ATMARP request containing host H2's IP address to the ATMARP server. The server will return host H2's ATM address to host H1, allowing host H1 to open a direct ATM connection to host H2.

Figure 8.9 shows the resulting connection to host H2 using its ATM address returned by the server:

8.3.3.2 Next Hop Resolution Protocol (NHRP)

As Figure 8.6 showed, only hosts in the same LIS may open direct ATM connections to each other; hosts in separate LIS must communicate via a router, even if it means the packets must leave the ATM network only to reenter the ATM network on the same physical access line to the router (but on a different VC, of course).

Figure 8.10 shows a more complicated example. The ATM network contains three LIS, and host H1 on LIS 1 needs to communicate with host H2 on LIS 3. Router R1 is logically on both LIS 1 and 2, and router R2 logically interconnects LIS 2 and 3. Because the hosts are in different LISes, they must indirectly communicate along the path provided by IP routing: from host H1 to router R1, then to router R2 on a different VC, and then to host H2 on a third VC.

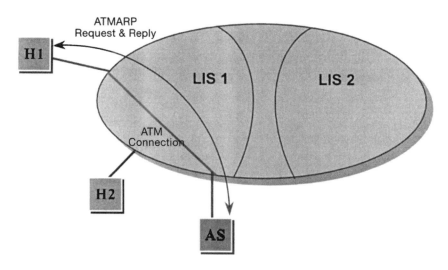

Figure 8.8 Host H1 Requesting Host H2's ATM Address

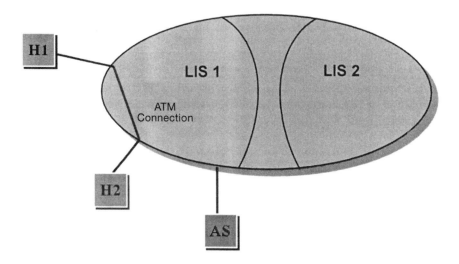

Figure 8.9 Resulting Connection Between H1 and H2

At the routers, the IP packets leave and reenter the ATM network. This is extremely inefficient, in that the packets must traverse each router's interface line twice, their cells must be reassembled into packets as they enter the router, each packet must be processed and switched by the router, and they must be resegmented into cells as they reenter the ATM network using the same ATM access line.

To prevent this inefficiency, the IETF developed the next hop routing protocol (NHRP)[12] to allow IP hosts in different LIS, but on the same ATM network, to be

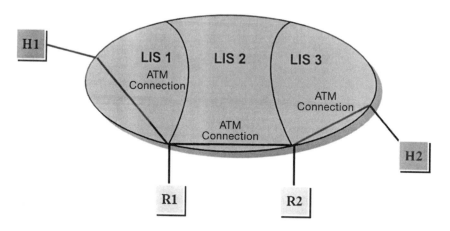

Figure 8.10 Routed Path From A to B

able to open direct VCs between each other (unless there are administrative reasons for this to be prevented).

NHRP can be viewed as an enhancement of ATMARP; ATMARP clients (end stations) become NHRP clients, and ATMARP servers are replaced by NHRP Servers (NHS). Because the client behavior is very similar to that of ATMARP, upgrading the client software is simple. Because NHRP is closely tied to IP routing, NHS are almost always implemented in IP routers (or ATM switches that include IP routing functionality).

When hosts H1 and H2 come up, they issue an NHRP registration request to their local NHS. In most cases, their default router will also be their NHS, so no extra configuration is required in the hosts. In Figure 8.11, router R1 is also the NHS for LIS 1, and router R2 is the NHS for LIS 3. While it is not germane to this example, LIS 2 could be served by either (or both) of the routers, or by a third router.

When host H1 needs to communicate with host H2, and it knows only H2's IP address, it issues an NHRP resolution request for H2's ATM address. Note that hosts H1 and H2 no longer need to be in the same LIS; host H1 can issue an NHRP resolution request for all hosts with which it needs to communicate.

As shown in Figure 8.12, if host H1 issues an NHRP resolution request to its server (router R1), and its server does not know the mapping between host H2's IP and ATM addresses, it forwards the request along the normal IP routed path towards host H2, until the request arrives at a router/NHS that knows the mapping (in this case, router R2). Note that the request stops at the first server that knows the answer, and is never sent to host B itself. (Keep in mind that ATM SVCs need to be opened to carry these signaling messages if there is not already a connection open.)

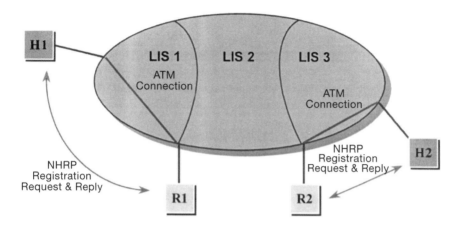

Figure 8.11 NHRP Registration

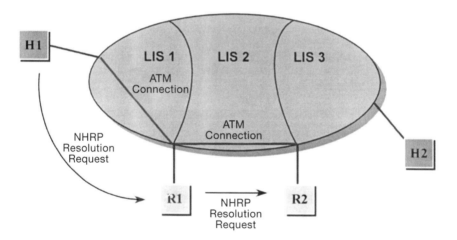

Figure 8.12 NHRP Query Processing

In Figure 8.13, the NHRP resolution reply is returned back along the same path to host H1. The answer is also cached by any NHSes on the way (router R1 in this example), so that future requests can be answered locally.

Finally, host H1 is able to open a direct ATM connection to host H2, bypassing the routers, as shown in Figure 8.14.

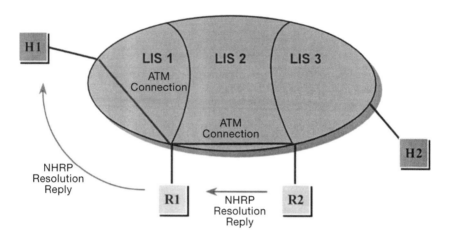

Figure 8.13 NHRP Resolution Reply

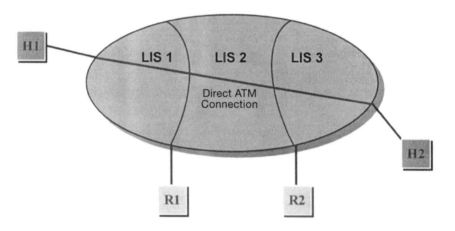

Figure 8.14 Direct Connection Between A and B

8.3.3.3   Multiprotocol Over ATM (MPOA)

The ATM Forum used NHRP as the basis for its multiprotocol over ATM (MPOA) [13] specification. MPOA, like Classical IP over ATM and NHRP, is based on the overlay model and relies on an NHRP server to provide address resolution services. In addition, it uses ATM Forum LAN Emulation (LANE)[14] to provide both bridging and routing support to MPOA clients. However, this functionality exacts a price; it uses multiple additional servers, including an MPOA server and several LANE servers and it requires ATM-level connections between the clients and multiple servers.

Classical IP over ATM, NHRP, and MPOA all work well in enterprise or local-area ATM networks. However, they all share serious limitations that do not allow them to scale for WAN use, such as on an Internet backbone network:

- They all open ATM connections to carry data based on detecting new IP traffic flows. This can lead to delays while connections open. In addition, many ATM switches and ATM router interfaces are seriously limited to the speed at which they can open ATM connections and maximum number of connections they can have open simultaneously. This is an extreme restriction because Internet backbone networks routinely carry 250,000 simultaneous flows.[15]
- There is excessive management complexity and overhead as a result of the large number of servers and the necessity to run both IP and ATM routing protocols.
- The lack of server redundancy in the LANE and MPOA specifications makes these protocols difficult to use as the basis of a public service network.

All of these restrictions led to the development of the integrated model of IP and ATM, which is discussed in the next section.

8.4 INTEGRATING IP AND ATM: THE INTEGRATED MODEL

The scaling limitation inherent in the overlay model of IP over ATM, whether for PVCs or SVCs, led to the development of a number of proprietary solutions by a number of IP router and ATM switch vendors. They are discussed in the following sections in the approximate order of their development. They have culminated in the development of MPLS as a multivendor, interoperable solution for integrating IP and ATM in the wide area.

8.4.1 IP SWITCHING

IP switching is a term invented by Ipsilon Networks, Inc. (which has since been acquired by Nokia Telecommunications, Inc.). Rather than using the classical IP over ATM approach of integrating IP and ATM, they chose to pair every ATM switch with a router that ran a new protocol, the Ipsilon Flow Management Protocol (IFMP).[16] IFMP was an alternate ATM SVC signaling protocol, which, like Classical IP, NHRP, and MPOA, would detect IP flows and open ATM connections to carry them. Once an IP flow was mapped to an ATM connection, it would be switched through the network at the ATM layer, rather than routed through the network at the IP layer. Ipsilon's main innovation was replacing the standard ITU and ATM Forum SVC signaling protocols with a much simpler protocol (IFMP), which allowed connections to complete faster and was much easier for them to implement. Also, not all flows were mapped to ATM connections; low bandwidth traffic continued to be routed over a default ATM connection over the trunk between two router/switches. However, Ipsilon's IP switching had several problems that restricted its suitability for use in wide-area Internet backbone networks:

- Because it opened ATM connections as a result of observing IP traffic flows, there was a delay that allowed packets to get out of order as they moved from the default ATM connection to a dedicated connection. Packet misordering slows down many TCP implementations.
- The switches were limited to the maximum number of flows they could support simultaneously. This problem was made worse by their flow detection algorithm, which was very fine-grained; it would open a connection based now on only the source and destination IP addresses, but also the particular application that was in use. This could result in multiple ATM connections to carry flows between the same IP source and destination hosts.
- Ipsilon refused to submit IFMP for multivendor standardization, preferring to publish it as a proprietary protocol. Most wide-area service providers prefer to use standardized protocols so they are not locked into a single vendor's products.

8.4.2 IP Navigator

IP Navigator,[17] developed by Cascade Communications Corporation (which was later acquired by Ascend Communications, Inc.), was the first method of integrating IP and ATM that based the ATM connections on IP routing. Cascade observed that in Internet backbone networks, IP routing is generally stable – it changes only during outages or as new IP networks are added to the Internet. In addition, they realized it would be easiest to interconnect to other parts of the Internet if no new protocols were required in external routers. Therefore they created IP Navigator, which has the following functionality:

- A network of Frame Relay and/or ATM switches running IP Navigator appear as a collection of routers, not switches, to outside routers. IP Navigator supports standard IP routing protocols, such as OSPF,[18] BGP, and so on. It presents standard IP router interfaces to its neighboring routers.
- Inside the network, the switches use OSPF for two functions. The first is to perform the standard IP routing functionality. The second is to use the switch-switch paths determined by OSPF's IP routing to set up switched connections at the Frame Relay or ATM layer. Because the connections are formed as a result of a network control protocol (OSPF) rather than actual traffic flows, IP Navigator is said to be *control-driven* rather than *flow-driven*, like IFMP or MPOA.
- These control-driven connections are used to carry IP traffic from the network ingress to egress. When an IP packet enters the network, it is

routed by the ingress switch, using a standard IP routing table, onto the proper Frame Relay or ATM connection to switch the packet to the correct egress switch. The packet is switched through the interior of the network, then routed again at the egress switch to find the proper egress interface.

The key to IP Navigator's scalability for the wide area is that it does not maintain a full mesh of connections from every ingress switch to every egress switch (after all, every switch with external interfaces is both an ingress and an egress). Rather, it uses point-to-multipoint trees (MPT) to aggregate the packets from the ingress switches to the proper egress switch.

An MPT is the opposite of a Frame Relay or ATM point-to-multipoint connection. Rather than traffic being sent from the root to the leaves of the connection, the IP packets are sent from the leaves of the tree to the root. There is one tree, and one root, for every egress switch. This allows the number of connections to scale with the number of switches, rather than square of the number of switches.

When the IP Navigator network is initialized, standard OSPF routing is used to determine the path from every ingress switch to each egress switch. IP Navigator then uses that information to create an MPT for every egress. This happens automatically, and prior to any user traffic being sent. If IP routing should later change, perhaps because of a power outage at a switch, then the network will automatically reform the MPTs to match. In each interior switch, packets are switched by their MPT.

Figures 8.15 and 8.16 show an example MPT. Figure 8.15 shows the physical network topology. As far as the external routers (R1-R5) are concerned, switches S1-S8 are actually routers and running standard routing protocols.

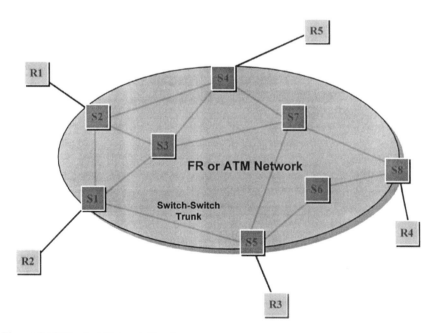

Figure 8.15 Physical Network Topology

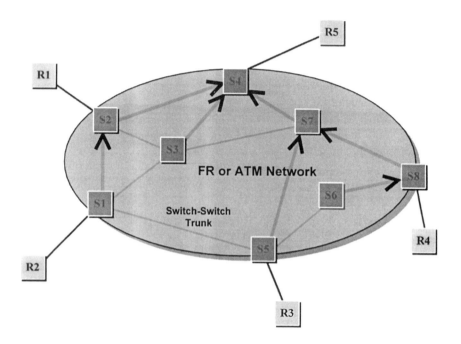

Figure 8.16 IP Navigator MPT with S4 as the Egress

Internally, however, the switches have set up MPTs to carry the IP traffic so that the packets do not have to be routed at each hop through the network.

Figure 8.16 shows an MPT with S4 as the egress switch.

In this example, IP packets from R2 to R5 will travel through switches S1, S2, and S4, while packets from R3 to R5 will travel though switches S5, S7, and S4. Note that S7 aggregates packets from S5, S6, S7, and S8 onto the MPT to S4.

IP Navigator implements MPTs over Frame Relay by assigning each egress switch a unique Frame Relay Data Link Connection Identifier (DLCI) at each ingress. This allows the interior switches to use normal FR label switching – at each interior switch, the ingress port and DLCI are switched to the proper egress port and DLCI.

IP Navigator over ATM is somewhat more complicated because the packets are actually contained in multiple cells, which have to be reassembled at the egress switch. Because AAL5 is used to segment and reassemble the cells, cells from different packets cannot be interleaved on the same ATM connection. To allow the packets to be reassembled, each egress switch is assigned a unique VPI and each ingress switch sending to that VPI is assigned a unique VCI. In the interior switches, only VP switching is used; the only use for the VCI field is to identify uniquely the source switch in order to reassemble the packets at the egress switch. This technique is called VP merging, and it again allows the number of VPs in the network to scale on the order of the number of switches.

Of course, this example shows only the MPT rooted at S4. Each of the other switches also has an MPT rooted at it, with leaves at every other switch. IP routing is used to determine the path each MPT uses through the network.

Once the MPTs have been established, IP Navigator uses the MPTs to route packets through the network, rather than IP routing. This allows a network operator to override routing when necessary to perform traffic engineering – to cause the packets to follow a particular route for reasons of policy. For example, in Figure 8.16, packets from S1 traveled through S2 to reach S4. However, the network administration could determine that the trunk from S2 to S4 is overutilized, and the trunks from S2 to S3 and S3 to S4 are underutilized. IP Navigator has facilities to allow the network administration to override IP routing, in this case to route the MPT branch from S1 to S3 rather than S1 to S2.

IP Navigator can also use MPTs to provide enhanced IP functionality, such as QoS support. By allocating multiple MPTs to a particular destination, with particular ATM or Frame Relay-layer QoS attributes associated with each MPT, the same QoS can then be imparted to the packets. So, best-effort packets can be sent along a best-effort MPT, while packets requiring a better QoS (such as for voice or video over IP) can be sent over a different MPT that has the required QoS at the switched layer.

IP Navigator uses OSPF as its IP routing protocol. OSPF is a link-state routing protocol which means that every switch knows the complete physical topology of the network. This makes it very easy for each MPT's root switch to determine the MPT's routing through the network, assign the necessary labels at each the leaves and at each hop, and ensure that routing loops will not occur. Routing loops must be avoided at all costs, because ATM cells and Frame Relay frames do not have any mechanism to timeout cells or frames if loops occur; they will travel through the loop for as long as it exists. Especially at higher speeds, this can cause looping cells or frames to consume considerable amounts of trunk and switch bandwidth. IP Navigator's use of OSPF makes stable loops impossible to form.

IP Navigator also includes IP multicast support by using normal ATM or Frame Relay point-to-multipoint connections to carry the multicast traffic. All required multicast replication takes places at the link layer in the switches.

As of this writing (July 1998), IP Navigator is running operationally in a number of production public IP networks. In addition, Ascend Communications is actively participating in the MPLS standardization effort, and has submitted much of IP Navigator's technology, especially MPTs, to the IETF for use in MPLS.

8.4.3 ARIS

IBM's Aggregate Route-Based IP Switching (ARIS)[19] is another control-driven method of integrating IP and ATM. ARIS took many of the concepts introduced in IP Navigator and generalized them for wider use and to remove the dependence on OSPF as the IP routing algorithm. Instead of piggybacking the MPT setup information into the OSPF routing updates, as in IP Navigator, ARIS has a separate setup algorithm that begins at the egress node of each MPT (for example, S4 in Figure 8.16) and spreads out, hop by hop, following IP routing backwards until it reaches the MPT's leaf nodes. It still depends on IP routing to set up the MPTs, but it is independent of any particular routing protocol. Because it cannot count on a link-state routing protocol, such as OSPF, to prevent routing loops, ARIS includes a separate loop prevention algorithm as a part of its MPT setup protocol. For use over ATM, ARIS also generalizes the VP-merging capability of IP Navigator's MPTs to allow VC merging as well, when used with ATM switches that include VC merge hardware. VC merging is very similar to the operation of IP Navigator over Frame Relay, in that only a single ATM VPI/VCI label is required to identify a MPT. However, this requires hardware in intermediate switches that can buffer the cells from a particular packet until they have all arrived at a switch, and then send them as an integral group of cells to the next switch. It should be noted that VC switching obviates all QoS support from the ATM network, thus should be used only for best-effort IP traffic.

As of this writing, IBM has canceled its ARIS project but has submitted ARIS' technology to the IETF for inclusion in MPLS, which it will then implement in its products.

8.4.4 TAG SWITCHING

Tag Switching,[20] from Cisco Systems is another control-driven label switching protocol (they use the term *tag* to represent labels used for switching). Tag switching further generalizes the concepts in IP Navigator and ARIS. New features in tag switching include the following:

- Routers as well as switches take part in tag switching, which was a natural step given that routers are Cisco's primary product. Tag switching in routers allows them to participate with switches in the formation of switched paths, to simplify their own operation as packets travel though them, and, for the first time, to support real traffic engineering.
- Cisco used TCP to carry its Tag Distribution Protocol (TDP), which is necessary since Cisco supports a wide range of IP routing protocols and wanted tag switching to work with all of them. Using TCP simplified TDP's operation because it could assume TCP's reliable transport service. In addition, it allowed TDP peers to be physically separate from each other.
- They added support for routing hierarchy by introducing the concept of a *tag stack*, which allows labels to be stacked in packets. The top-most label is always the one used for tag switching through a router or switch, but they have lower-level tags for routing them through the various Internet routing domains, and within each domain, the top-most tag routes the packets through the local routers and switches. When packets enter a routing domain, a new tag is pushed on the stack, and when they leave a routing domain the top-most label is popped from the stack.
- They added the concept of the *forwarding equivalency class* (FEC), which is a set of packets that follow the same path to a particular destination.

Normal IP routing has one equivalency class, which is the best match of its IP address to a network identifier in a router's routing tables. They identified a number of additional FECs, from the egress point from a network used by a number of IP routes to the set of packets belonging to a particular application at one particular destination host.

• In addition to using TDP to distribute tags, they also allow tag distribution to be piggybacked onto other protocols, such as the Resource ReSerVation Protocol (RSVP)[21] for flows that require a particular QoS, or onto the Border Gateway Protocol (BGP)[22] to support interdomain routing.

One particular criticism of tag switching is that it does not include a loop prevention algorithm when setting up tags. Cisco felt that loops, when they do form, are of short enough duration that looping packets would not consume excessive network resources. This may be true for routers, but loops lasting even several seconds at ATM speeds could be disastrous.

As of this writing, Cisco has deployed tag switching in its routers and is testing tag switching in its ATM and Frame Relay switches. Cisco has also contributed heavily to the MPLS effort in the IETF.

8.4.5   MULTIPROTOCOL LABEL SWITCHING (MPLS)

The IETF's MPLS working group began its work in April 1997. Their primary goal is to produce an interoperable, multivendor approach to using label swapping to integrate IP with ATM and Frame Relay, to speed up switching in routers, and to provide unified traffic engineering functionality in both switched and routed networks. The primary technology inputs to the process are, as mentioned above, IP Navigator, ARIS, and tag switching. To quote from the working group's charter, "The working group is responsible for standardizing a base technology for using label swapping forwarding paradigm (label switching) in conjunction with network layer routing and for the implementation of that technology over various link level technologies, which may include Packet-over-SONET, Frame Relay, ATM, Ethernet (all forms, such as Gigabit Ethernet, etc.), Token Ring, and so on. This includes procedures and protocols for the distribution of labels between routers, encapsulations, multicast considerations, use of labels to support higher layer resource reservation and QoS mechanisms, and definition of host behaviors."[23]

As of this writing, all of the work is in the draft stage, which means that any and all of the details given here are subject to change before the documents are published as RFCs. However, the group has reached consensus on the broad outline of the solution.

First, it has produced draft framework[24] and architecture[25] documents. The framework document discusses MPLS technical issues and requirements and provides a broad survey of the different approaches and solutions that have been considered by the working group, including the three major inputs to the group described above in Sections 8.4.2, 8.4.3, and 8.4.4. In contrast, the architecture document describes the technical approaches that have been agreed upon to this point in the working group; it winnows through the range of possible solutions and mechanisms presented in the framework document to make specific choices. The highlights of the architectural agreements to date are provided below:

- The working group has agreed upon the basic terminology to be used, which is actually a very important step. The *label* is the basic information that is used to switch an incoming packet to an outgoing interface, possibly with a new label. The label is the same as the tag in tag switching, and can be an ATM VPI/VCI, Frame Relay DLCI, or a field in an MPLS header on point-to-point lines or LANs. Routers and switches that participate in MPLS are *Label-Switched Routers* (LSRs). They set up *Label-Switched Paths* (LSPs) to carry the IP data packets (an IP Navigator MPT is an example of an LSP). LSRs cooperate to set up LSPs by using a *Label Distribution Protocol* (LDP). LSPs are unidirectional, and, as a result, LSRs are either *upstream* or *downstream* from their neighbor, depending on the direction of traffic flow on an LSP.
- MPLS will be control-driven but will have options for being flow-driven, as well, for use in enterprise networks. MPLS' use in the Internet backbone will always be control-driven.
- As in IP Navigator, LSPs use merging in order to scale for use over wide-area Internet backbone networks.
- As in ARIS, loop prevention mechanisms are used to prevent loops from forming in LSPs.
- As in tag switching, labels can be stacked for hierarchical routing and can represent different forwarding equivalence classes.

Figure 8.17 shows an example LSP to R5. Note that the major difference between this and Figure 8.16 is that the routers are now also participating in the LSP, which saves an IP routing hop when the packets enter the FR or ATM network. In Figure 8.17, switch S2 is downstream from S1 and upstream to S4.

The working group has also produced a first draft of the Label Distribution Protocol specification,[26] which is largely a combination of ARIS' setup algorithm and tag switching's TDP. Like TDP, it also runs over TCP.

The working group is considering several quite different proposals to provide advanced traffic engineering and bandwidth management services. One provides the functionality as a part of LDP itself. The other, which proposes modifying the RSVP to carry labels, similar to tag switching's use of RSVP, is somewhat controversial because of doubts about the suitability of RVSP in WANs.

A number of additional technical issues remain to be settled before the MPLS work in the IETF can be considered complete enough for vendors to implement and attempt interoperability testing. These include encapsulation of IP packets over ATM and Frame Relay networks, the ATM classes of service for best-effort and QoS LSPs crossing ATM networks, and how to provide virtual private network functionality in MPLS, among others. However, it does promise the potential to improve greatly, in an interoperable manner, the operation of IP both across the Internet backbone and in enterprise networks.

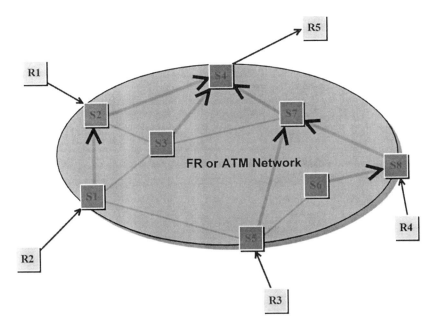

Figure 8.17 MPLS LSP to R5

REFERENCES

1. Heinänen, J., RFC 1483, Multiprotocol Encapsulation over ATM Adaptation Layer 5, July 1993, www.ietf.org.
2. ITU-T Recommendation I.363.5, B-ISDN ATM Adaptation Layer specification: Type 5 AAL, August 1996, www.itu.ch.
3. ISO/IEC Technical Report 9577 (also published as ITU-T Recommendation X.263), Protocol Identification in the Network Layer, November 1995, www.itu.ch.
4. International Standard, Information Processing Systems – Local Area Networks - Logical Link Control, ISO 8802-2: 1989 (E), IEEE Standard 802.2-1989, December 1989, www.ieee.org.
5. Postel, J., Reynolds, J., RFC 1700, Assigned Numbers, October 1994, www.ietf.org.
6. Perez, M., et al., RFC 1755, ATM Signaling Support for IP over ATM, February 1995, www.ietf.org.
7. Laubach, M., Halpern, J., RFC 2225, Classical IP and ARP over ATM, April 1998, www.ietf.org.
8. Maher, M., RFC 2331, ATM Signalling Support for IP over ATM — UNI Signalling 4.0 Update, April 1998, www.ietf.org.
9. ATM Forum af-uni-0010.001, ATM User-Network Interface Specification V3.0, September 1993, www.atmforum.com.
10. ATM Forum af-uni-0010.002, ATM User-Network Interface Specification V3.1, 1994, www.atmforum.com.

11. ATM Forum af-sig-0061.000, UNI Signaling 4.0, July 1996, www.atmforum.com.
12. Luciani, J., et al., RFC 2332, NBMA Next Hop Resolution Protocol (NHRP), April 1998, www.ietf.org.
13. ATM Forum af-mpoa-0087.000, Multi-Protocol Over ATM Specification v1.0, July 1997, www.atmforum.com.
14. ATM Forum af-lane-0021.000, LAN Emulation over ATM 1.0, January 1995, www.atmforum.com.
15. Thompson, K., et al., Wide-Area Internet Traffic Patterns and Characteristics, *IEEE Network*, November/December 1997.
16. Newman, P., et al., RFC 1953, Ipsilon Flow Management Protocol Specification for IPv4 Version 1.0, May 1996, www.ietf.org.
17. H. Ahmed, et al., IP Switching for Scalable IP Services, *Proc. IEEE*, December 1997.
18. Moy, J., RFC 2328, OSPF Version 2, April 1998, www.ietf.org.
19. Viswanathan, A., et al., ARIS: Aggregate Route-Based IP Switching, Work in progress, March 1997, www.ietf.org.
20. Rekhter, Y., et al., RFC 2105, Cisco Systems' Tag Switching Architecture Overview, February 1997, www.ietf.org.
21. Braden, R., et al., RFC 2205, Resource ReSerVation Protocol (RSVP) — Version 1 Functional Specification, September 1997, www.ietf.org.
22. Rekhter, Y., et al., RFC 1771, Border Gateway Protocol 4 (BGP-4), March 1995, www.ietf.org.
23. IETF MPLS Working Group charter, http://www.ietf.org/html.charters/mpls-charter.html.
24. Callon, R., et al., A Framework for Multiprotocol Label Switching, Work in progress, November 1997, www.ietf.org.
25. Rosen, E., et al., Multiprotocol Label Switching Architecture, Work in progress, March 1998, www.ietf.org.
26. Andersson, L., et al., Label Distribution Protocol, Work in progress, March 1998, www.ietf.org.

9 Designing Multipoint Logically Switched Optical Networks

Eric Bouillet

CONTENTS

9.1 INTRODUCTION

The huge potential of optical networks for satisfying the skyrocketing needs of broadband telecommunications services while meeting rigid quality of service requirements has long been acknowledged. However, although fiber has become the medium of choice in telecommunication networks, its vast resources are severely underused because of the much slower electronics that are interfaced with the optical medium. For instance, transceivers operate at speeds that are several orders of magnitude below the actual usable capacity of the fiber (several Gbps versus hundreds of Gbps). In order to achieve higher rates, wavelength division multiplexing (WDM) techniques have been widely suggested. The concept behind WDM is to partition the optical spectrum into multiple nonoverlapping λ channels, each assigned a wavelength and modulated at electronic speed. In parallel to WDM transmission, recent studies have enabled the realization of photonic routing and switching devices that perform elementary functions such as wavelength routing and switching. With those devices in hand we can envision all-optical transparent networks that carry traffic between distant users on single beams of light from end-to-end. The optical network has the capability to route the signal on specified fiber paths, thereby allowing wavelength reuse, in addition to wavelength multiplexing. Furthermore, as in the radio frequency paradigm, sources can multicast signals to multiple destinations.

While purely optical networks have very large capacity, they lack the processing power useful for many network applications, especially the ability to support high connectivity. For this reason, an electronic layer must be superposed on the optical layer. This hybrid combination of electronics with optics is depicted in Figure 9.1. We consider general transparent optical networks called linear lightwave networks (LLN) as the optical infrastructure for the hybrid network. LLNs are all-optical networks whose nodes are generalized switches called linear divider combiners (LDC). The LDCs are controllable photonic switches that can create multipoint optical connections in a waveband-selective manner. In the sequel a waveband is a bunch of adjacent λ-channels and its width depends on the discrimination capabilities of the LDC. More precisely, the waveband is the smallest segment of the optical

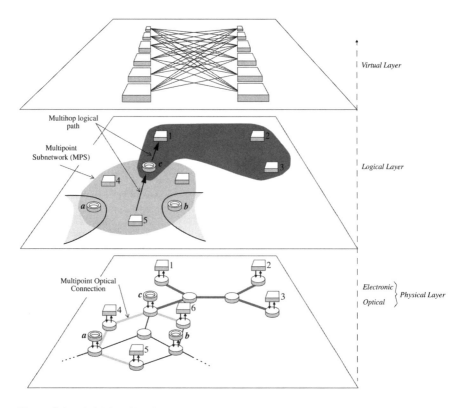

Figure 9.1 Hybrid Combination of Electronics with Optics

spectrum that is distinguishable in the LDC, while the λ-channel is the smallest unit resolvable in the access station by a tunable receiver. Typically, the spectral width of a waveband is much larger than a λ-channel. The LLN is accessed through network access stations (NAS) containing optical transceivers. Electronic equipment (either end user equipment or switches) accesses the network via external ports on the NAS. The connections between these external ports are called logical connections. The NAS are connected through access fibers to the LDCs, which in turn are set to create end-to-end paths between the access stations, either point-to-point or multipoint. In the latter the optical path is, de facto, a tree. Once an optical connection is set, it can carry one or more logical connections on it, using multiplexing protocols such as WDM, TDM, or CDM to avoid interference. In order to distinguish logical connectivity and optical connectivity, we lay a logical layer on top of the physical topology (PT). The role of the logical layer is to conceal the details of the optical paths and center our view on the logical connectivity.

There are four possible types of logical connectivities:

- *one-to-one* — one connection from one source to one destination
- *one-to-many* — multiple connections from one source to many destinations
- *multicast* — one connection from one source to many destinations
- *many-to-one* — multiple connections from many sources to one destination.

A fifth possibility, the many-to-many, combines one-to-many and many-to-one. We will call a set of NAS fully connected by a many-to-many connection a multipoint subnet (MPS). A MPS acts like a broadcast-and-select subnetwork of a larger network.

Even though the physical support is all-optical, it would be futile to attempt to establish full optical connectivity among all the NAS (or, in other words, to set up a MPS that covers the whole network). There are two reasons for this:

- The electromagnetic signal degrades as it propagates through a large transparent network. The degradation becomes even more evident when the signal is split for multicasting purposes in the LDC. Thus, a purely optical connection between distant NAS is sometimes impossible.
- To achieve a full connectivity among n NAS, $n(n - 1)$ simultaneous unidirectional connections must be established, and the purely optical approach soon reaches its limits when n becomes large.

Therefore, in order to maintain the connectivity between all NAS, the optical signal will sometimes have to be converted into an electronic form, processed, switched, and converted back to an optical signal, to continue on another optical path. This is the basic idea of multihop optical networks. In the networks proposed here, the key idea is to cluster the NAS into several MPS and let the electronic logical switches relay the traffic between stations that belong to different subnets, while the stations belonging to the same subnet communicate optically. Following this idea, we end up with a multipoint logically switched network (MSN) characterized by mutihop paths in a logical topology (LT). The difference between our

MSN and a multihop network is that the latter has an LT in the form of a graph, while the LT of the MSN is a hypergraph. The whole concept is illustrated in Figure 9.1, which shows a LT composed of two MPS and three NAS each, with the NASs numbered from 1 to 6. The link between the two MPSs is assured by the ATM switch c, which includes its own NAS to access the optical network. The figure also shows a multihop logical path from NASs 5 to 1. The capacity of the links in the virtual layer superimposed on top of the LT, must correspond to some prescribed traffic requirement between the NAS. The model for the traffic should include the notions of multicasting, time-scales (call level and bit rate), service classes, and Quality of Service (QoS).

The MSN architecture presented above relies on the optical resources to provide high throughput and electronic techniques to achieve higher connectivity. There is no free lunch, however, and by coupling optical switching with electronic switching, electronic bottlenecks are introduced. Nevertheless, intelligent electronics is essential to compensate for the lack of processing power in the optical layer.

9.1.1 DESIGN OF THE NETWORK

Broadband network design is usually aimed at satisfying a traffic requirement between NASs, taking into account their geographical location and without exceeding a maximum cost. In our case, we will assume that the virtual layer connectivity and capacity requirement and the PT of the optical infrastructure, described above, are given. The objectives, then, will be first to design the LT, and second to embed it into the PT. The design of the logical topology involves three operations:

- Pick an LT
- Map the NAS in the LT
- Route the traffic in the LT

Next, the LT is embedded in the PT, which requires two operations:

- Set up the optical path in the LLN such that the NAS in each MPS can see each other.
- Assign appropriate wavelengths to the connections among NAS.

In the next section, we will explain the value of a multipoint optical network. We then present the LT design and the embedding problems in Sections 9.3 and 9.4.

9.2 MOTIVATION

The properties of MSN come from the ability of the LLN backbone to support multipoint optical connections, thereby enabling higher connectivity with fewer electro-optic interfaces than in electronically switched networks connected by point-to-point fiber links. This is shown in Figure 9.2 with a comparison between a multipoint optical network and a standalone ATM network. The physical connectivity layer corresponds to the fiber deployment on the ground. This layer is fixed because

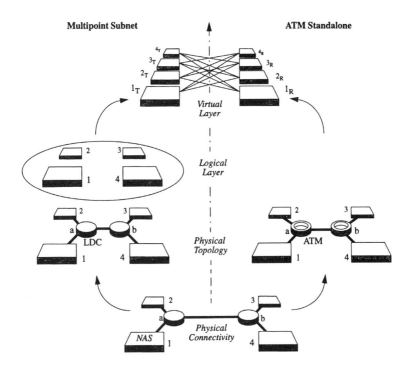

Figure 9.2 Comparison of Multipoint Optical Network and Standalone ATM

it is usually bound to geographical constraints. We consider bidirectional links with a unidirectional fiber for each direction. The two networks to be compared lie on top of the physical connectivity and must satisfy a traffic requirement represented in the figure by the virtual layer. We assume a uniform traffic with one unit of flow (*uof*) in either direction of the NAS pairs. In addition, we assume that all the transceivers have a same capacity *Cuof* that represents the maximum bit rate that can be carried through one λ-channel.

9.2.1 STANDALONE ATM NETWORK

The standalone ATM network reposes entirely on a point-by-point connectivity. There is one optical transceiver per NAS and ATM port, thus a total of 10 transceivers (two per link). A trivial computation tells us that the traffic on any fibers from/to the NAS is $3uof$, whereas the traffic on the inter-switch fibers is $6uof$. Thus the utilization of the NAS fibers is $\rho = 3/C$, and $6/C$ for the inter-switch fibers. Furthermore, each ATM switch has to process $10uof$.

9.2.2 MULTIPOINT OPTICAL NETWORK

In the multipoint network version, the connections are realized on a broadcast and select basis. All the transceivers are tied together by one multipoint optical path which in turn supports the 12 logical connections. Thus all stations are included in a single MPS, and no logical switching is required in this case. A time wavelength division multiple access (T-WDMA) protocol is then used to multiplex the logical connections, with one optical transceiver in each station. In one version of T-WDMA the transmitter of each station is permanently tuned on a dedicated wavelength which is broadcast to all other stations. Each wavelength carries a one-to-many (one-to-three in the example) connection using TDM in the transmitters. The receivers select the information from a given transmitter by tuning on that transmitter's wavelength. In this scheme, an appropriate schedule is pre-arranged so that the information can be retrieved at the destination side. Figure 9.3 shows one appropriate T-WDMA schedule for the present example.

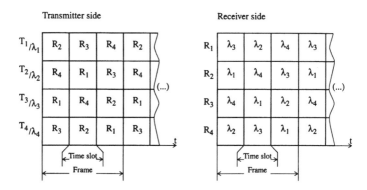

- Each transmitter T_i has a capacity C, and transmits on wavelength λ_i.
- The capacity C is divided into 3 time slots (one frame), one slot per logical connection.
- Capacity per logical connection is C/3.
- During a time slot transmitter T_i transmits to receiver R_j, and the receiver R_j is tuned to wavelength λ_i.

Figure 9.3 T-WDMA Schedule

9.2.3 STANDALONE ATM VERSUS MULTIPOINT OPTICAL

In the ATM version, a signal from station 1 to station 2 goes through 2 links with utilization 3/C, one link with 6/C, and 2 ATM switches that must process 10uof. Using the schedule of Figure 9.4, the same signal needs to go through only one logical connection with utilization 3/C. Also, one transceiver per NAS is sufficient to achieve a full connectivity. Thus, there is a gain of 4 to 10 in terms of transceivers (an expensive component). A difference, however, is that the receivers must be rapidly tunable (order of microseconds) in the multipoint network.

In conclusion, the multipoint optical approach provides us with a higher throughput for a lower cost. This is the reward of a more efficient management of the fiber's resources. The standalone ATM uses only one λ-channel with a capacity limited by the electronic interface, whereas the optical approach uses 4 separate λ-channels. Another important feature of the multipoint architecture is its ability to dynamically assign the capacity among the logical links inside an MPS. In the example above, a nonuniform capacity assignment would be simply accomplished by allotting more time slots for some logical connections in the T-WDMA schedule. In the standalone ATM the capacity is fixed by the hardware.

9.3 DESIGN OF THE LOGICAL LAYER

The design of the LT involves three steps. The first step is to find a suitable target topology for the LT. This target topology is bound to the following constraints:

- The number of transceivers per NAS — This determines the degree of connectivity between the MPS
- The maximum admissible number of logical hops between any pair of NAS — Determines the diameter of the topology
- The maximum admissible number of logical connections per MPS — This determines the maximum number of NAS per MPS
- Optional constraints such as self-routing properties, k-connectivity for fault tolerances, etc.

In the second step, terminals are mapped into the target topology, with respect to the following:

- The physical locations, or more precisely the geographical distances between the NAS — We want to prevent optical connections between nodes that are too far apart.
- The traffic requirements — Fast purely-optical channels are reserved for the virtual connections with stringent requirements, whereas virtual connections with relaxed requirements can be routed through mutihop logical paths.
- Feasibility of embedding the LT into the PT.

The last step addresses the routing problem in the LT. This corresponds to the embedding of the virtual layer into the LT. In other words, we want to find the routes which satisfy the traffic requirements and minimize the delays and congestion. The details of what we have already done in this direction are given in the Preliminary Work section below.

9.3.1 PRELIMINARY WORK

In Section 9.2, we explored the potential of the dynamic capacity allocation offered by the MSN architecture, and compared two identical LTs. Both are regular Kautz hypergraph topologies,[12] with 252 NAS clustered into 42 MPS of 12 NAS each, giving a total of $12 \times 11 = 121$ logical one-to-one connections per MPS. One topology has static MPS with a fixed capacity for each logical connection, and the other has dynamic MPS thereby enabling dynamic capacity sharing among the logical channels of the MPS.

We evaluate the routes that maximize the throughout in each LT, using a flow deviation algorithm.[11] Briefly, this algorithm iteratively evaluates the direction of a gradient on the surface of some objective function and recomputes a new metric for the next iteration. Advancing step by step, and always following the steepest descent/ascent, the algorithm stops when it reaches a minima/maxima. We demonstrated that the congestion and the delay in a dynamic MPS depends only on the total capacity and the aggregate flow loading the MPS. Thus the dynamic MPS can be viewed as an atomic object with a given capacity. This approach allows a simplified view of the network and reduces the complexity of the problem by a factor of 121 (the number of logical links per MPS).

The results of the comparison are reproduced in Figure 9.4, which shows that the dynamic MPS converges to an optimal solution 100 times faster than the static MPS, and that it achieves a better throughput.

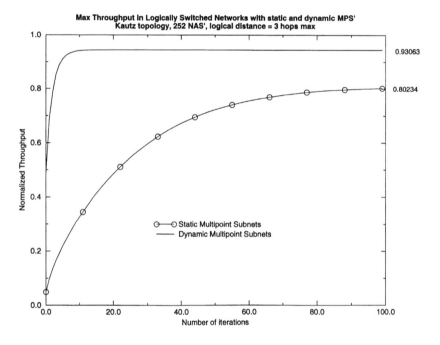

Figure 9.4 Dynamic MPS Compared to Static MPS

9.3.2 FUTURE WORK

The two first steps of the LT design problem, i.e., find a suitable target topology and map the NAS, have to be solved. Much has already been done in this area for graph topologies [8-9, 13] and we want to extend the results for the case of hypergraphs with the constraints presented above.

In the previous version of the algorithm, only point-to-point traffic requirements have been addressed. However, the MSN architecture is a natural candidate for multicast services, and we would like to modify the algorithm so that it also takes advantage of this additional ability.

9.4 THE EMBEDDING OF THE LOGICAL LAYER INTO THE PHYSICAL TOPOLOGY

Once a valid LT is found, it must be embedded in the physical topology. This is achieved by setting up optical paths in the form of trees for each MPS, so the NASs in each MPS can see each other by tuning their transceivers appropriately. All λ-channels used by NAS in the same MPS are grouped into a common waveband for ease of routing through the LDC. An example of embedding is illustrated in Figure 9.5. The combined action of setting the optical tree and tuning the transceivers is bound by some constraints, however, which are

- The optical trees must either by routed on separate fibers or WDM techniques must be used to avoid interference (different wavebands must be assigned to trees sharing the same fiber).
- The number of wavelengths is limited.

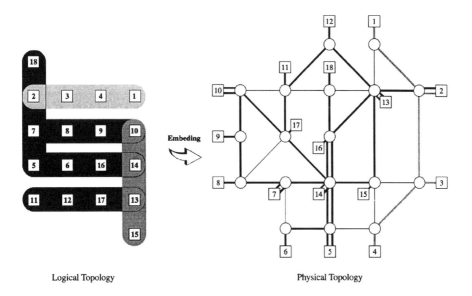

Logical Topology   Physical Topology

Figure 9.5 Embedding Logical Layer into Physical Topology

9.4.1 PRELIMINARY WORK

We have developed various Integer Programming (IP) formulations to solve this problem, with some variants in the objectives to be optimized. Typically, they concern the average point-to-point distance and the number of wavelengths. A self-explanatory formulation of the IP used to solve the embedding shown in Figure 9.2 can be found in the appendix.

9.4.2 Future Work

- The IP formulation becomes intractable when the network becomes large (a hundred nodes or more). We therefore need to devise a heuristic that solves the embedding in large networks, with possible modifications of the LT when the embedding is unfeasible.
- We need to adapt the design techniques to networks carrying logically multicast traffic.

9.5 CONCLUSION

We have presented the concept of multipoint logically switched networks, a hybrid architecture that marries the speed of optics and the intelligence of electronics. This architecture is justified by the many advantages it offers:

- higher connectivity than either electronically switched networks or purely optical networks
- higher throughput
- logical multicasting is naturally supported
- lower cost
- dynamic capacity allocation is possible

In order to concretise this architecture, we will have to solve the challenging problems of designing a suitable logical topology and embedding this topology into a given physical topology.

REFERENCES

1. Stern, T.E., A linear lightwave MAN architecture, *Proc. of NATO workshop on High Speed Networks*, 1990.
2. Stern, T.E., Linear Lightwave Networks: How far can they go? *Proc. Globecom* 1990.
3. Stern, T.E., Linear Lightwave Networks, CTR Technical Report No. 184-90-14, Columbia University.
4. Bala, K. and Stern, T.E., Algorithms for Routing in Linear Lightwave Network, *Proc. Infocom*, 1991.
5. Bala, K., *Routing in Linear Lightwave Networks*, Ph.D thesis, Columbia University, 1992.
6. Jiang, S., Stern, T.E., and Bouillet, E., Design of Multicast Multilayered Lightwave Networks, *Globecom* 1993, Houston.
7. Sharony, J., *Architectures of Dynamically Reconfigurable Wavelength Routing/ Switching Networks*, Ph.D thesis, Columbia University, May 1993.

8. Jiang, S., *Multicast Multihop Lightwave Network: Design and Implementation*, Ph.D thesis, Center for Telecommunications Research, Columbia University, New York, 1995.

9. Ramaswami, R. and Sivarajan, K. N., Design of Logical Topologies for Wavelength-Routed All-Optical Networks, *Proc. Infocom*, 1995.

10. Ramaswami, R. and Sivarajan, K. N., Routing and Wavelength Assignment in All-Optical Networks, *Transactions on Networking*, October 1995.

11. Fratta, L., Gerla, M., and Kleinrock, L., *The Flow Deviation Method: An Approach to Store-and-Forward Communication Network Design.*

12. Bermod, J.C. and Peyrat, C., de Bruijn and Kautz networks: a competitor for the hypercube? In *Hypercube and Distributed Computers*, Elsevier Science Pub., North Holland, 1989.

13. Bienstock, D. and Günlük, O., *Computational Experience With a Difficult Mixed-Integer Multicommodity Flow Problem*, IEOR, Columbia University, 1993.

APPENDIX: A MIXED-INTEGER PROGRAMMING FORMULATION

In this appendix, we provide a mixed-integer programming formulation for the embedding problem given in Section 9.4. The objective of this formulation is to minimize the optical capacity requirement and the average size of the optical trees. Each optical connection Λ_w supports K logical connections $\lambda_{w,k}$, $k \in (1, \ldots, K)$, that connect K NAS-pairs tied to Λ_w. A boolean variable Λ_w^{xy} is set to 1 if Λ_w goes through a fiber between optical switches x and y, or 0 otherwise. Similarly, a boolean variable $\lambda_{w,k}^{xy}$ is set to 1 if $\lambda_{w,k}$ belongs to Λ_w^{xy}. $\overline{\Lambda}$ denotes the maximum number of optical connections in a fiber. The overall formulation is:

Minimize

$$m\overline{\Lambda} + \sum_{xy} \sum_{w} \Lambda_w^{xy} \tag{9.1}$$

Such that

$$\sum_{y \neq x} \lambda_{w,k}^{xy} - \sum_{y \neq x} \lambda_{w,k}^{yx} = \begin{cases} 1 & \text{if } x \text{ is the source of } \lambda_{w,k} \\ -1 & \text{if } x \text{ is the destination of } \lambda_{w,k} \\ 0 & \text{otherwise} \\ \forall w, k, x \end{cases} \tag{9.2}$$

$$\Lambda_w^{xy} = \max_k (\lambda_{w,k}^{xy}) \, \forall w, x, y$$
$$\sum_{w} \Lambda_w^{xy} \leq \overline{\Lambda} \, \forall x, y \tag{9.3}$$

$$\Lambda_w^{xy} \in \{0, 1\}, \lambda_{w,k}^{xy} \in \{0, 1\} \forall w, x, y, k \qquad (9.4)$$

In the objective function 9.1, m is an appropriately large quantity. Equation 9.2 is a flow conservation equation. Equation 9.3 indicates that any logical connection must be supported by an optical connection, and Equation 9.4 is the optical capacity constraint.

10 Multiwavelength Optical Networks (WDM)

Krishna Bala

CONTENTS

This chapter discusses the evolution of high-capacity wavelength division multiplexed (WDM) networks or multiwavelength optical networks from point-to-point configurations to more advanced flexible and survivable network architectures. It discusses architectures that enable a flexible and survivable WDM optical layer for a data-centric network. In particular, it proposes three architectures: WDM optical cross-connect mesh (W-Mesh), WDM unidirectional path switched ring (W-UPSR) and WDM bidirectional line switched ring (W-BLSR). These architectures provide a cost-effective evolution to a high capacity data-centric environment.

10.1 INTRODUCTION

Traffic growth in the Internet and other new data communications have resulted in new business opportunities and challenges for telecommunications network operators. The number of Internet hosts has grown hundred-fold from 1990 to 1998, and the resulting demand is straining the capacity of the telecommunications network infrastructure. In the past, telecommunications network design and economics have been determined by voice traffic considerations. It is now apparent that the dominant traffic in the network will be data. This results in a fundamental change in principles underlying network architecture choice and technology. It is envisioned that the wavelength division multiplexed (WDM) optical networking layer will form the core of this data-centric network architecture and provide the network operator with functionality similar to that which SONET (synchronous optical network) TDM (time division multiplexing) has provided in a voice-centric environment. SONET TDM terminals support several rates of OC-3 (corresponds to 155 Mbps), OC-12 (corresponds to 622 Mbps), OC-48 (corresponds to 2.5 Gbps), OC-192 (corresponds to 10 Gbps), etc.

10.2 POINT-TO-POINT WDM NETWORKS

WDM optical systems combine several optical wavelengths (e.g., separated by 100 GHz in the 1550 nm band of the spectrum) in a single fiber. These multiplexed wavelengths are then amplified by erbium-doped fiber amplifiers (EDFAs) that boost signals to overcome losses in transmission (see Sternard Bala,[8] Bala,[1] and Brackett[3]). As shown in Figure 10.1, these point-to-point WDM systems are mostly deployed in open architectures with 1.3 μm standard short-reach SONET interfaces. These interfaces interconnect directly to equipment like SONET TDM equipment, or Internet Protocol (IP) routers and asynchronous transfer mode (ATM) switches that support the standard SONET interfaces. The part of WDM systems that converts a 1.3 μm signal into an ITU wavelength signal in the 1550 nm band is called a transponder. Also, the WDM multiplexers and demultiplexers along with EDFAs in a typical long-haul network configuration are shown in Figure 10.1.

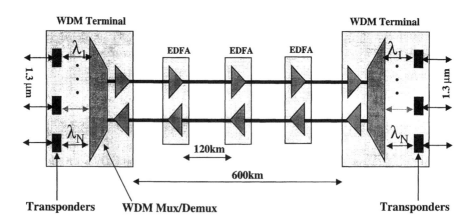

Figure 10.1 Open WDM Architecture

WDM is a proven method of increasing bandwidth by a factor of 30-50% of the cost of alternate methods. These cost advantages are particularly significant in cases where new fiber builds are avoided by using WDM equipment. In cases where a route between central offices runs out of capacity because of fiber exhaust, it can cost up to a $100,000/mile to lay new fiber facilities. In this case, the ability to increase the capacity of existing fibers by multiplexing several wavelengths adds tremendous economic value. Several long-distance network and local-exchange carriers have less than 50% fiber available in their cables. In addition, the long haul carriers have a small number of fibers per cable. This exacerbates the fiber exhaust problem and has brought about a mass deployment of WDM into the network.

The capacity of point-to-point WDM systems is increasing rapidly. The first systems were deployed with 8 wavelength channels. The number of wavelengths deployed in the network has increased to 40 wavelengths recently. Vendors have already announced the availability of up to 128 wavelengths per fiber. The total span of these systems varies from tens of km (local) to 1000 km (long haul) with optical amplifiers at intermediate locations for boosting the signal level. Several network operators have already deployed these systems and are looking to increase the number of wavelengths per fiber to even higher counts. The International Telecommunications Union has specified a 100 GHz spacing for the wavelengths. However, several equipment providers have announced products based on 50 GHz spacing and are considering 25 GHz spacings for adjacent channels. At the same time, the EDFA band has been flattened and increased by using dual-stage amplifiers with intermediate stage filters. New advancements in optical amplifiers[8] and lasers suggest there is still much room for growth in the number of wavelengths carried on a single fiber.

Now consider advanced network architectures that introduce configurability, flexibility, and survivability into the WDM optical layer. Figure 10.2 shows the overall network architecture vision for WDM optical networks. Traditional TDM networks demultiplex all the traffic down to sub-rates (e.g., DS3 at 45 Mbps) and perform capacity routing and assignment at the lower rates. This strategy makes sense in a voice-centric network where the fundamental unit is the 64 Kbps channel. In a data-centric network, equipment such as IP routers and ATM switches support optical interfaces (e.g., OC-Nc concatenated SONET interfaces). This requires that the traffic not be demultiplexed to lower rates but be routed and managed at the higher optical concatenated rates. Any demultiplexing operation is both unnecessary and costly. The optical layer allows such pass-through traffic at a node to traverse the node optically without any demultiplexing down to the lower sub-rates. However, the optical layer now has to provide the flexibility and survivability that the network operators are accustomed to obtaining from the SONET layer.

Consequently, the WDM optical networking layer must have an open architecture with interfaces that enable it to carry signals originating from equipment such as IP routers, ATM switches, and SONET terminals. This implies that the WDM layer must be configurable, survivable, and manageable. This chapter discusses and compares several architectures that will enable a flexible optical networking layer by providing survivability and dynamic wavelength provisioning while minimizing manual intervention.[1] These architectures are enabled by two key network elements: the reconfigurable wavelength-add/drop multiplexer (WADM)[7] and the optical cross-connect.[1]

10.3 WDM OPTICAL CROSS-CONNECT MESH (W-MESH)

The large deployment of point-to-point systems has resulted in numerous wavelengths terminating at the central offices. This in turn has resulted in a strong need to manage these wavelengths at the optical level without sub-rate demultiplexing of the individual connections. Figure 10.2 shows a central office that has several WDM systems deployed in a mesh network architecture. Traffic patterns in data networks

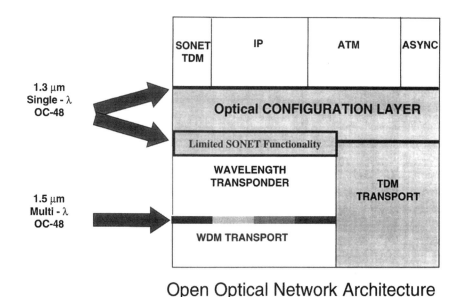

Open Optical Network Architecture

Figure 10.2 WDM Optical Networking Architectural Vision

are expected to be more arbitrarily mesh-like in their behavior. This is accentuated by the emergence of distributed applications on the web. As a result, mesh architectures for WDM networks need careful consideration.

Figure 10.3 shows a traditional digital crossconnect system (DCS) that has been optimized for telephony applications deployed in support of the WDM mesh network. Each incoming signal (e.g., 2.5 Gbps optical OC-48 SONET) at the DCS is demultiplexed down to smaller tributaries (e.g., STS-1 or 50 Mbps, VT1.5 or 1.5 Mbps) which are switched individually to output ports. The output ports re-multiplex the signals back for transmission (e.g., 2.5 Gbps optical OC-48 SONET). Thus, the traditional DCS switching allows for grooming and capacity allocation at lower rates. However, this lower rate processing becomes unnecessary (and very uneconomical) in an environment in which large volumes of traffic (e.g., several optical signals at OC48 rate or 2.5 Gbps from WDM transport systems) pass through a node (e.g., > 25%) without requiring any lower-rate grooming. This situation is faced by carriers that have deployed large numbers of WDM point-to-point systems in mesh.

Another problem with current DCS-based mesh networks is that sub-rate switching is an impediment to fast restoration from catastrophic events such as link and node failures. The restoration algorithms must restore at the sub-rate level. This situation results in slower restoration times on the order of seconds or minutes in these mesh networks. There is also a limit on the total number of sub-rate connections that can be restored in such a network.

Also, several equipment providers have announced ATM switches and IP routers with OC-Nc concatenated SONET interfaces where N = 3 to 192. The

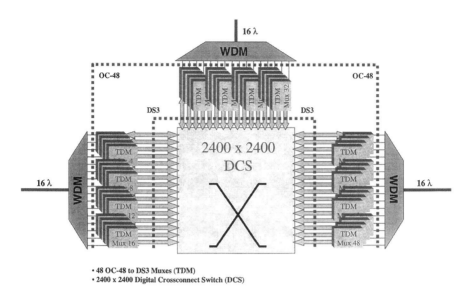

- 48 OC-48 to DS3 Muxes (TDM)
- 2400 x 2400 Digital Crossconnect Switch (DCS)

Figure 10.3 Traditional Digital Crossconnect System (DCS) in a WDM Point-to-Point Network

interfaces are specified as concatenated, which means that the entire bit stream is required to stay together from source to destination. In this case, demultiplexing the signal down to lower rates is unnecessary and uneconomical. However, these signals are unnecessarily demultiplexed and remultiplexed to lower rates, such as 1.5 Mbps, as they traverse the DCS. Since access to these sub-rate tributaries are not required in these data networks, it is expensive to take these OC-Nc signals apart and put them back together.

Figure 10.4 shows a WDM mesh network node that uses WDM transport systems in conjuction with OCS to provide dynamic wavelength assignment and restoration at the optical layer in an opaque network architecture.[2] The opaque architecture clearly separates switching from transport and eliminates any cascaded impairments that accumulate during transmission by providing signal regeneration between the WDM transport systems and the OCS. Furthermore, it enables wavelength interchange functionality between WDM transport systems by using the 1.3 μm interface as a common intermediate frequency between two WDM systems. In this fashion, it enables multivendor interoperability and allows different wavelength sets from different vendors to be interconnected by bringing them all to a common denominator, the 1.3 μm standard interface. The OCS will be deployed to operate in conjunction with existing WDM point-to-point systems providing interconnection and interoperability between them. In such a scenario, the point-to-point systems and the OCS might be provided by independent suppliers with interoperability achieved using a standard open interface, e.g., the 1.3 μm short reach SONET interface.

In this architecture, optical signals pass through the OCS without demultiplexing the signal down to the tributary sub-rate level. Besides significant cost and space

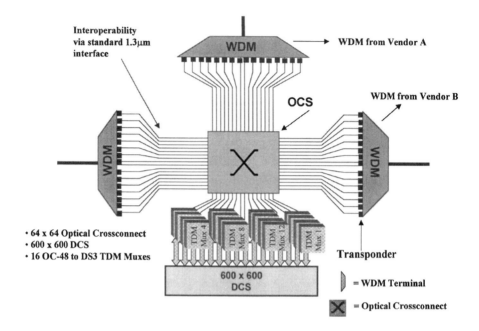

Figure 10.4 Optical Crossconnect (OCS) in a WDM Mesh Architecture

savings (at a factor of 5-10 over a traditional DCS solution), the OCS allows a direct interface to IP routers and ATM switches. Support for OC-Nc concatenated interfaces comes about naturally using the OCS. Furthermore, since the restoration is done at the optical layer (not at the lower rate tributaries), restoration times are expected to approach that of SONET rings (50 msecs).

Figure 10.3 shows three WDM terminals supporting 16 wavelengths each at OC-48 rates (2.5 Gbps). Each of these OC-48 signals is demultiplexed to 48 individual DS3 signals (approximately 50 Mbps) using a SONET terminal. A total of 48 TDM SONET terminals are required (one/wavelength). These DS3s are then switched through a DCS that has a maximum size of 2400 DS3 ports. Figure 10.4 shows the same scenario using one 64-port OCS and a smaller DCS with 600 ports resulting in a significant savings over the traditional DCS, assuming that 75% of the traffic passes through the node without requiring demultiplexing.

10.4 WAVELENGTH BIDIRECTIONAL LINE-SWITCHED RINGS (W-BLSR)

A 2-fiber W-BLSR architecture comprises two counter-rotating rings as shown in Figure 10.5. The figure shows a connection between two nodes on a wavelength channel. In this architecture only the working path is set up without dedicating the protection. The protection is shared among all the working paths on the ring.

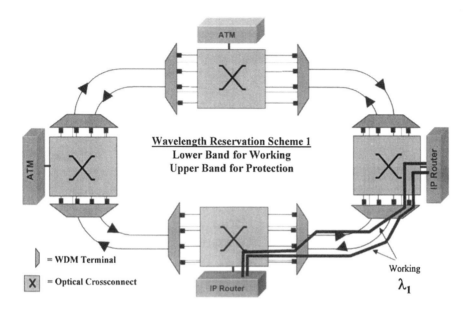

Figure 10.5 2-fiber WDM Bidirectional Line Switched Ring (2F W-BLSR) with λ Reservation Scheme 1

Automatic protection switching (APS) against failures is achieved by reserving half the wavelengths on each fiber for working traffic and the other half for protection on each fiber. Two wavelength reservation schemes are possible:

1. The lower band of wavelengths is used for working traffic in each fiber and the upper band of wavelengths is used for protection traffic in both directions on the two counter-rotating fibers.
2. In one direction the upper band is used for working traffic and the lower band for protection. In the opposite direction the lower band is used for working traffic and the upper band is used for protection traffic.

Both cases require configurable WADM. However, in the first case above, when there is a fiber cut a wavelength interchange operation is required at the two ends of the cut in order to move working traffic in one direction onto protection wavelengths in the opposite direction. On the other hand, case 2 above does not require the use of wavelength interchange because of the unique wavelength reservation strategy employed.

Figure 10.5 shows a 2-fiber W-BLSR ring with a bidirectional connection between two nodes on wavelength λ_1. In this case, the lower band of wavelengths (λ_1 through $\lambda_{w/2}$) are reserved for working and the upper band of wavelengths ($\lambda_{w/2}$ through λ_w) are reserved for protection. Figure 10.6 shows the same ring with a link failure. The two nodes adjacent to the failure perform a wavelength interchange

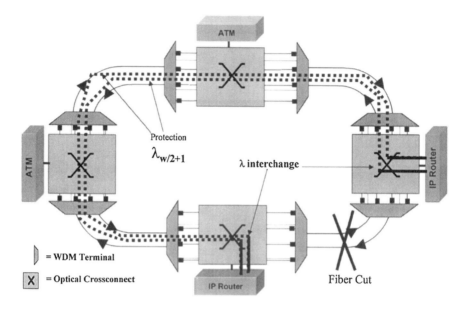

Figure 10.6 2-fiber W-BLSR with λ Reservation Scheme 1 (Fiber Cut)

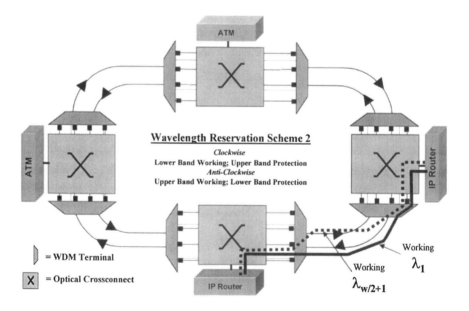

Figure 10.7 2-fiber WDM Bidirectional Line Switched Ring (2F W-BLSR) with λ Reservation Scheme 2

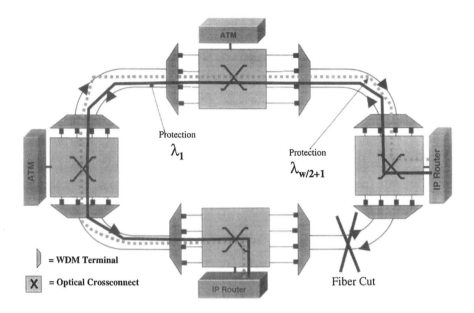

Figure 10.8 2-fiber W-BLSR with λ Reservation Scheme 2 (Fiber Cut)

operation to move the signal on λ_1 to wavelength $\lambda_{w/2+1}$ in the opposite direction for protection. In the near term, the only practical way to perform the required wavelength interchange is via optoelectronic conversions. However, in core networks, these optoelectronic conversions might be needed anyway to clean up the transmission impairments accumulated over long distances.

Figure 10.7 shows the 2-fiber W-BLSR ring with wavelength reservation scheme 2. This scheme results in the allocation of different wavelengths for the two directions of each bidirectional connection. Figure 10.8 shows the operation of the W-BLSR ring under a link failure condition. Note that in this case a wavelength interchange is not required during ring switching. Wavelength reservation scheme 1 is simpler to use from the point of view of network operations. It allows the assignment of the same wavelength to both directions of a two-way connection. Furthermore, this scheme reduces the amount of spare inventory by reducing the number of wavelengths that are assigned to connections.

Figure 10.9 shows a 4-fiber WDM bidirectional line switched ring (W-BLSR). This architecture is especially useful in long distance networks where span switching is important. Since the 4-fiber W-BLSR has dedicated fibers for working and protection, the protection fibers can be used to span-switch from a failure (e.g., transmitter). In span-switching, the failed traffic is moved to the protection fiber using the shortest path without needing to switch the traffic around the ring on the longer path. This also has the advantage of allowing testing, maintenance, and equipment upgrade on the protection fibers without service interruption.

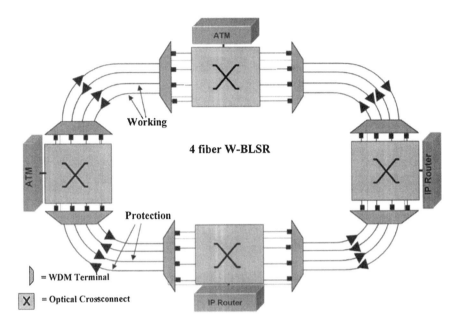

Figure 10.9 4-fiber WDM Bidirectional Line Switched Ring (4F W-BLSR)

10.5 WAVELENGTH UNIDIRECTIONAL PATH PROTECTION RING (W-UPSR)

The W-UPSR has one fiber dedicated for working and one for carrying protection traffic. Reconfigurable WADMs along the ring dynamically add and drop individual wavelengths from the working and protection fibers. As shown in Figure 10.10, in the W-UPSR, a wavelength path (e.g., an OC-48 signal on a wavelength) is bridged at the transmit end to create a working path and a dedicated protection path which are sent along opposite directions on the ring. Compare this architecture to the W-BLSR schemes discussed earlier that used shared, as opposed to dedicated, protection. The receiving end monitors both the working and the protection signals continuously. In case of a fiber cut, each receiving end switches independently to its protection path to provide Automatic Protection Switching (APS). Figure 10.11 shows the operation of the W-UPSR ring under a link failure condition. The WADM switches to the protection path and recovers from the link failure condition. The WADMs pass through traffic without demultiplexing the signals down to the lower speed tributary level. In this manner they offer significant cost and space savings compared to traditional TDM networks.

10.6 COMPARISON OF NETWORK ARCHITECTURES

We now compare the different network architectures from the standpoint of their capacity.

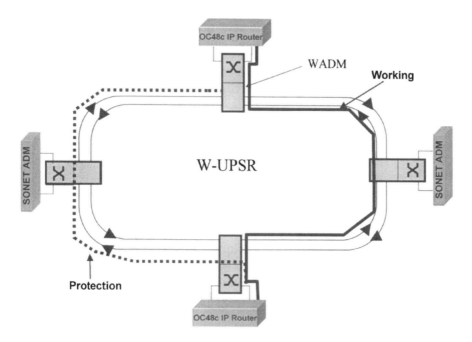

Figure 10.10 2-fiber WDM Unidirectional Path Switched Ring (W-UPSR)

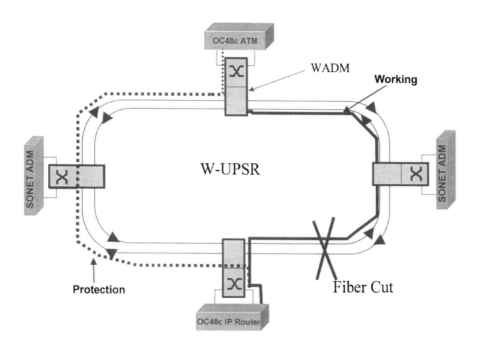

Figure 10.11 2-fiber W-UPSR (Fiber Cut)

Figure 10.12 shows a table comparing these three architectures with 10 network nodes each. The formulas used for the capacity calculations can be found in Stern and Bala.[8] It shows the number of wavelengths required in each architecture to support the required traffic pattern.

Architecture (10 nodes)	Number of Wavelengths	
	Hub Traffic *1 λ to each node from Hub*	**Mesh Traffic** *1 λ between every pair of nodes*
W-UPSR	9	45
2F W-BLSR	10	25
W-Mesh (degree 3)	3 + 2 (protection)	9 + 5 (protection)

Figure 10.12 Comparison of Architectures: Network Capacity

Figures 10.13 and 10.14 compare the amount of protection capacity in a ring versus a mesh network. In a ring network the amount of redundant capacity reserved for protection is 50%. In a mesh network, the number of wavelengths/fiber reserved for protection is much less and is equal to W/d because in a mesh network there are (d-1) different routes along which a group of signals can be protected in case of a fiber cut.

In general, mesh networks offer significantly better network utilization by using a smaller number of wavelengths to support the same amount of traffic than do ring networks. On the other hand, mesh networks require more complex restoration algorithms for protection switching as compared to the UPSR case for ring networks. See Figure 10.15 for a comparison of the manageability of the architectures.

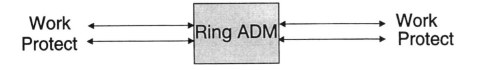

	Number of Wavelengths reserved for Protection
W-UPSR	**W**
2- fiber W-BLSR	**W/2 per direction**

Figure 10.13 Comparison of Architectures: Ring Protection Capacity

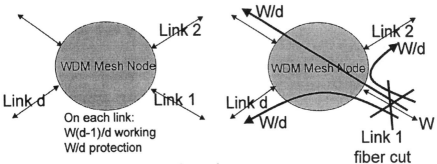

On each link:
W(d-1)/d working
W/d protection

Required Protection Capacity:
W/d wavelengths per direction reserved on each link

e.g. For W = 12 and d = 3;
Working Capacity = 8 λ s, Protection Capacity = 4 λs

Figure 10.14 Comparison of Architectures: Mesh Protection Capacity

Control & Management	
W-UPSR	Simple - Tail End Switch
W-BLSR	Complex - Ring Coordination
W-Mesh	Complex - Network Coordination

Figure 10.15 Comparison of Architectures: Manageability and Network Protection

10.7 CONCLUSION

This chapter discussed the evolution of WDM networks from point-to-point systems to more complex architectures like rings and meshes. It introduced three WDM optical network architectures: W-Mesh, W-UPSR, and W-BLSR. The W-Mesh architecture has significantly better utilization than the W-BLSR and the W-UPSR rings; it uses the least number of wavelengths to route the same traffic. Also, the W-BLSR architecture results in more efficient utilization of network resources than does the W-UPSR. However, the W-UPSR is much simpler to manage for protection switching than are the W-BLSR and the W-Mesh networks.

REFERENCES

1. K. Bala, "WDM Network Architectures for a Data-centric Environment," *Natl. Fiber Optic Engineers Conf.*, Orlando, FL, September 1998.
2. K. Bala, R. R. Cordell, and E. L. Goldstein, "The Case For Opaque Multiwavelength Optical Networks," *IEEE LEOS Summer Topical Meeting on Optical Networks*, Keystone, CO, August 1995.
3. C. A. Brackett, "Is there an emerging consensus on WDM Networking?" *IEEE/OSA J. Lightwave Techol.*, Volume 14, Number 6, June 1996.
4. P. V. Hatton, F. Cheston, "WDM Deployment in the Local Exchange Network," *IEEE Commun. Mag.*, Volume 36, Number 2, January 1998.
5. K. McCammon, V. Cacal, A. Eriksen, M. Esfandiari, S. Koehler, L. Lamb, and G. Pearson, "High Bandwidth Transport Technology Introduction at Pacific Bell," *Natl. Fiber Optic Engineers Conf.*, San Diego, CA, September 1997.

6. J. M. Simmons, E. L. Goldstein, and A. A. M. Saleh, "On the Value of Wavelength Add/Drop in WDM Rings with Uniform Traffic," *IEEE/OSA Optical Fiber Comm. Conf.*, San Jose, CA, February 1998.

7. W. J. Tomlinson, "Comparison of Approaches and Technologies for Wavelength Add/Drop Network Elements," *Natl. Fiber Optic Engineers Conf.*, Orlando, FL, September 1998.

8. T. E. Stern and K. Bala, *Multiwavelength Optical Networks: A Layered Approach*, Addison-Wesley, December 1998.

11 IP and Integrated Services

Saleem N. Bhatti

CONTENTS

11.1 INTRODUCTION

During the nineties applications became increasingly reliant on Internet protocols to provide data communications facilities. The use of the Internet protocols seems likely to increase at an extremely rapid rate and the Internet Protocol (IP) will be the dominant data communications protocol in the next decade. IP is being used for a huge variety of traditional applications, including e-mail, file transfer, and other general non-real-time communication. However, IP is now also being used for real-time applications that have quality of service (QoS) sensitive data flows*. Applications such as conferencing, telephony – voice-over-IP (VoIP) – as well as streaming audio and video are being developed using Internet protocols. The Internet and IP were not designed to

---

* A flow is a stream of semantically related packets which may have special QoS requirements, e.g., an audio stream or a video stream.

0-8493-9594-1/00/$0.00+$.50
© 2000 by CRC Press LLC

handle such traffic, so the Internet community must evolve the network and enhance the Internet protocols to cater to the needs of these new and demanding applications. Users want access to a plethora of telecommunication and data communication services via the Internet; they want an Integrated Services Network (ISN).

In this chapter, we consider the evolution of the changes that will be occurring in the Internet to support the ever-increasing demand of applications that populate it, and we look at how to evolve the Internet to an ISN. We examine the requirements for provision of QoS awareness and QoS mechanisms within the network, as well as looking at the trends for the future.

We give an overview of a set of technologies that must work and evolve together in order to allow integrated services provision over IP. Some of these technologies (e.g., RSVP, IP multicast) are described in other chapters. Sections 11.2 and 11.3 consider the state-of-the-art and likely near-term developments for service deployment. Section 11.4 considers technology that is likely to be deployed in the midterm while Section 11.5 looks at some longer term technology issues.

11.2 INTEGRATED SERVICES

People today use the Internet for many different applications. Some of these applications already exist on specific network technologies, e.g., voice on POTS, data on X.25. As the ability to use a more diverse range of applications becomes available from the desktop to an increasing number of people, demand for these applications increases. To provide access to such a diverse range of applications, it would be impractical to maintain access to each of the application-specific networks for each user. So, over the past two decades, there has been a move to provide a single ISN that can support the provision of any and all of these applications, e.g., N-ISDN (narrowband-ISDN), B-ISDN (broadband-ISDN). Although, in principle such a network should be able to provide very good QoS guarantees, the notion of a single, ubiquitous (sub)-network technology is not realistic (in fact, today's Internet services are provided across networks that consist of many different technologies). Internet protocols are widely available, generally easy to use, have well-defined software APIs, and can operate on many network technologies. Consequently, the Internet is being seen as a means for allowing access to integrated services [DT97].

Internet users have increasing demands to use a range of multimedia applications with QoS-sensitive data flows. These applications may require different QoS guarantees to be provided by the underlying network. An e-mail application can function with a best-effort network service. Interactive or real-time voice and video applications require delay, jitter, loss, and throughput guarantees in order to function. Web access can also work with a best-effort service, but it typically requires low delay and may require high throughput depending on the content being accessed. The Internet was not designed to cope with such a sophisticated demand for services [Cla88] [RFC1958]. Today's Internet is built upon many different underlying network technologies, of different age, capability, and complexity. Most of these technologies are unable to cope with such QoS demands. Also, the Internet protocols themselves are not designed to support the wide range of QoS profiles required by the plethora of current and future applications. This deficiency is currently being addressed by the IETF INTSERV (Integrated Services) WG* [RFC1633]. The explosive growth in the use of the Internet has resulted in much of the network being heavily loaded or overloaded so there is a need to allow controlled use of resources. Clark, Shenker and Zhang [CSZ92] speak of the Internet evolving to an integrated services packet network (ISPN), and identify four key components for an integrated services architecture for the Internet:

- **service-level** — the nature of the commitment made, e.g., the INTSERV WG has defined guaranteed and controlled-load service-levels (discussed later) and a set of control parameters to describe traffic patterns
- **service interface** — a set of parameters passed between the application and the network in order to invoke a particular QoS service-level, i.e., some sort of signalling protocol plus a set of parameter definitions
- **admission control** — for establishing whether or not a service commitment can be honored before allowing the flow to proceed
- **scheduling mechanisms within the network** — the network must be able to handle packets in accordance with the QoS service requested

A simple description of the interactions between these components is as follows:

- A service-level is defined (e.g., within an administrative domain or, with global scope, by the Internet community). The definition of the service-level includes all the service semantics: descriptions of how packets should be treated within the network, how the application should inject traffic into the network as well as how the service should be policed. Knowledge of the service semantics must be available within routers and within applications.
- An application makes a request for service invocation using the service interface and a signalling protocol. The invocation information includes specific information about the traffic characteristics required for the flow, e.g., data rate. The network indicates if the service invocation is successful or not, and may also inform the application if there is a service violation, caused either by the application's use of the service or a network failure.
- Before the service invocation can succeed, the network must determine if it has enough resources to accept the service invocation. This is the job of admission control that uses the information in the service invocation, plus knowledge about the other service requests it is currently supporting, and determines if it can accept the new request. The admission control function will also be responsible for policing the use of the service, making sure that applications do not use more resources than they have requested. This will typically be implemented within the routers.
- Once a service invocation has been accepted, the network must employ mechanisms that ensure that the packets within the flow receive the service that has been requested for that flow. This requires the use of scheduling mechanisms and queue management for flows within the routers.

---

* http://www.ietf.org/html.charters/intserv-charter.html

We examine how the realization of these key components has developed and discuss how research and development may progress in the next decade in order to move the Internet towards an ISN.

11.2.1 QoS Service Definitions and Service Invocation

The IETF INTSERV working group has proposed an architecture for evolving the Internet to an ISN. To support the architecture, INTSERV has produced a set of specifications for specific QoS service-levels based on a general network service specification template [RFC2216] and some general QoS parameters [RFC2215]. The template allows the definition of how network elements should treat traffic flows. With the present IP service enumerated as best-effort, currently, two service-level specifications are defined:

- **controlled-load** service [RFC2211] — the behaviour for a network element required to offer a service that approximates the QoS received from an unloaded, best-effort network
- **guaranteed** service [RFC2212] — the behaviour for a network element required to deliver guaranteed throughput and delay for a flow

Also specified is how to use a signalling protocol, RSVP [RSVP], to allow the use of these two services to be signalled through the network [RFC2210]. INTSERV also defines SNMPv2 extensions [RFC2213 and RFC2214] to allow remote monitoring and management of network elements that support these network services. Part of the INTSERV work is the definition of an architecture for a QoS manager (QM) entity that coordinates flow activities and resource usage at the end system [INTSERVQM]. Note that this architecture requires that the network elements and applications have semantic knowledge about the service-levels for the application flows, as specified in the service templates.

RSVP is used by applications to make a resource reservation, by asking the network to provide a defined quality of service for a flow. The reservation request consists of a FlowSpec identifying the traffic characteristics and service-level required. One part of the FlowSpec is a TSpec, a description of the traffic characteristic required for the reservation. Consequently it is possible for the same traffic characteristic to be used with different service-levels. This difference in QoS service-level could, for example, act as a way for offering cost differentials on the use of a particular application or service.

To invoke a particular service, the application uses a signalling protocol, RSVP, for a particular communication session (which may consist of one or more flows). To make a resource reservation, an appropriate FlowSpec is used along with the session IP destination address, the protocol number in the IP packet, and, optionally, the destination port number in the service invocation. The reservation procedure is as follows. The sender transmits a Path message advertising the session QoS requirements towards the destination IP address. All RSVP routers forwarding the Path message hold soft-state – information about the resource reservation required – until one of the following happens: a PathTear is sent from the sender cancelling the reservation, a Resv message is transmitted from a receiver effectively confirming the reservation, or the soft-state times-out. A Resv message from a receiver is sent back along the same route as the Path message*, establishing the reservation, then the application starts sending data packets. Path and Resv messages are sent by the sender and receiver, respectively, during the lifetime of the session to refresh the soft-state and maintain the reservation. A PathTear or ResvTear message explicitly tears down the reservation and allows resources to be freed. It is possible for the reservation to be changed dynamically during the lifetime of the session. RSVP can be used for unicast or multicast sessions.

RSVP allows the reservation to be made using filters that control how the reservation is applied. A fixed-filter (FF) is used to make a distinct reservation (cannot be shared by other flows along the path) with an explicit sender selection criteria (similar to a closed user group in telephony). A shared-explicit (SE) filter is used to request a reservation that is a union of all the requirements of the senders but still with explicit user selection. A wildcard-filter (WF) is a shared reservation with an open sender selection, i.e., an open group. FF would typically be used for unicast reservations or reservations for a lecture-type multicast session. FF or SE would be used for closed user groups, such as virtual meeting rooms. SE and WF would be used for open multicast groups, such as public seminars or conferencing.

Note that RSVP

- provides end-to-end signalling (between applications)
- sets up unidirectional reservations
- is specific to one session
- requires the applications and the network to be RSVP and INTSERV aware

11.2.2 QoS Service Provision

In the last subsection we considered how the QoS services are defined and how they are invoked. We now consider how they are implemented in the network. The INTSERV WG does not mandate any particular algorithms or mechanisms for the provision of a particular service. The INTSERV WG defines the behaviour required in the network elements and only suggests ways in which this behaviour might be implemented at the current time, albeit with reference to existing implementation techniques. The philosophy behind this approach is that as technology matures or as new, better technology is produced, it can be used to provide the same service as long as the service behaviour is honoured.

In the wide area, the behaviour implemented depends on the underlying network technology (the bearer service used by IP) to provide the links between the routers, and the behaviour of the routers connecting the various IP links. The bearer technology may be asynchronous transfer mode (ATM), frame relay (FR) or various point-to-point technologies, e.g., SONET/SDH (synchronous optical

---

* It is assumed that routes are symmetrical and relatively stable, but this is not always true in the wide area [Pax97a].

network/synchronous digital hierarchy). In some cases, it may be possible to exploit the properties of the underlying technology in order to achieve some traffic engineering goals at the IP level, for example, use of ATM virtual circuits for traffic segregation. Work is in progress within the Integrated Services for Specific Link layers (ISSLL)* and the Interworking over Non-Broadcast Multiple Access networks (ION)** working groups of the IETF in order to provide solutions specific to certain bearer technologies. In particular, at the time of writing, the ISSLL WG work on mappings of INTSERV onto ATM was approaching publication. This includes the mapping of RSVP onto ATM signalling.

The ISSLL WG is also tasked with examining the provision of IP integrated services within an IEEE 802-based LAN environment, and this work is currently in progress. The IEEE has recently extended the definition of IEEE 802.1D to support priority classes for traffic (the work was carried out under a working group that was originally labelled IEEE 802.1p). Work is in progress within the Internet community to map INTSERV and differentiated services (DIFFSERV – see below) onto such mechanisms.

However, the main issue concerning integrated services provision is the handling of the individual packets that make up a flow in order to honour the QoS requirements of that flow. The router has a nontrivial forwarding process for each packet:

- classify the packet in order to identify its QoS requirements (classification)
- determine when the packet should be forwarded (scheduling)
- manage the output queues under congested conditions (queue management)

Note these activities are logically distinct from the *routing* functions that all routers must be able to perform in order to determine in which direction to forward a packet (i.e., which output interface should be used). Several schemes have been developed within the Internet community for performing classification, scheduling, and queue management tasks, and they are currently undergoing experimentation and development. The most popular mechanisms establish a class-based hierarchy that allows sharing of resources in some way, for example sharing of the link capacity. The mechanisms are refined in order to incorporate scheduling mechanisms that ensure that packets are transmitted within a given timeframe. The mechanisms are based around analysis presented in Parekh and Gallagher [PG93 and PG94], which defines a model that allows fair sharing of resources. However, the realisation of this model is subject to some practical constraints in implementation, due in part to the computational complexity of the algorithms involved. Three models currently receiving attention within the Internet community are weighted fair queuing (WFQ) [DKS90], class-based queuing (CBQ) [FJ95 and WGCJF95], and worst-case fair-weighted fair queuing (WF^{2}Q+) [BZ96]. Note that *fair* does not necessarily imply *equal*, and all of the techniques that have been developed allow for different users***

---

to have different shares of resources. For example, CBQ was designed to allow sharing of link capacity within a class-based hierarchy, and, in Figure 11.1, we see an example showing the link capacity as a root node in a tree at 100%. Organisations X, Y, and Z that share the link are assigned 40%, 30%, and 30% of the link capacity, respectively. Within their own allocations of capacity, the organisations can choose to partition the available capacity further by creating sub-classes within the tree. Organisation X decides to allocate 30% to real-time traffic and 10% to all non-real-time traffic. Within the real-time allocation, X decides to allocate capacity to individual applications. Organisation Y also divides its allocation into real-time and non-real-time, but with a different share of the available link capacity. Organisation Z decides not to further refine its allocation of link capacity. The percentages indicate the minimum share of the link capacity a node in the tree will receive. Child nodes effectively take capacity from their parent node allocation. If some sibling nodes are not using their full allocations, other siblings that might be overshooting their own allocation are allowed to borrow capacity by interacting with the parent node. With an appropriate scheduling mechanism, this allows support for QoS-sensitive flows. Classifications in Figure 11.1 could be made per application, per flow, per IP source address, etc., as dictated by the policy chosen by the individual organisations in conjunction with their Internet connectivity provider.

WFQ, CBQ, and WF^{2}Q+ have different capabilities and different levels of computational complexity, depending on the policy used to define the granularity of the flow and the exact nature of the resource sharing implemented. However,

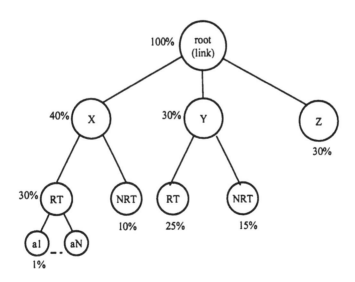

RT   real-time
NRT non-real-time

Figure 11.1 Example class hierarchy for link-sharing

they all have their advantages and drawbacks and, at the time of writing, are still generally in experimental use, though products will soon be appearing incorporating these schemes.

When a resource reservation is invoked, one of the functions that may be applied is admission control. Given a suitable description of the resource reservation requirements for a flow, admission control determines whether or not it is currently possible to provide the service required for the flow. This also requires knowledge of other flows that are currently sharing any resources along the network path of the flow. The nature of the admission control algorithm is dependent on the type of service that is being invoked; controlled-load service admission control will be handled in a different manner from guaranteed service admission control. Where a network element supports both controlled-load and guaranteed services for different flows, careful engineering must ensure that the service commitments undertaken by the network element are maintained. There is work in progress within the IETF to address admission control, and schemes have been proposed by the research community (e.g., [BFMM94 and JDSZ95]).

11.3 DIFFERENTIATED SERVICES

We have said that resource reservation with RSVP is a useful mechanism for applications with QoS-sensitive data flows. However, as IP cannot rely on any particular network technology-specific mechanisms, RSVP uses a soft-state technique with a two-pass protocol. We summarize the main problems with RSVP below [SB95 and WGS97]:

- During reservation establishment, if the first pass of each of two separate reservation requests are sent through the same network element, where one request is a superset of the other, the lesser one may be rejected (depending on the resources available), even if the greater one eventually fails to complete (of course it is possible to retry).
- If the first pass does succeed, the router must then hold a considerable amount of state for each receiver that wants to join the flow (e.g., in a multicast conference).
- The routers must communicate with receivers to refresh soft-state, generating extra traffic, otherwise the reservation will time out.
- Complete heterogeneity is not supported, i.e., in a conference everyone must share the same service-level (e.g., guaranteed or controlled-load), although heterogeneity within the service-level is supported.
- If there are router failures along the path of the reservation, IP route changes, so the RSVP reservation fails and the communication carries on at best-effort service, with the other routers still holding the original reservation until an explicit tear-down, the reservation times-out, or the reservation can be re-established along the new path.
- The applications must be made RSVP-aware, which is a nontrivial goal to realize for the many current and legacy applications, including multimedia applications with QoS-sensitive flows.

Resource reservation could be expensive on router resources, and adaptation capability is still required within the application to cope with reservation failures or lack of end-to-end resource reservation capability. Indeed, the Internet community has acknowledged the shortcomings of RSVP, especially with respect to scalability, and it is now recommended for use only in restricted network environments [RFC2208]. Such concerns about resource reservation have directed the Internet community to consider alternatives, specifically differentiated services [DIFFSERV]. Without resource reservation, we require some mechanisms to allow service differentiation within the network, but also we require a more flexible and dynamic adaptation capability within the application.

11.3.1 SERVICE DIFFERENTIATION

The IETF DIFFSERV (Differentiated Services) WG* takes a different view of using network resources to that of the INTSERV WG. At the time of writing, this work is still at very early stages, so several schemes are being discussed. The general model is to define a class-based system where packets are effectively marked with a well-known label. This label identifies the aggregate service-level the packet will receive much like a letter can be marked as registered, first class, or second class delivery. This is a much coarser granularity of service, but it reflects a well understood service model used in other commercial areas. The DIFFSERV model is different from RSVP. A key distinction of the DIFFSERV model is that it is geared to a business model of operation, based on administrative bounds, with services allocated to users or user groups. Whereas RSVP can act on a per-flow basis, the DIFFSERV classes may be used by many flows. Any packets within the same class must share resources with all other packets in that class, e.g., a particular organization could request a premium (low delay) quality with an assured (low loss) service-level for all their packets at a given data rate from their provider. The packets are treated on a per-hop basis by traffic conditioners, routers that determine the way a packet should be treated based on a policy that is selected by examining the value of the packet's class marking. The policy could be applied to all the traffic from a single user (or user group) and could be set up when subscription to the service is requested or on a configurable profile basis. The DIFFSERV mechanisms would typically be implemented within the network itself, without requiring runtime interaction from the end-system or the user, so they are particularly attractive as a means of setting up tiered services, each with a different price to the customer.

The RSVP mechanism seeks to introduce well-defined, end-to-end, per-flow QoS guarantees by use of a sophisticated signalling procedure. The DIFFSERV work seeks to provide a virtual pipe with given properties, in which the user may require adaptation capability or further traffic control if there are multiple flows competing for the same virtual pipe capacity. Additionally, the DIFFSERV architecture means that different instances of the same application throughout the Internet could receive different QoS, so the application needs to be adaptable.

---

* http://www.ietf.org/html.charters/diffserve-charter.html

The service itself will be defined in terms of a service level agreement (SLA) that embodies the contract between the service user and service provider. The policy implemented by the SLA may include issues other than QoS that must be met, e.g., security, time-of-day constraints, etc. Figure 11.2 highlights the main difference between INTSERV and DIFFSERV scope. INTSERV tries to provide, per application, end-to-end resource reservation. DIFFSERV aims to provide a SLA-based contract between service networks. One very attractive feature of DIFFSERV is that it can be introduced into existing networks in a piece-wise manner, without having to modify current or legacy applications. The packets leaving a network are marked for DIFFSERV handling by DIFFSERV-capable routers that sit at administrative boundaries. Therefore, only the routers need to be updated, and the applications themselves can remain unchanged. (However, this does not preclude individual hosts or individual applications being DIFFSERV-aware and marking packets accordingly as they leave the host.) The DIFFSERV-capable routers could be at the edge of the customer network or part of the provider's network. If the DIFFSERV-marking is performed within the customer network, then policing is required at the ingress router at the provider network in order to ensure that the customer does not try to use more resources than allowed by the SLA.

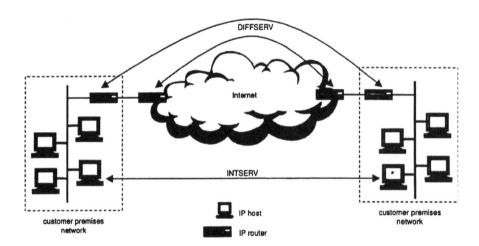

Figure 11.2 Scope of INTSERV and DIFFSERV

11.3.2 PROVIDING DIFFERENTIATED SERVICES

The DIFFSERV work is aimed at providing a way of setting up QoS using policy statements that form part of a SLA. The policy may use several packet header fields to classify the packet, but the classification marking is a simple identifier (currently a single byte) within the packet header. The classification is by way of a special value for a single header field, the DS (differentiated services) byte, which will be used in place of the ToS (Type of Service) field in IPv4 packets or the traffic-class field in IPv6 packets. The DS byte will have the same syntax and semantics in both

IPv4 and IPv6. There are likely to be some global values – DS codepoints – agreed for the DS field within the IETF but the intention is that the exact policy governing the interpretation of the DS codepoints and the handling of the packets is subject to some locally agreed SLA. SLAs could exist between customer and Internet Service Provider (ISP) as well as between ISPs. The DS codepoints are used to identify packets that should have the same aggregate per-hop behaviour (PHB) with respect to how they are treated by individual network elements. The PHB definitions and the DS codepoints used may differ between ISPs, so translation mechanisms between ISPs will be necessary.

The meaning of the DS codepoints and the content of SLAs are established at subscription time, and, although there will be scope for change by agreement between customer and provider, the kind of dynamic and flexible resource reservation that is described above for using RSVP is not envisioned for DIFFSERV.

The mechanisms for classification and handling of packets within the network can be the same as for INTSERV – WFQ, CBQ and WF^{2}Q+ could be used. The big gain is that the end-to-end signalling and the maintenance of per-flow soft-state within the routers that is required with RSVP is no longer required. This makes DIFFSERV easier to deploy and more scalable than RSVP and INTSERV services. However, this does not mean that INTSERV and DIFFSERV services are mutually exclusive. Indeed, it is likely that DIFFSERV SLAs will be set-up between customer and provider for general use, then RSVP-based per-flow reservations may be used for certain applications as required, for example, an important video conference within an organisation. This concept is shown in Figure 11.3. The DIFFSERV capability provides the aggregate service to the provider while individual applications with special needs can use RSVP to setup INTSERV reservations within this aggregate pipe, as required.

Note that while INTSERV is based on the notion of receiver-generated control messages for confirming the resource reservation, DIFFSERV requires that the ISPs for the receiver and the sender have a way of allowing the PHB definition to be honoured across the network. This requires cooperation between many ISPs. So, it

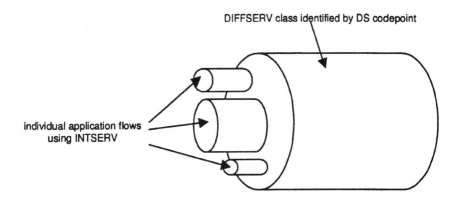

Figure 11.3 Conceptual view of INTSERV reservations within a DIFFSERV class

is expected that the DIFFSERV facilities will be used initially to offer individual customers of single ISPs the ability to establish virtual private network (VPN) scenarios, with the network of that single provider enabling the wide-area connectivity. Of course, individual ISPs (or backbone providers) may form peering agreements to enable wide-area connectivity based on DIFFSERV. Such connectivity could be used to provide enterprise intranet services, as well as conferencing, group-working, and software distribution based on use of IP-multicast across the VPN.

11.4 PERFORMANCE ENHANCEMENTS FOR IP

With the evolution to integrated services provision with IP, one thing is certain: the amount of IP traffic will increase so the networks must be able to handle this increased load. Over the past few years, particular emphasis has been placed on developing techniques to allow increased performance of IP-based networks. It should be noted that performance issues are not necessarily the same as QoS issues. Performance issues are concerned with getting packets from A to B across the network as fast as possible. QoS issues are concerned with making sure that as packets traverse the network, they receive appropriate handling at the routers to ensure that QoS performance criteria (such as delay, jitter, data rate) are met. Nevertheless, as IP traffic increases so the networks must be able to handle large volumes of IP packets or the QoS criteria may not be met.

In this section we examine three of the issues affecting performance that may impact QoS – the use of IP over high-speed bearer services, enabling fast forwarding mechanisms within the network, and the evolution of IP routers.

11.4.1 HIGH SPEED BEARER SERVICES

A suitable subnetwork technology to provide integrated services capability might be ATM, which is itself designed to be an integrated service bearer. Consequently, IP and ATM might be seen as competing technologies. However, the evolution is such that it seems there will probably be few native ATM applications, while many IP applications already exist and many more are being created. Therefore there has been much activity within the Internet community to make IP work effectively over ATM networks [RFC1821]. Basic connectivity mechanisms for IP over ATM have been proposed by the IETF [RFC2225]; however, other solutions, not designed specifically for IP but with the advantage of providing support for other layer 3 protocols, have been proposed by the ATM Forum* – LAN Emulation [LANEv2] and multi-protocol over ATM [MPOA]. These technologies all provide an encapsulation mechanism for IP and a set of rules for establishing ATM-level connectivity for the transportation of IP packets.

In general, there is a need to allow IP to be carried in a whole range of non-broadcast multiple access (NBMA) scenarios including ATM, Frame Relay, and other point-to-point technologies, and this need is being addressed by the ION WG within the IETF. However, in certain backbone scenarios, the use of ATM is seen

---

* http://www.atmforum.com/

as an overhead for carrying IP, and many network operators are now investigating carrying IP packets directly in SONET/SDH frames. Indeed, an encapsulation method for IP in SONET/SDH [RFC1619], IP over ISDN [RFC1618], and IP in Frame Relay [RFC1973] have all existed for some time, based on the use of the standard point-to-point protocol (PPP) [RFC1661]. However, these mechanisms say nothing of how high performance can be achieved with IP. In general, there are likely to be relatively few problems with allowing IP to be carried within a particular network bearer service – IP works over anything.

The mapping of IP onto any lower layer should be as simple as possible so protocol overhead will not become a performance bottleneck. An example of protocol inefficiency is seen in the protocol stack of [RFC2225] for classical IP over ATM (CIPA). The main goal for CIPA is connectivity, and Figure 11.4 shows the protocol stack used to attain connectivity with CIPA. An IP packet must be framed in a LLC frame, then within an AAL5 protocol data unit, and then shredded into ATM cells. This process must be reversed at every IP-level routing hop within the ATM network (to reform the IP packet) and then the IP packet must be re-encapsulated with the same process if forwarding onto another ATM interface from the router. CIPA does not allow direct ATM-level communication between IP-nodes at the ATM-level if they are on *different IP sub-networks*, even if they are on the *same ATM network*.

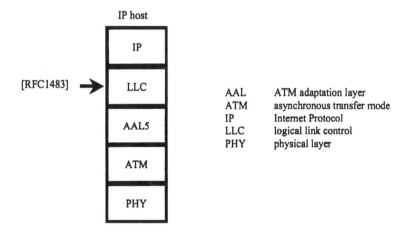

Figure 11.4 Protocol stack for Classical IP over ATM (CIPA) [RFC2225]

Therefore, the main issue for performance is to find an efficient forwarding scheme for transporting IP packets over NBMA network connections.

11.4.2 FAST FORWARDING MECHANISMS

The problem of making fast forwarding decisions is inherent in IP networking. A description of the task is quite simple: move a packet from an input port to an output port as fast as possible. To make a forwarding decision for an IP packet, the following steps take place at a router:

1. A packet arrives at an input port and the packet may need to be buffered.
2. The router must read the destination address of the packet.
3. Based on the destination address, the router selects candidate routing table entries, and, for each candidate entry, saves the next hop address, the address mask for the address, and the output port for that entry.
4. After all the candidate entries have been found, an entry must be selected by using the longest prefix match using the routing entry address mask and the destination address in the packet.
5. When the appropriate candidate entry has been selected, the packet is placed on the appropriate output queue.

Steps 3 and 4 in this process may require the consideration of other information, such as routing metrics, policy-based routing, and security information. In general, this may slow the forwarding process, although clever caching and recent developments in table lookups can help (more on this below). Where there are many packets in a flow, it seems that this process need be executed only once as all packets for the flow will be subject to the same forwarding decision. This is the main principle behind multi-protocol label switching (MPLS). MPLS expedites the forwarding process by using simple, fixed-length labels to identify packets within a flow. The labels may be set up using network management tools or other administrative measures, or they may be generated dynamically as a flow is detected. The label acts as a selector, just like a virtual circuit identifier (VCI), with only local significance, to allow switching of IP packets based on labels rather than on IP address information. This situation is sometimes called short-cut routing or cut-through routing, reflecting the fact that the concatenation of the locally generated labels along a path describe the route for a packet along that path. As the labels are of a short, fixed length (currently 20 bits), they are easy to look up using tables. Note that this is not a new *routing* mechanism – it is simply a way of making *forwarding* decisions easier. In fact, it still uses standard IP routing information and relies on standard IP routing protocols to establish the forwarding table. The label sits as part of a short (32 bit) shim-header between the link-layer header and the IP-header. In fact, as the name suggests, MPLS is designed to work for any layer 3 packet switched protocol, not just IP, but most of the effort is currently around the implementation of IP solutions. Use of labels in this way introduces its own requirements: labels must be generated, distributed, and maintained throughout the network. The MPLS technology is in progress and covers all aspects of label distribution and handling.

Different vendors have already produced products that use different flavours of cut-through routing to exploit subnetwork-specific technology features in place of the generic label of MPLS, for example, the use of ATM VCI/VPI to switch ATM cell streams containing IP packets.

11.4.3 SMARTER, NOT JUST FASTER

Making networks faster means more than just increasing the line speed connecting the routers; it means improving the performance of the routers themselves [KS98] [KLS98]. Current technology is already at the point where memory access speeds and the execution of the software algorithms implementing the forwarding code within routers are becoming the bottleneck. The design and implementation of router hardware and software is now an art and can determine the overall performance of a network more than the line speed of the individual links can. In dealing with the provision for INTSERV/DIFFSERV mechanisms, router manufacturers and the research community are working hard, with some success, to produce smarter routers with better software, and not just running existing routing/forwarding software on faster hardware platforms.

Also, for high-speed packet processing, routers must be able to process and classify packets at line speed, i.e., there should be no queuing before packets have been classified, lest this delay contribute to the violation of the QoS requirements for the flow to which this packet belongs. Additionally, routers must not rely on any knowledge of possible traffic patterns as history shows the traffic patterns are hard to model and predict [PF95, PF97 and Pax97b].

There is much progress in devising new algorithms for performing fast routing table look-ups [BCDP97] [VTP97] and packet classification [LS98]. There are moves to integrate the hardware and software as much as possible and to devise algorithms that are as simple as possible so that they can be implemented in hardware.

This aim of hardware-friendliness is also visible with the evolution of the IP protocol itself. In Figure 11.5 we can compare the IPv4 packet header [IPv4] and the (currently proposed) IPv6 header. We see that the latter is much simpler in nature. We see that the IPv6 header is much more amenable to hardware processing than the IPv4 header, and as fragmentation and re-assembly have now become an end-to-end issue in IPv6, this simplifies the router's task in handling IP packets. Additionally, IPv6 potentially has better support for QoS support by including a flow-label and the traffic-class field within the first word of the IP packet header; however, the exact use of both these fields has not yet been fully defined.

Consequently, current work suggests it is wise to take into account QoS and performance issues when considering hardware, software, *and* protocol design, and this trend seems likely to continue and be an important factor in promoting IP as an integrated services bearer.

11.5 TECHNOLOGY COMPONENTS FOR FUTURE IP INTEGRATED SERVICES

So far we have considered current research or developments that are likely to be deployed within the next decade. In this section we consider three technology issues that are likely to affect integrated services in the coming decade and may change the way in which applications and services are used and deployed.

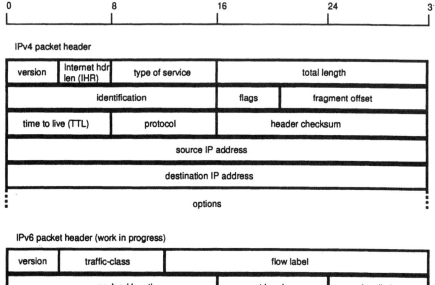

Figure 11.5 IPv4 packet header (top) and IPv6 packet header (bottom)

11.5.1 DYNAMICALLY ADAPTABLE APPLICATIONS

We discussed above that the original INTSERV work using RSVP may not scale and that it is likely that DIFFSERV and INTSERV mechanisms will be used together. This situation has the potential to allow different users of the same applications to have different QoS. In general, the QoS experienced by a particular application instance may vary due to a number of factors:

- variations in network behaviour because of network traffic from other sources
- variations in network paths because of the behaviour of routing functions

- the application resides on a mobile host
- the (human) user selects different user preferences depending on the costs of a particular service or the QoS required for a particular use of an application

The first three of these are based on QoS observed by an application instance during operation. In that case, the application must be able to detect QoS changes and adapt its operation to match the QoS available at the current time. Additionally, differences in QoS may be because different users of the same application subscribe to different service-levels from their ISP. Most ISPs currently provide only a single best-effort service, but this is sure to change in the near future.

The last factor in the list above is a distinct choice made by the user. For example, consider that the user has a video-telephony application. When that application is used to contact family, the user may select high-quality, full-screen video and high-quality audio, but when the same application is used to contact the office, the user may select slow-scan and small-size video and phone quality audio. Therefore, the application must be able to adapt in response to changes in the network QoS as well as to a change in user preferences. This adaptation is dynamic and involves changes in the application's configuration. Such configuration changes are currently handled manually, and they must rely on a knowledgeable user being able to determine the correct application configuration for a particular network QoS scenario. We need mechanisms that can provide summaries of QoS information that allow either the application to dynamically adapt (reconfigure) itself automatically (taking into account network QoS factors as well as user preferences) or at least to allow the user to make an informed decision using simple feedback to the user [BK98a]. Some work is in progress to design mechanisms that can enable dynamic adaptation [BK98b].

11.5.2 ACTIVE NETWORKS

Another way to try to capture the adaptation capability required for integrated services provision is to make the networks themselves adaptable. Such networks could consist of active network components and elements that are effectively programmable by the application and that can adapt their behaviour in response to changes in the network QoS or in the application behaviour (the latter occurring due to interaction with the user). Campbell [Cam97], Lu, Lee, and Bharghavan [LLB97], and Campbell, Coulson, and Hutchison [CCH96] discuss issues concerning the provision of adaptation capability within the network itself. In such situations the QoS requirements for the flow, including adaptation capability, are submitted to the network which must manage resources to maintain the service for the user.

In more general terms, active networks allow the deployment of new services into the network in an incremental fashion as required. In theory, therefore, a new application with its own sophisticated QoS requirements could effectively download the code for the processing mechanisms for its packets to all the relevant network elements along its communication path as required [TW96] [WLG98]. The application would not be concerned with the availability or deployment of definitions such as those currently being specified by INTSERV or DIFFSERV.

In order to enable such active networking, we need to have common application programming interfaces (APIs) that allow applications to interact with the network components. The task of establishing such APIs has been undertaken by the IEEE application Programming Interfaces for Networks (PIN) Working Group (P1520).* The goal of this group is to define a reference model and set of APIs to allow access to network elements. Making the network elements programmable allows a much more flexible and dynamic approach to deployment of service and facilities in the network. The approach is heavily based on a distributed systems model, with resources and network entities modelled as objects that are accessed via well-defined interfaces. The intention is that the hardware and software aspects of service development and deployment are separated to the extent that the service may have some level of independence from the hardware substrate. At the time of writing, this work is at a very early stage, with no draft standards produced.

In general, there is currently great momentum behind the idea that the network can be made more active, and platforms such as Java** and CORBA (Common Object Request Broker Architecture)*** are seen as enabling technologies in this arena.

11.5.3 SECURITY

One of the biggest issues raised by the use of the Internet for carrying media flows such as voice and video (as well as other more sensitive data, such as credit card numbers!) is security. IP has a security architecture [RFC1825] as well as some specific defined security standards that allow encryption of packets to provide privacy in communication [RFC1827 and RFC1829], as well as to allow per-packet authentication [RFC1826 and RFC1828]. There are some technical issues concerning the use of security, for example the performance loss when security mechanisms such as encryption are used. However, most of the major security issues are currently concerned with national and international political activities and the provision of trusted third parties (TTPs). Various governments around the world see that provision of strong cryptographic techniques will make it almost impossible for them to monitor the communications of criminals. This has led to legislation where the use of strong cryptography is permissible only if security agencies or other authorised government agencies have the ability to access the encrypted information. The proposed implementation of this is the use of TTPs that will securely store the cryptographic keys that are used and make them available to an authorised body as required. The argument against such a mechanism is that it is highly unlikely that criminals will register their keys with the TTP and will continue to use the strong cryptographic techniques that are available. TTPs acting as certification authorities (CAs) are also required in order to provide verification of electronic credentials – signatures for electronic identifiers.

Other security issues arise when considering active networking. Who has the right to program the network elements? How does a network element know the code is safe? What would happen if a network virus were to infect active network components? The security requirements of active networking have yet to be clearly identified.

---

* http://www.ieee-pin.org/
** http://www.javasoft.com/
*** http://www.omg.org/

11.6 SUMMARY

Internet protocols and APIs are widely used to develop applications that have particular QoS requirements. However, the Internet was never designed to offer QoS guarantees for applications. There is a need to provide support for QoS-sensitive applications within the Internet using additional mechanisms. Work is in progress within the INTSERV WG of the IETF to produce an integrated services model and to develop QoS mechanisms that can reserve resources for applications. However, the current developments, based around the use of RSVP for application-network-application signalling are not fully deployed, and, indeed, it is considered that at the current time they may not scale for use across whole of the Internet.

The DIFFSERV WG of the IETF proposes a different model based on classifying packets according to specific QoS requirements and implementing special packet handling criteria based on this packet classification. The DIFFSERV approach is more coarse-grained than the INTSERV approach and is based on providing administratively controlled service differentiation, rather than fine-grained, per-application, dynamically requested QoS. This will allow ISPs to offer a tiered service on a per-customer or per-application basis.

As increasing numbers of people and applications make use of the Internet, the core network must be capable of handling large amounts of traffic. In order to ease the congestion that is currently seen across the Internet, new protocols and mechanisms to provide performance enhancements in the network elements are needed, and not just faster transmission capability. One of the major bottlenecks in the Internet is the capability of the routers. To support INTSERV and DIFFSERV, routers must be enhanced with controlled scheduling, classification, queue management, and fast-forwarding mechanisms.

Applications need to be dynamically adaptable so that they can be easily reconfigured (under user control) to make the best use of the resources available to them in a particular situation. Networks themselves may become active and programmable to support the diverse range of applications, QoS options, and user preferences that may be available. Lastly, security mechanisms are required to allow protected real-time communication such as person-to-person voice and video flows and conferencing.

REFERENCES

[BCDP97]   A. Brodnik, S. Carlsson, M. Degermark, S. Pink, "Small Forwarding Tables for Fast Routing Lookups," *Proc. ACM SIGCOMM'97*, Sep 1997, 3–14.

[BFMM94]   A. Banerjea, D. Ferrari, B. A. Mah, M. Moran, "The Tenet Real-time Protocol Suite: Design, Implementation, and Experiences," Technical Report TR-94-059, University of California at Berkeley, Berkeley, California, Nov 1994.

[BK98a]   S. N. Bhatti, G. Knight, "QoS Assurance vs. Dynamic Adaptability
               for Applications," *Proc. 8th Int. Workshop on Network and Operating
               Systems Support for Digital Audio and Video* (NOSSDAV'98), New
               Hall, Cambridge University, Cambridge, UK, 8-10 Jul 1998.
[BK98b]   S. N. Bhatti, G. Knight, "Notes on a QoS Information Model for
               Making Adaptation Decisions," *Proc. 4th Int. Workshop on High
               Performance Protocol Architectures* (HIPPARCH'98), 1998.
[BZ96]   J. C. R. Bennett, H. Zhang, "Hierarchical Packet Fair Queue Algo-
               rithms," *Proc. Acm SIGCOMM'96*, Sep 1996, 143–156.
[Cam97]   A. T. Campbell, "Mobiware: QoS-Aware Middleware for Mobile
               Multimedia Networking," *Proc. IFIP 7th Int. Conf. High Perfor-
               mance Networking*, Apr 1997.
[CCH96]   A. Campbell, G. Coulson, D. Hutchison, "Supporting Adaptive
               Flows in Quality of Service Architecture," *ACM Multimedia Systems
               J.*, May 1996.
[Cla88]   D. D. Clark, "The Design Philosophy of the DARPA Internet Pro-
               tocols," *Proc. ACM SIGCOMM'88*, Aug 1988, 106–114.
[CSZ92]   D. D. Clark, S. Shenker, L. Zhang, "Supporting Real-Time Appli-
               cations in an Integrated Services Packet Network: Architecture and
               Mechanism," *Proc. ACM SIGCOMM'92*, Aug 1992, 14-26.
[DIFFSERV]   K. Nichols, S. Blake (Eds), "Differentiated Services Operational
               Model and Definitions," IETF DIFFSERV WG, work-in-progress,
               1998.
[DKS90]   A. Demers, S. Keshav, S. Shenker, "Analysis and Simulation of a
               Fair Queuing Algorithm," *Internetworking Res. Exp.*, vol. 1, 1990,
               3-26.
[DT97]   M. Decina, V. Trecordi, "Convergence of Telecommunications and
               Computing to Networking Models for Integrated Services and Appli-
               cations," *Proceedings of the IEEE*, vol. 85 no. 12, Dec 1997.
[FJ95]   S. Floyd, V. Jacobson, "Link-sharing and Resource Management
               Models for Packet Networks," *IEEE/ACM Trans. Networking*, Vol.
               3 No. 4, Aug 1995, 365–386.
[INTSERVQM]   D. Clark, "The Quality Management Interface," slides presented at
               the 31st IETF meeting, Jan 1995.
[IPv4]   J. Postel, "Internet Protocol," RFC791, Sep 1981.
[JDSZ95]   S. Jamin, P. Danzig, S. Shenker, L. Zhang, "A Measurement-based
               Admission Control Algorithm for Integrated Services Packet Net-
               works," *Proc. ACM SIGCOMM'95*, Sep 1995, 2–13.
[KLS98]   V. P. Kumar, T. V. Lakshman, D. Stiliadis, "Beyond Best Effort:
               Architectures for the Differentiated Services of Tomorrow's Inter-
               net," *IEEE Communications*, no. 5, vol. 36, May 1998, 151–164.
[KS98]   S. Keshav, R. Sharma, "Issues and Trends in Router Design," *IEEE
               Communications*, no. 5, vol. 36, May 1998, 144–151.
[LANEv2]   "LANE v2.0 LUNI Interface," ATM Forum, af-lane-0084.000, Jul
               1997.
[LLB97]   S. Lu, K.-W. Lee, V. Bharghavan, "Adaptive Service in Mobile Com-
               puting Environments," in *Building QoS into Distributed Systems*, (A.
               Campbell, K. Nahrstedt, Eds), Chapman & Hall, 1997, 25–36.
[LS98]   T. V. Lakshman, D. Stiliadis, "Packet Classification Algorithms for
               Gigabit Internet Routers," *Proc. ACM SIGCOMM'98*, Sep 1998.

[MPOA]	"Multi-Protocol Over ATM Specification v1.0," *ATM Forum*, af-mpoa-0087.000, Jul 1997.
[Pax97a]	V. Paxson, "End-to-End Routing behaviour in the Internet," *IEEE/ACM Trans. Networking*, vol. 5 no. 5, Oct 1997, 601–615.
[Pax97b]	V. Paxson, "End-to-End Internet Packet Dynamics," *Proc. ACM SIGCOMM'97*, Sep 1997, 139–152.
[PF95]	V. Paxson, S. Floyd, "Wide-Area Traffic: The Failure of Poisson Modeling," *IEEE/ACM Trans. Networking*, vol. 3 no. 3, 1995, 226–244.
[PF97]	V. Paxson, S. Floyd, "Why We Don't Know How to Simulate the Internet," *Proc. 1997 Winter Simulation Conf.*
[PG93]	A. Parekh, R. Gallagher, "A Generalised Processor Sharing Approach to Flow Control in Integrated Services Networks – The Single Node Case," *ACM/IEEE Trans. on Networking*, vol. 1 no. 3 1993.
[PG94]	A. Parekh, R. Gallagher, "A Generalised Processor Sharing Approach to Flow Control in Integrated Services Networks – The Mutiple Node Case," *ACM/IEEE Trans. on Networking*, vol. 2 no. 2, 1994.
[RFC1483]	J. Heinanen, "Multiprotocol Encapsulation over ATM Adaptation Layer 5," RFC1483, Jul 1993.
[RFC1618]	W. Simpson, "PPP over ISDN," RFC1618, May 1994.
[RFC1619]	W. Simpson, "PPP over SONET/SDH," RFC1619, May 1994.
[RFC1633]	B. Braden, D. Clark, S. Shenker, "Integrated Services in the Internet Architecture: An Overview," RFC1633, Jun 1994.
[RFC1661]	W. Simpson (Ed), "The Point-to-Point Protocol (PPP)," RFC1661 Jul 1994.
[RFC1821]	Borden, Crawley, Davie, Batsell, "Integration of Real-time Services in an IP-ATM Network Architecture," RFC1821, Aug 1995.
[RFC1825]	R. Atkinson, "Security Architecture for the Internet Protocol," RFC1825, Aug 1995.
[RFC1826]	R. Atkinson, "IP Authentication Header," RFC1826, Aug 1995.
[RFC1827]	R. Atkinson, "IP Encapsulating Security Payload (ESP)," RFC1827, Aug 1995.
[RFC1828]	P. Metzger, W. Simpson, "IP Authentication using Keyed MD5," RFC1828, Aug 1995.
[RFC1829]	P. Karn, P. Metzger, W. Simpson, "The ESP DES-CBC Transform," RFC1829, Aug 1995.
[RFC1958]	B. Carpenter, "Architectural Principles of the Internet," RFC1958, Jun 1996.
[RFC1973]	W. Simpson, "PPP in Frame Relay," RFC1973, Jun 1996.
[RFC2208]	A. Mankin, F. Baker, B. Braden, S. Bradner, M. O'Dell, A. Romanow, A. Weinrib, L. Zhang, "Resource ReSerVation Protocol (RSVP) – Version 1 Applicability Statement Some Guidelines on Deployment," RFC2208, Sep 1997.
[RFC2210]	J. Wroclawski, "The Use of RSVP with IETF Integrated Services," RFC2210, Sep 1997.
[RFC2211]	J. Wroclawski, "Specification of the Controlled-Load Network Element Service," RFC2211, Sep 1997.
[RFC2212]	S. Shenker, C. Partridge, R. Guerin, "Specification of Guaranteed Quality of Service," RFC2212, Sep 1997.

[RFC2213]   F. Baker, J. Krawczyk, A. Sastry, "Integrated Services Management Information Base using SMIv2," RFC2213 Sep 1997.

[RFC2214]   F. Baker, J. Krawczyk, A. Sastry, "Integrated Services Management Information Base Guaranteed Service Extensions using SMIv2," RFC2214, Sep 1997.

[RFC2215]   S. Shenker, J. Wroclawski, "General Characterization Parameters for Integrated Service Network Elements," RFC2215, Sep 1997.

[RFC2216]   S. Shenker, J. Wroclawski, "Network Element Service Specification Template," RFC2216, Sep 1997.

[RFC2225]   M. Laubach, J. Halpern, "Classical IP and ARP over ATM," RFC2225, Apr 1998.

[RSVP]   R. Braden, Ed., L. Zhang, S. Berson, S. Herzog, S. Jamin, "Resource ReSerVation Protocol (RSVP) – Version 1 Functional Specification," RFC2205, Sep 1997.

[SB95]   S. Shenker, L. Breslau, "Two Issues in reservation Establishment," *Proc ACM SIGCOMM'95*, Sep 1995, 14–26.

[She95]   S. Shenker, "Fundamental Design Issues for the Future Internet," *IEEE J. Selected Areas in Commun.*, no.13, 1995, 1141–1149.

[TW96]   D. L. Tennenhouse, D. J. Wetherall, "Towards an Active Networking Architecture," *ACM Comput. Commun. Rev.*, no. 2 vol. 26, Apr 1996.

[VTP97]   G. Varghese, J. Turner, B. Plattner, "Scalable High Speed IP Routing Table Lookups," *Proc. ACM SIGCOMM'97*, Sep 1997, 25–36.

[WGCJF95]   I. Wakeman, A. Ghosh, J. Crowcroft, V. Jacobson, S. Floyd, "Implementing Real Time Packet Forwarding Policies using Streams," *Proc. USENIX'95*, Jan 1995, 71–82.

[WGS97]   L. Wolf, C. Gridwodz, R. Steinmetz, "Multimedia Communication," *Proc. IEEE*, vol. 85 no. 12, Dec 1997, 1915–1933.

[WLG98]   D. Wetherall, U. Legedza, J. Guttag, "Introducing New Internet Services: Why and How," *IEEE Network*, no. 3 vol. 12, May/Jun 1998.

12 Satellite Communication Systems

Ahmad Khanifar

CONTENTS

0-8493-9594-1/00/$0.00+$.50
© 2000 by CRC Press LLC

This chapter portrays the evolutionary path of satellite communication systems into the new millenium. It emphasizes the new services and their effect on what is perceived to be the established technology. System designers will be faced with the enormous task of integrating satellite-based mobile communications with the existing terrestrial network.

The geostationary orbital latency associated with satellite-based communication networks can be an inconvenience on voice transmissions and will be untenable for real-time applications, such as video conferencing, as well as many other standard data protocols. Using satellites in low earth-orbit (LEO) appears to be the obvious remedy to this problem, as a typical LEO distance is an order of magnitude closer to earth than the geostationary orbit.

Just as networks on the ground have evolved from centralised systems built around a single mainframe computer to distributed networks of interconnected personal computers, space-based satellite networks are evolving from centralised networks relying on a single geostationary satellite to distributed networks of interconnected LEO satellites. With a distributed network, reliability can be built into the network rather than into the individual unit. In such a distributed architecture, the dynamic routing and the robust scalability should be comparable to the Internet, while adding the benefits of real-time capability and location-insensitive access.

The fixed satellite services (FSS) arena will continue to provide global voice communications as well as witness a widespread incorporation of business service applications. The technology envisioned to fulfill the forthcoming service demands is very small aperture antennas (VSAT).

The choice of orbits and the access schemes are the two areas that have a decisive impact on the performance of systems discussed in the preceding section. These systems are discussed in greater detail in the following pages.

12.1   INTRODUCTION

Over the next ten years, space communications will play an integral role in the development of a global information infrastructure, allowing video, audio, and data to be distributed irrespective of borders and restrictions. Satellites are the key element to intercontinental traffic, as well as to direct user connection, bypassing terrestrial means. The ever-growing need to provide higher capacity at lower cost is the main driving force in this field.

It has been nearly 50 years since trans-horizon communication via satellites was proposed as a theoretical possibility. Since then, the science fiction-like idea has come a long way to manifest itself as an established technology. Despite this tremendous progress, 1998 will be remembered as another milestone in the evolutionary path of this industry. It was a year when a number of systems offering seamless worldwide satellite telecommunication using a portable handset were commissioned. Before we discuss these complex systems and other counterparts, we will take a look at where it all started.

The concept of global telecommunication coverage using satellites was first suggested in an article in *Wireless World* by science fiction author Arthur C. Clark.[1]

> All these problems can be solved by the use of a chain of space-stations with an orbital period of 24 hours, which would require them to be at a distance of 42,000 Km from the centre of the earth. There are a number of possible arrangements for such a chain but that shown figure (1) is the simplest. The stations would lie in the earth's equatorial plane and would thus always remain fixed in the same spots in the sky, from the point of view of terrestrial observers. Unlike all other heavenly bodies they would never rise or set. This would greatly simplify the use of directive receivers installed on the earth.

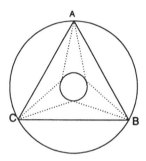

Figure 12.1 Arthur Clark's view of a Global Communications System

12.2 ORBITS

For the telecommunication coverage using a space vehicle, there are basically three orbital options: equatorial, polar, and inclined. Arthur Clark's vision of satellite communications was based on the circular equatorial orbit, but Figure 12.2 shows two more options, the circular polar orbit and the elliptically inclined orbit. There are merits for each in terms of earth coverage and the intended services.

12.2.1 CIRCULAR EQUATORIAL ORBIT (GEOSTATIONARY)

A satellite in a circular orbit at 35,800 km has a period of 24 hours and appears stationary over a fixed point on the earth's surface, hence, it is referred to as geostationary. Such a satellite is visible from one third of the earth surface. Examples of telecommunication services using these orbits are the INTELSAT* and Eutelsat** satellites. Figure 12.3 shows a typical coverage area for different geostationary satellites.

The geostationary operation requires in-orbit stabilisation because the earth is not a perfect sphere and because of other perturbations, such as the earth's tidal

---

* http://www.intelsat.int/
** http://www.eutelsat.com/home.html

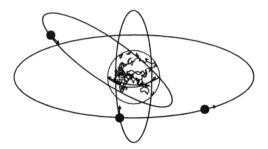

Figure 12.2 Three basic orbits

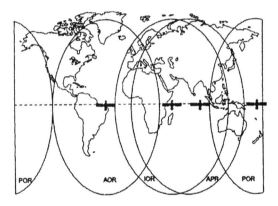

Figure 12.3 INTELSAT Global Network and Coverage Areas

motion and the gravitational forces of the moon and sun. An orbit that is inclined towards the equatorial plane produces a sinusoidal variation in the longitude. This is seen from earth as a motion around an ellipse with the period of 24 hours. An incorrect orbital velocity results in inaccurate altitude and drifts to the east or west. The east-west position of a satellite has to be corrected continuously over its operational lifetime, which typically is 10–15 years. Positioning is regularly corrected to within ±0.1°. The north-south positioning poses a greater demand on fuel reserves; to save fuel some satellite operators, including INTELSAT, allow their satellites to drift, resulting in an inclination angle of up to 3°. This, in turn, can typically extend the operational life of the satellite by as many as 3 years.

12.2.2 CIRCULAR POLAR ORBIT

The circular polar orbit is the only orbit plane that can provide full global coverage, including the polar regions, by a single satellite. For nearly all telecommunication systems where the instantaneous transfer of information is required, full global connectivity can be achieved with a series of satellites in circular polar orbit phased in an appropriate order. Until recently, this type of orbit has been used primarily for military, navigation, or meteorologic applications, but in the past few years a number of polar orbiter systems have been proposed and are being deployed for telecommunication applications. The Iridium constellation (to be discussed later) is a prime example and is due to become operational shortly after the time of this writing.

12.2.3 ELLIPTICALLY INCLINED ORBIT

The elliptically inclined orbit offers unique properties that have been used by some communication satellite systems, notably a former Soviet Union domestic system called Molniya. Inclined orbits can provide visibility to higher northern and southern latitudes, but they require the earth station to track the satellite continuously. In addition, inclined orbits usually require at least three satellites, suitably phased to be spaced along the orbit, to provide continuity of service. The former USSR preferred to put its communication satellites into an eccentric orbit with an inclination of about 63° and a period of 12 hours. The main reason for this was geographical. Because all of the Soviet Union land area is in the northern hemisphere, extending from mid-latitude to the poles, and all of its launch sites are at relatively high northern latitudes, it takes considerably less energy to put a satellite into inclined orbit than into an equatorial one. A typical Molniya orbit may have an eccentricity of about 7, making an apogee of about 40,000 km and a perigee of about 500 km so that on alternate orbits, the apogee is above the former Soviet Union states. Because the height at apogee is close to a geosynchronous altitude, the satellite moves relatively slowly through the sky, spending a large fraction of its orbital period in this part of the sky. A typical Molniya satellite spends up to about 8 hours over the eastern part of the former Soviet Union republics.

The problem with the highly eccentric orbit is that the perigee is at a relatively low altitude. Consequently the drag due to the earth's atmosphere slows the satellite, thus reducing the height of the perigee. The satellite's lifetime is therefore reduced, and the spacecraft re-enters the atmosphere and burns up. As a result, over the years the Soviet Union has had to launch in excess of 100 Molniya satellites to maintain coverage. This type of orbit will be used in upcoming satellite systems such as Archimedes, a pan-European satellite broadcast system.

12.3 FREQUENCY

A communication satellite may be considered a distant repeater whose function is to collect the minute impinging electromagnetic field and retransmit the amplified frequency-converted carriers in the downlink towards the intended parts of the earth. With limited output power and a path loss on the order of 200 dB, the signal levels from the satellite are expected to be very weak at the ground receivers, hence any natural phenomena to facilitate the reception must be exploited.

Figure 12.4 shows that sky noise is a minimum in the 2–10 GHz range. Consequently this band is favoured for satellite communications. As demand for capacity increased, higher frequencies of 14 GHz and 11/12 GHz (Ku-band) have also been utilised. Table 12.1 summarises the frequency allocations for various civilian applications.

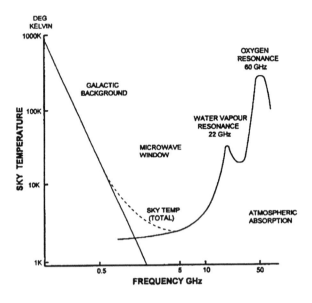

Figure 12.4 Sky-Noise and Frequency Bands

TABLE 12.1
Communications Satellite Frequency Allocations

Application	Downlink frequency (GHz)	Uplink Frequency (GHz)
Fixed Services		
C-band (Commercial)	3.7–4.2	5.925–6.425
X-band (Military)	7.25–7.75	7.9–8.4
K-band (Commercial) Domestic	11.7–12.2	14–14.5
International	10.95–11.2	27.5–31
Mobile Services		
Maritime	1.535–1.5425	1.635–1.644
Aeronautical	1.5435–1.58888	1.645–1.660
Broadcast Services		
	2.5–2.535	11.7–12.75
Telemetry, Tracking and Command		
	0.137–.138, .401–.402	1.525–1.54

12.4   MULTIPLE ACCESS

Multiple access is the method by which a number of ground stations may use a repeater (in this case a satellite) simultaneously. Frequency division multiple access (FDMA) and time division multiple access (TDMA) are widely used in commercial satellite applications. Code division multiple access (CDMA) has been used in military applications in the past because of its innate immunity to jamming, but, in recent years, it is being considered for both satellite and terrestrial mobile applications.

12.4.1   FREQUENCY DIVISION MULTIPLE ACCESS (FDMA)

FDM/FM/FDMA stands for frequency division multiplex (of base-band signals), frequency modulation (of carriers), and frequency division multiple access (in RF band). The transponder can be shared among several earth stations, which has been used in satellite telecommunication networks such as the INTELSAT series. Typically, the available bandwidth is in the order of 500 MHz shared between several transponders (nominally 12 × 36 MHz transponders). Each earth station is assigned a segment of this bandwidth, with sufficient guard band allocated between segments to ensure that one user will not interfere with an adjoining user. In C-band, the uplink band of frequencies transmitted to the satellite is 5925–6425 MHz; thus, a 6000 MHz carrier received by the satellite is retransmitted to the earth at 3775 MHz. The local oscillator (L.O.) frequency onboard the satellite in this case uses a frequency of 2225 MHz to down-convert the uplink frequency.

In such a network, although an earth station may transmit only one carrier to the satellite, it must be equipped to receive at least one carrier from each location with which it wishes to communicate. Some earth stations intercommunicate with dozens or more distant stations, so they require an equal number of receiver equipment. On the uplink, depending on the required capacity, they may transmit to all stations using only a single carrier. FDMA was widely used in the earlier INTELSAT network, and its main attractive feature is the simplicity; however, FDMA does not permit demand assignment, and it is not compatible with the widely used digital switching and multiplexing infrastructure.

12.4.1.1   SPADE

The SPADE (single-channel per carrier PCM multiple access demand assignment equipment) system was conceived to address the lack of flexibility in the FDMA networks. The SPADE system is capable of the following:

- providing efficient service to light traffic links
- handling overflow traffic from medium capacity pre-assigned links
- allowing the establishment of a communications link from any earth station to any other earth station within the same zone of command
- utilizing the satellite capacity efficiently by assigning circuits individually
- making optimum use of existing earth station equipment

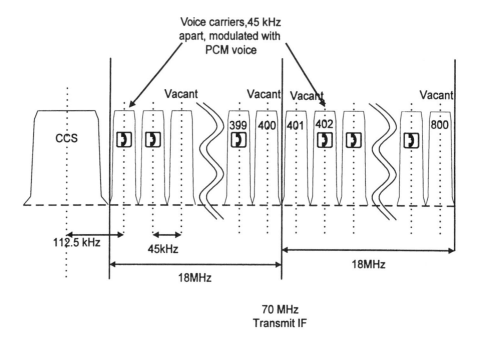

Figure 12.5 SPADE frequency plan

Figure 12.5 shows the frequency allocations of SPADE. At the center of the transponder bandwidth, a pilot frequency is used as a reference in receiving stations for automatic frequency control (AFC). On either side of the pilot, there are 400 carrier channels spaced 45 kHz apart, with a bandwidth of 45 kHz. At the extreme left a carrier, with a wider allocated bandwidth than the others, is used to carry the common signaling channel at 128,000 b/s. This channel is used to control the allocation of voice carriers to the earth stations.

The 36 MHz transponder frequency plan shown in Figure 12.5 is a pool of 397 usable two-way voice channels. A free channel can be taken by any earth station on demand. The 128 kb/s signaling channel is shared by all earth stations and uses TDMA. Each earth station is assigned a 1-ms slot every 50 ms, permitting it to transmit a burst of 128 bits.

In general, more information can be transmitted via a transponder if only one carrier is used. The more carriers to share a transponder, the lower the overall capacity.

As the number of carriers increases, more guard bands are needed. Furthermore, with more carriers, the level intermodulation noise is greater, because carriers tend to modulate one another caused by the non-linear characteristics of the travelling wave tube amplifiers (TWTAs), resulting in reduced output, as shown in Table 12.2.

Despite the outline presented in the above table, in certain applications, such as thin routes, or to accommodate traffic spillover, SPADE offers a higher efficiency than the fixed systems. This efficiency is because there is one voice

TABLE 12.2
FDMA System Throughput

No. of carrier / transponder	Carrier band MHz	No. of channels / Carrier	No. of channels / transponder
1	36	900	900
4	3 of 10 MHz	132	456
–	1 of 5 MHz	60	–
7	5	60	420
14	2.5	24	336

channel/carrier, and the carrier is switched off when no one is speaking (voice-activated channel). So, even when the channel is occupied, the carriers can be switched off half the time. In fixed systems, the carriers cannot be switched off when there is speech inactivity, because they are modulated by master groups or other blocks of speech channels.

12.4.2 TIME DIVISION MULTIPLE ACCESS (TDMA)

Time division multiple access (TDMA) circumvents many problems associated with the FDMA, and, because of its maturity and compatibility with digital switching, it is the most widely used access system in satellite and terrestrial mobile systems. Consequently, it is discussed in detail here.

In the TDMA scheme, each earth station is allowed to transmit a high-speed burst of bits for a brief period of time. The transmission time of the bursts are controlled so that no two bursts overlap. For the period of the burst, the full transponder bandwidth is available.

In its simplest form, each station is allocated in turn an equal length burst. To be efficient, however, stations must be able to vary their transmission rate, so either the bursts will be of variable length or the scheme must permit some stations to transmit more often than others. TDMA also avoids the intermodulation problem by using a single carrier, and it is highly efficient in using satellite power. The uplink power amplifiers can be operated at full power which in turn means a more efficient use of uplink resources.

The input to the transponder thus consists of a set of bursts originating from a number of earth stations. This set is called the TDMA frame. There are two reference bursts (RB1 and RB2), traffic bursts, and guard time between bursts. The TDMA frame is the period between two reference bursts. Figure 12.6 shows this arrangement.

Every earth station receives the entire bit stream and extracts the bits addressed to it, therefore the entire system needs to be synchronized. Maintaining synchronization is a complex issue because several factors contribute to the need for frequent resynchronization in a TDMA-based satellite communication network.

The propagation channels to different earth stations are of different path lengths, so there is a variation in propagation times. Moreover, in each path the delay is not

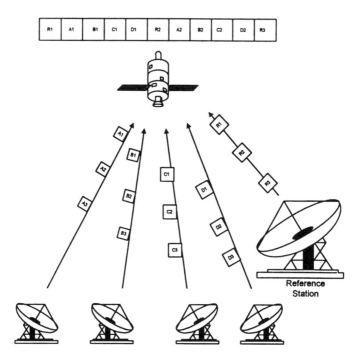

Figure 12.6 TDMA system using a reference station for burst synchronization

constant, as satellites tend to drift from their orbital station. In addition, the sun and moon create an oscillation in the satellite position (tidal oscillation) that is superimposed onto the long term drift. To circumvent these problems, each burst carries its own means of synchronization so that it can be transmitted and received in isolation. A burst starts with a synchronization pattern that permits the receiving modem to carry out the carrier and clock recovery.

12.4.2.1 Reference Burst

There are normally two reference bursts, RB1 (primary reference burst) and RB2 (secondary reference burst) which is added for reliability. The TDMA traffic stations take their timing reference from the primary reference burst.

12.4.2.2 Traffic Burst

Each station may transmit one or more traffic bursts / TDMA frame and may position them in the frame according to a burst time plan that coordinates traffic between stations. The length of the traffic bursts depends on the information it carries and can be changed if required. The location of the burst in the frame is referenced to the primary reference burst's time of occurrence. By detecting the primary reference burst, a traffic station can locate and extract the traffic burst intended for it. It can also derive the exact transmit timing of its own bursts.

12.4.2.3   Guard Time

A short guard time is used between bursts originating from several stations that access a common transponder to ensure that bursts never overlap. The guard time must be long enough to allow for transmission timing inaccuracies and the range variation of the satellite distance.

The TDMA frame length is normally selected to be in the 0.75–20 ms range, and it is usually a multiple of 0.125 ms, which is the sampling period of PCM (8 kHz). The frame length is chosen at the outset and remains constant for a TDMA system.

The structure of the traffic and reference bursts is shown in the Figure 12.7. The traffic burst contains the actual information bits and is preceded by the preamble. The reference burst contains only the preamble and no traffic data. The preamble normally consists of the carrier and clock recovery sequence (CCR), the unique word (UW), and the signaling channel.

12.4.2.4   Carrier and Clock Recovery Sequence

Each burst begins with a sequence of bits enabling the earth station demodulator to recover the carrier phase and regenerate the timing clock. The number of bits in the CCR depends on the carrier to noise ratio (C/N) and the acquisition range (carrier frequency uncertainty). Therefore, a high C/N and a small acquisition range require a short CCR and vice versa; typically, 300 to 400 bits for a120 Mbps TDMA system.

12.4.2.5   Unique Word

There are two unique words in TDMA bursts:

- the unique word that follows the CCR bits in the RB and is used to provide receive-frame timing that allows a station to locate the position of a TB in the frame.
- the unique word in the traffic burst that marks the time of occurrence of the traffic burst and provides receive-burst timing that allows the station to extract only the wanted sub-bursts within the traffic bursts.

The unique word is a sequence of ones and zeros selected to have good correlation properties to enhance its detection. At the demodulator, the unique word enters the unique word detector, which is a digital correlator (as discussed in Appendix A), where it is correlated with a strict pattern of itself. The maximum number of errors allowed in the unique word detection is called detection threshold. The unique word detection occurs at the instant of reception of the last bit of the UW. It marks receive-frame timing if the unique word belongs to the primary burst, or it marks receive-traffic burst timing if the unique word belongs to the traffic burst.

The position of every burst in the frame is defined with respect to receive-frame timing, and the position of every sub-burst in the traffic burst is defined with respect to receive-burst timing. Clearly, the accurate detection of the unique word is of utmost importance.

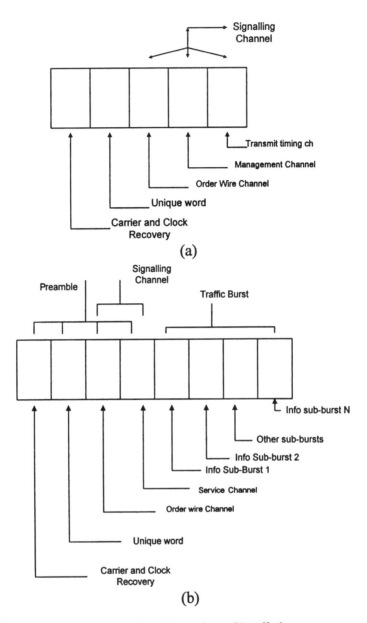

Figure 12.7 TDMA burst structure, (a) reference burst, (b) traffic burst

12.4.2.6   Traffic Data

The traffic data follows the preamble in a burst. The traffic data sub-burst is further subdivided into time slots addressed to the individual destination stations. The length of the traffic sub-burst depends on the type of services. The greater the fraction of frame time that is given over to traffic, the higher the efficiency.

12.4.2.7 Frame Efficiency

The TDMA efficiency η is defined as

$$\eta = 1 + \frac{T_x}{T_f}$$

where T_x is the overhead portion of the frame. For n bursts in the frame,

$$T_x = n \cdot T_g + \sum_{i=0}^{n} T_{p,i}$$

where T_g is the guard time and $T_{n,i} =$ preamble of burst i.

Therefore, η can be improved by increasing the frame length. However, this increase in turn increases the size of buffer memory required. Furthermore, the frame length must be kept small compared with the maximum roundtrip delay of 274 ms (maximum path distance at 5° antenna elevation angle) to avoid adding a significant delay to the transmission of voice traffic. The frame length for voice traffic is normally selected as less than 20 ms.

12.4.2.8 TDMA Super-frame Structure

The burst position control is carried out by the reference station using the transmit timing channel, while coordination of traffic is carried out by the reference station's burst using the management channel. This, in turn, means that to control n stations in the network, there will be n messages in the transmit timing channel and n messages in the management channel of the reference burst. Furthermore, to improve the reliability of this operation, these control messages are sent using the 8:1 redundancy-coding algorithm (repeated 8 times). This procedure further reduces the frame efficiency. The same applies to the service channel of the traffic bursts.

To reduce the length of the preamble of the reference and traffic bursts, the reference station can send one message to each station per frame instead of n messages to the n stations per frame. To address n stations, therefore, the process takes n frames. In this way, n frames can be put into one group, called a super-frame.

12.4.2.9 Advanced TDMA

In multiple beam satellites, the antenna beam interconnections are normally fixed in the basic TDMA systems discussed so far. However, it is possible to increase the system capacity if a number of spot beam antennas are employed to provide spatial division, hence reusing the same frequency band. Narrow beam antennas provide high gain and power saving in the uplink and downlink channels. Satellite-switched TDMA (SS-TDMA) will then be required for the interconnection of upbeams with the downbeams. This is accomplished by dynamic switching using a microwave switch matrix onboard the satellite. During a SS-TDMA frame, the satellite switch is controlled by a sequence of states of various duration. The duration of a given

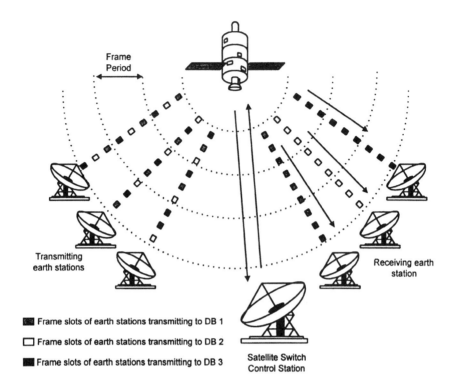

Figure 12.8 Satellite-switched TDMA

switch state is selected to accommodate a segment of the total traffic between the two earth stations. This arrangement is shown in Figure 12.8.

INTELSAT VI series of satellites have a number of transponders that are inter-connectable using either static switch matrices or a subsystem which provides SS-TDMA. Two SS-TDMA networks are in operation, one in the Atlantic Ocean Region (AOR) and one in the Indian Ocean Region (IOR). The C-band frequencies are reused six times through two hemispherical beams and four zone beams using dual circular polarization and spatial isolation. The Ku-band frequencies are reused twice by the two spatially isolated spot beams, using orthogonal linear polarization.

Another advanced feature of SS-TDMA system is beam hopping, which is capable of offering satellite services to sparsely populated areas. Beam hopping is implemented by using phase array antennas onboard the satellite. The antenna steers its beam towards a particular spot for a new TDMA burst and dwells for the duration of the burst and the guard band. The information in the burst is stored in the uplink memory, then the beam is steered in the direction of the second burst, and the second burst is stored, and so on. The stored uplink bursts are demodulated and reconfigured for downlink transmission. A combination of SS-TDMA and beam-hopping TDMA allows the flexibility and provision of service for low, medium, and high bit rate requirements applicable to different traffic requirements.

12.4.3   CDMA

Unlike the FDMA or TDMA systems that have been widely used in fixed satellite services so far, the interest in a commercial application of Code Division Multiple Access (CDMA) started with the mobile satellite systems. In FDMA and TDMA, the user terminal discrimination is achieved through frequency or time separation of the relevant channel, whereas in CDMA the system relies on the signature sequence assigned to the individual user (terminal) to ensure signal separatability. This is particularly suitable for mobile satellite services. The individual user signature is used to spread the spectrum of the modulated carrier in a direct-sequence spread-spectrum (DS-SS) arrangement. The user signals are modulated with approximately orthogonal signature sequences. The modulation scheme used in the DS-CDMA is normally binary phase-shift keying (BPSK) or quaternary phase-shift keying (QPSK). Since the signature signal usually is at a much higher rate than the transmitted data, the transmission bandwidth is much higher. The spread-spectrum user signals are simultaneously transmitted using the same frequency band. In the receiving end, the composite incoming signal is correlated with the exact signature sequences of each user, thus reconstructing the information signals of the individual user. However, the residual correlation of the signatures produces interference.

The noise power in a correlation receiver can be calculated from Lutz[6]:

$$N_{tot} = N_0 + \alpha(1 + k)\frac{(N_u - 1)E_b}{G}$$

where N_u is the number of active users in the satellite beam, G is the chip rate of the spread sequence. Further assumptions are Gaussian channels, no shadowing, asynchronous user signals, and discontinuous transmission (i.e., during speech pauses the signal transmission is interrupted). N_0 represents thermal noise and the second term describes the interference caused by other users, the signal of which is received with E_b. The interference is inversely proportional to G, the processing gain with typical values of 100 to 1000. α is speech activity of users and is normally taken as 0.5. The factor $(1 + k)$ accounts for the additional interference produced by users in the neighbouring cells, with k defined as the other cells' interference power divided by the user's cell interference power. In terrestrial CDMA, k is taken as 0.44. However, in satellite mobile systems, k may vary between 0.75 to 1.25 based on the type of orbit and the antenna beam contours. The CDMA links are normally interference-limited, i.e., the term N_0 can be ignored and the number of user per beam can be approximated to

$$N_u \approx \frac{G}{\alpha(1 + k)} \cdot \frac{1}{\left(\dfrac{E_b}{N}\right)_{req}}$$

The term $(E_b/N)_{req}$ designates the required signal-to-noise for the link. Assuming $(E_b/N)_{req}$ = 5 dB and a Gaussian channel, for BER of 10^{-6}, α = 0.5 and k = 1.2, the number of user is given as

$$N_u \approx 0.29G = 0.29\frac{B_s}{Rb}$$

A similar treatment of a TDMA system with a bandwidth of B_s and a user bit rate of R_b using M-level modulation results as in Lutz[6]:

$$N_u = \frac{1}{C} \cdot \frac{B_s}{R_b} \cdot \frac{IdM}{1 + \frac{H + G}{I + P}IdM}$$

where H is the burst header, I and P are the user information and parity, and G is the guard time. Assuming a four-cell cluster, $P + I$ = 1000 bits, H = 100 symbols, G = 30 symbols, and M = 2, the number of user within a cell becomes

$$N_u = 0.22\frac{B_s}{R_b}$$

A comparison of the above two systems suggests that CDMA may be more efficient than TDMA. However, the CDMA efficiency may not be fully achieved because of imperfect power control.

12.4.4 SATELLITE PACKET COMMUNICATIONS

The access systems introduced so far are designed primarily for voice and data traffic with fixed or demand assignment. These systems use multiple access protocols such as FDMA, TDMA, or CDMA, and they are based on circuit-switched networks, which are efficient for voice/data transmission when messages are long in comparison to the time required to establish the link. Data traffic is somewhat different from voice traffic as its message length ranges from a few characters to hundreds of megabytes. This characteristic is referred to as bursty and results in a large peak-to-average ratio. Therefore, if fixed capacity allocation is used, each user must be assigned enough capacity to meet the peak transmission rate. Consequently, the resulting channel utilization will be low. To circumvent this problem, the data is formatted into fixed length packets, which are routed through shared communication resources by a sequence of node switches. Packet switching does not store packets for a prolonged period of time, and packets are discarded if difficulty arises in delivery, in which case they have to be retransmitted. As an implementation example, a shared satellite global beam offers full connectivity between users, eliminating the routing and switching. Each user can listen to its own message and receive automatic acknowledgement. This allows the implementation of a special multiple access protocol for the dynamic assignment of the satellite capacity.

12.4.4.1   Statistical Channels

Sharing of a common communication medium requires protocols governing the behavior of the group of users. Protocols enable the users to gain access to the central resource (transponder capacity). An example is the Aloha protocol, a random access scheme pioneered at the University of Hawaii for interconnection of terminals and computers via radio and satellite. Aloha has several refinements:

- Pure Aloha *(P-Aloha)*
- Slotted Aloha *(S-Aloha)*
- Aloha with Capture Effect *(C-Aloha)*
- Aloha with Capacity Reservation *(R-Aloha)*

P-Aloha — The P-Aloha scheme is simple and needs no sophisticated hardware. In essence, the users address the message packets in a TDMA frame in a purely random fashion. As a result of this unrestricted access freedom, message packets do collide and need to be retransmitted, reducing the throughput and transmission efficiency. Figure 12.9 shows the arrangement for the P-Aloha multiple access.

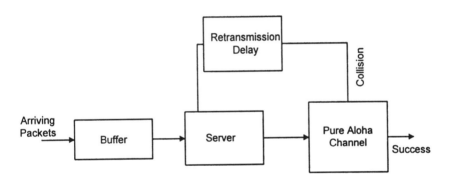

Figure 12.9 The Aloha multiple access protocol

S-Aloha — The S-Aloha technique decreases the probability of collision between packets by requiring that users transmit only at the beginning of discrete time intervals (slots). The S-Aloha channel has two disadvantages: the potential complexity of the synchronization and the limited packet length. Also, for users with small traffic load, the time between the end of a user transmission and the beginning of the next slot is wasted. The maximum throughput of S-Aloha is twice that of the P-Aloha.

C-Aloha — Improvement in capacity can be achieved if each user transmits at slightly different levels. If two packets with different signal levels collide, the stronger of the two is likely to capture the receiver and be transmitted without an error. The C-Aloha can yield a three-fold increase in capacity over P-Aloha. The performance of such a technique in satellite transponders may suffer from transponder nonlinearities.

R-Aloha — In a system where there are a few large and frequent users, it is possible to dedicate a portion of the channel on a fixed-assignment basis, leaving the remaining portion of the channel open to contention among many sporadic users. For example, if half of the channel is dedicated to a fixed-assignment basis of 80% utilization and the remaining half at 36.8 efficiency on S-Aloha mode, the overall channel efficiency becomes 58.4%. There are a variety of reservation protocols in use.

12.4.4.2   Very Small Aperture Terminals (VSAT)

The VSAT is the physical layer realization of an access system described in the preceding section. The system enables a large number of remote microterminals to have communication access to a central HUB, share a computing system, have access to a common database, and exchange e-mail, voice, and fax services. The network comprising a central HUB and remote terminals is characterized by its star topology. The network architecture is shown in Figure 12.10. The satellite links can be implemented in C-band, but they are mostly implemented in Ku band because of the wide bandwidth availability.

At the central HUB site, the inbound satellite links are typically at 56 Kb/s and the outbound link is typically at 256 Kb/s. The outbound carrier is different from the inbound link in two fundamental ways: it is at higher rate, allowing for multiple VSATs to receive a common outbound channel, and the outbound carrier uses continuous modulation enabling the VSAT terminals to use a low cost demodulator.

Direct VSAT-to-VSAT communication is not usually supported; however, VSAT-to-VSAT communications can be supported using a two–hop technique (VSAT-to-HUB-to-VSAT).

The VSAT terminals send data in packet form through a random access/time division multiple access (RA/TDMA) satellite channel with transmission delay of τ seconds. After processing, the host acknowledges the successful reception of the packet data via a broadcast TDM channel. A failure due to packet error or packet collision requires retransmission. This extra delay due to the retransmission loop introduces significant complexity in system design. The TDM frame from the HUB-station can be a combination of variable length data messages multiplexed at the HUB-site and broadcast to all remote stations in the network. A synchronization pattern is sent every frame to synchronize remote stations in the network. The synchronization pattern provides the start of the TDMA frames to all terminals.

The TDM frame structure is shown in Figure 12.11. Each message contains an address field that identifies the microstation for which the message is intended. All microstations receive the TDM stream and filter out the messages not intended for their own use. By using an appropriate addressing scheme, it is possible to broadcast a single message to all stations or to a specific station.

For inbound transmission from the microstations to the HUB, the TDMA carriers can be a S-Aloha shared by many microstations. The timing on the TDMA channel is divided into a series of contiguous frames and slots. Each frame is

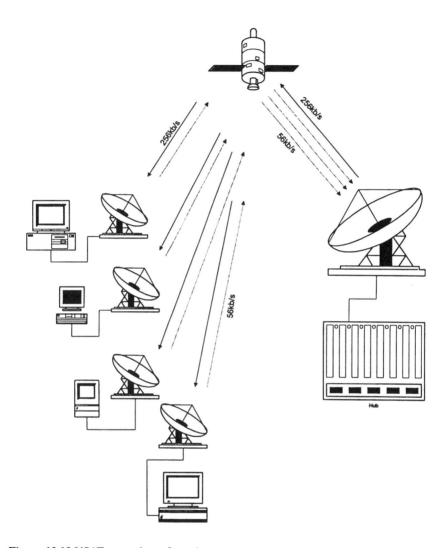

Figure 12.10 VSAT network configuration

comprised of N slots as shown in Figure 12.12. Remote terminals may transmit packets only within the slots, and a packet can never cross the slot boundaries. The size of a packet may vary, but the maximum size of a packet can never be greater than the size of slot. Packets are transmitted as bursts. The size of a slot and the number of slots in the frame will depend on the type of application. The slot period is software selectable.

The RA/TDMA channel is a contention channel and needs to be synchronized to the start of each frame (SOF) and the start of each slot instant. Each slot on the TDMA channel may be either a random access TDMA, or a demand assigned (DA/TDMA) slot. A RA/TDMA slot is a contention slot and is available to all

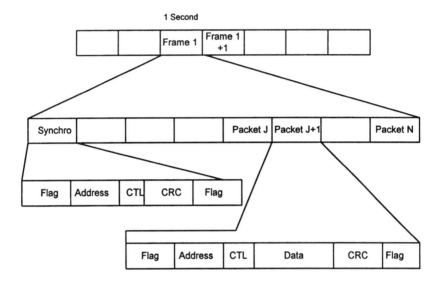

Figure 12.11 TDM frame structure

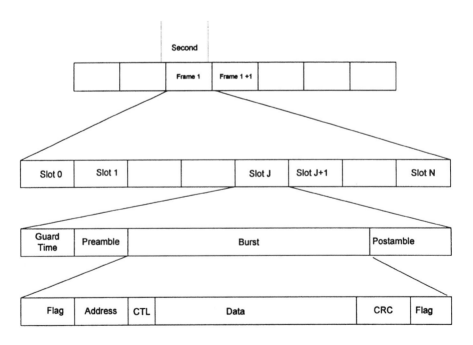

Figure 12.12 TDMA frame description

microstations for transmission of their packet. AD/TDMA slot is a slot dedicated to a single remote station. Normally there is no contention on DA/TDMA slots. The HUB broadcasts a slot map that defines the RA/TDMA and the DA/TDMA slots in each TDMA frame.

All microstations must receive this map and may transmit in the permitted slots. The VSAT terminal is characterized by its low cost and small sized antenna, normally 0.6 to 1.8 meters in diameter for Ku band systems. Other characteristics include ease of installation and maintenance.

12.5 MOBILE SATELLITE COMMUNICATIONS

Fixed-satellite services (FSS) have been the primary application of communication satellites since commercial operation was started by INTELSAT. More recently, FSS has been used for the implementation of large numbers of VSAT networks throughout the globe. However, since 1980, the interest in the use of satellite communications for mobile applications has grown dramatically. The growing interest in mobile satellite systems (MSS) has been driven by the same factors that led to changes in the entire telecommunications industry: availability of the technology and deregulation. The main features of MSS include the following:

- large service area and limited capacity
- global coverage
- appropriate for rural service provision
- appropriate for layered architectures
- complementary to terrestrial networks

MSS has proved to be the largest new growth area in satellite communications. Table 12.3 shows some representative MSS that are in operation or will be in operation over the next 5 years.

TABLE 12.3
The Satellite-Mobile Systems

Name	Organisation	Features	Start of Service
ICO	Inmarsat	MEO	–
Iridium	Motorola	66 LEO Sat	1998
GlobalStar	Loral/Qualcomm	48 LEO Sat	1999
Teledesic	Motorola	288 LEO Sat	2003

The current generation of MSS have geosynchronous satellites with ground mobile terminals varying in size from portable briefcase-sized transceivers to a standard 'A' INMARSAT station with dish antennas of one meter in diameter. Future MSS will use handheld telephone or laptop ground terminals and the satellite networks will be made up of a fleet of LEO (low-earth circular orbiters, 400–1000 miles in altitude) or MEO (medium-earth orbits, 5000–7000 miles in altitude) with either polar or inclined orbits. In addition to handheld terminals, these MSS will also serve mobile ground terminals such as aircraft, shipboard, and land vehicles.

The next few years will see increasing levels of integration between satellite and terrestrial cellular communication networks. The primary purpose of MSS is not to substitute but to complement the terrestrial mobile communication networks; indeed, most ground terminals have the capability of functioning with both networks. Iridium[7] is currently operational, and Globalstar [8] and ICO[9] are at an advanced stage of development and due to start commercial mobile communication services in 1999 and 2000, respectively. With these systems fully deployed, business travellers can circumvent the incompatibility between terrestrial mobile networks in operation around the globe. The incompatibility problem appears to endure even in the third generation of mobile networks that are currently being developed, hence the necessity for the MSS complementary role.

Figure 12.13 shows the basic system architecture. Each satellite covers a circular area on the earth's surface that increases with increasing orbit height and a decreasing minimum elevation angle. The choice of orbital planes and satellite phasing within the orbits must guarantee continuous coverage of the service area. The number of required satellites is determined by the orbit height and minimum satellite elevation angle.[10]

Direct communication via satellite, using a handheld terminal with low transmit power and omni-directional antenna, requires a high antenna gain or spot-beam antenna onboard the satellite. In each constellation, the coverage area of each satellite is composed of a large number of beams, allowing frequency reuse within the coverage area and, therefore, increasing the bandwidth efficiency of the system. The gateway to the fixed terrestrial network is comprised of fixed earth stations and mobile switching centres (MSC). Call routing and mobility management is achieved by use of databases, such as the home location register (HLR) and the visitor location register (VLR). The network control centre (NCC) allocates, among other tasks, spot-beam frequencies and devises routing tables to the satellites in the network.

12.5.1 FREQUENCY

The spectrum allocated by WARC-92 and 95 for the MSS falls mainly in the L-band and divided into the following categories:

- aeronautical mobile satellite services (AMSS)
- maritime mobile satellite services (MMSS)
- land mobile satellite services (LMSS)

Frequencies around 140 and 400 MHz can be used for MSS data systems. After the year 2000, additional frequency bands at 1.980–2.025 GHz (uplink) and

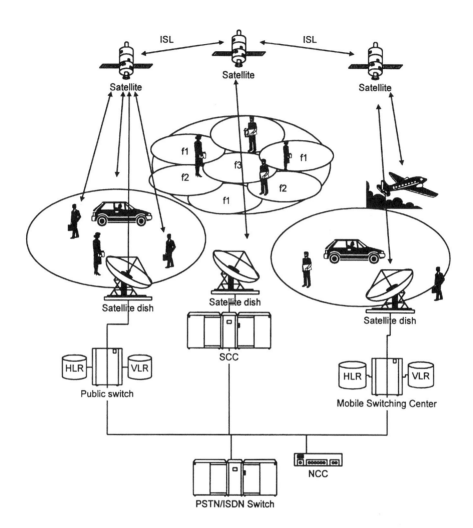

Figure 12.13 Satellite-mobile system architecture

2.160–2.200 GHz (downlink) may be used as intended by the ICO[9] system. Higher frequencies at 15 GHz and 20/30 GHz are foreseen for the feeder links. Intersatellite links may operate at 23 GHz, 60 GHz, or at optical bands. Table 12.4 shows the allocated bands.

12.5.2 Low Earth Orbit (LEO)

System Concept: Low earth orbits at 700–1500 km avoid the large propagation loss and delay associated with geostationary orbits. However, many LEO satellites are needed for continuous coverage of the earth's surface.

TABLE 12.4
The Allocated Frequency Band for Mobile Satellite
Communications

1.6/2.4 GHz and 1.6/2.4 GHz bands approved by WARC 92		
Uplink	Bandwidth	Downlink
1610 MHz	16.5 MHz	2483.5 MHz
1626.5	34 MHz	1500
		1525
1660.5 MHz		1559 MHz

12.5.2.1 Globalstar (Commercial Service Planned for September 1999)

The Globalstar network is based on a joint venture between Loral and QUALCOMM. The access system is CDMA, and the system uses 48 satellites in 8 inclined orbits at 1414 km. Figure 12.14 shows the system configuration.

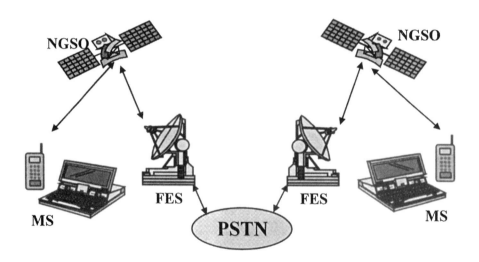

Figure 12.14 GlobalStar network configuration

The service is meant to be totally transparent to the user, whether provided by the Globalstar or the terrestrial network and is fully integrated into the terrestrial network. A Tri-mode handset, which will automatically switch between the terrestrial analogue system (AMPS), the digital cellular system (DAMPS), or the Global satellite network is being developed for the North American market. The system specifications are given and summarised in Table 12.5.

TABLE 12.5
GlobalStar System Configuration

Orbital characteristics		Frequency	
Orbit	Inclined circular	Mobile uplink	1610–1625 MHz
Altitude	1414 km	Mobile downlink	2483–2500 MHz
Orbital period	≅ 2 hours	Feeder uplink	6484–6541.5 MHz
Inclination	52°	Feeder downlink	5158.5–5216.0 MHz
No. of orbital planes	8		
No. satellites per plane	6		

12.5.2.2 Iridium (Operational)

The Iridium system is based on 66 satellites that are grouped in 6 orbital planes, each containing 11 active satellites and one spare satellite. The orbits are circular at the height of 783 km at an inclination angle of 86°. The separation between satellites in each orbit is 32.7°. The satellites in adjacent planes travel out of phase and are located at half-satellite spacing. Collision avoidance is built into the orbital planing, and the closest approach between the satellites is 223 km.

Satellites in planes 1, 3, and 5 cross the equator in synchronisation, while satellites in planes 2, 4, and 6 also cross in synchronisation, but out of phase with those in planes 1, 3, and 5. The first and last planes are counter-rotating. The separation between the corotating planes is 31.6°, which allows 22° separation between the first and last planes. This closer separation is needed because the earth's coverage under the counter-rotating planes is not as efficient as corotating planes.

Two-way communication links exist between each satellite and its nearest neighbour ahead and behind and to the nearest satellites in the adjacent planes.

The satellite antenna beam footprints are similar to the cells encountered in terrestrial cellular mobile systems with one major difference: unlike the fixed cellular network, the beams (cells) move relative to the mobile subscriber.

The orbital period for the Iridium satellites is approximately 100 minutes. Taking the average earth radius of 6371 km, the surface speed is

$$\frac{2 \times 6371 \times \pi}{100} \cong 400km/\min \ or \ over \ 15000 \ miles/h$$

Therefore, satellites are travelling at a much greater velocity than that normally encountered by the terrestrial mobiles (cars included) which may be considered as stationary relative to the satellite.

A 48-beam antenna pattern is used from each satellite, with each beam under a separate control. This arrangement is shown in Figure 12.15.

At the equator, for example, overlap of patterns will be minimal and all beams may be on, while at high latitudes considerable overlap occurs and certain beams can be switched off. It is also possible to switch off the beams where their operation

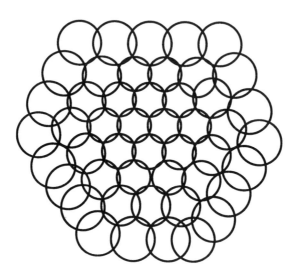

Figure 12.15 Iridium antenna foot-print

is prohibited by the telecommunication administrations. The beam switching is referred to as cell management. The access system for Iridium is TDMA, and dual mode handsets compatible with terrestrial DAMPS are commercially available.

As a result of the small antenna foot print of the LEO satellite, a large number of gateway stations are required in systems without intersatellite links (ISL). Globalstar will use approximately 150 gateways all over the world. If ISLs are used, the number of gateways can be reduced, and their position can be chosen freely. The Iridium system, which uses ISLs, has plans for 11 gateway stations. Moreover, ISLs allow routing long distance calls within the satellite network, saving cost for terrestrial lines. The use of ISLs necessitates more signal processing and switching onboard the satellites to route the calls within the ISL network.

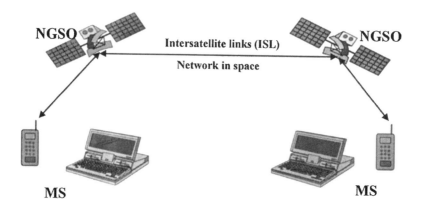

Figure 12.16 Iridium system architecture

12.5.2.3 Teledesic (Service Targeted for 2003)*

The Teledesic system* is proposed by a consortium formed by the major players in the satellite communications industry** and is building a global, broadband network using a constellation of LEO satellites. The commercial service is targeted for 2003. This system will create a network capable of worldwide access to telecommunications services, such as broadband Internet access, videoconferencing, high-quality voice, and other types of digital data.

The Teledesic Network will consist of 288 operational satellites, divided into 12 planes, each with 24 satellites. To make efficient use of the radio spectrum, frequencies are allocated dynamically and reused many times within each satellite footprint. Within any 100-km radius area, the Teledesic Network can support over 500 Mbps of data to and from the user terminals. The Network supports bandwidth-on-demand, allowing a user to request and release capacity as needed. The frequency of operation is at the high end of Ka-band (28.6–29.1 GHz uplink and 18.8–19.3 GHz downlink). The use of a high frequency band and a large number of LEO satellites (resulting high elevation look angles) enables the use of low-power terminals and small antennas.

The Teledesic Network is designed to support a very large number of simultaneous users. Most users will have two-way connections that provide up to 64 Mbps on the downlink and up to 2 Mbps on the uplink. Broadband terminals will offer 64 Mbps of two-way capacity. With such a bandwidth it would be possible to transmit a set of X-rays over the Teledesic Network in a few seconds.

Satellites in adjacent planes travel in the same direction except at the constellation seams, where ascending and descending portions of the orbits overlap.

12.5.3 MEDIUM EARTH ORBIT (MEO) SYSTEM CONCEPT

Systems with satellites in medium earth orbits (10,000 km) provide links that avoid the large signal attenuation and delay associated with the geostationary orbits and still allow global coverage with a few (10–15) satellites. Since the required state-of-the-art satellite antennas are now available, and no ISLs are necessary, the technical risks of MEO systems are acceptable. An example of a MEO system is the ICO (Intermediate Circular Orbits)[9] system based on 10 satellites at an orbital height of 10,354 km.

REFERENCES

1. Clark A C, "Extra-terrestrial Relays," *Wireless World*, Vol. 51, No. 10, Oct. 1945, 305–308.
2. Gagliardi R M, *Satellite Communications*, 2nd Ed., Pub.Van Nostrand Reinhold, 1991.
3. Calcutt D, Tetley L, *Satellite Communications Principle and Applications*, Edward Arnold, 1994.
4. Gilhousen K S, et al., "Increased Capacity Using CDMA for Mobile Satellite Communication," *IEEE J. on Selected Areas in Comm.*, Vol.8, No. 4, May 1990, 503–514.

---

* www.teledesic.com
** Motorola, Boeing, and Matra Marconi Space form an international team who will develop the technology.

5. De-Gaudenzi R, et al., "Advances in Satellite CDMA Transmission for Mobile and personal Communications," *Proc. IEEE* Vol. 84, No. 1, Jan. 1996, 18–37.
6. Lutz E, "Issues in Satellite Personal Communication Systems," *Wireless Networks*, No. 4, 1988, 109–124.
7. Hartleid J E, "The Iridium System Personal Communications Anytime, Anyplace," *Proc. 3rd Int. Mob. Sat. Conf., IMSC'93*, 1993, 285–290.
8. Wiedeman R A, Viterbi, A J, "The Global Mobile Satellite System for Personal Communications," *Proc. 3rd Int. Mobile Satellite Conf., IMSC'93*, 1993, 291–296.
9. Poskett P, "Satellite System Architecture," *Proc. 2nd European Workshop on Mobile/Personal Satcoms, EMPS'96*, 1996, 485–500.
10. Werner M, et al., "Analysis of System Parameters for LEO/ICO-Satellite Communications Networks," *IEEE J. on Selected Area in Comm.*, Feb 1995, 371–381.

APPENDIX

The unique word is a sequence of ones and zeros selected to have good correlation properties to enhance its detection. At the demodulator the unique word enters the unique word detector, which is a digital correlator as shown in the Figure 12.17, where it is correlated with a strict pattern of itself. The correlator consists of two *n*-stage shift registers where *n* is the length of the unique word. There are also *n* modulo-2 adders, a summer, and a threshold detector. Each stage in the shift register is applied to a modulo-2 adder whose output is a logical zero when the data matches the unique word in the same position. All the modulo-2 adder outputs are summed, and the sum is compared to a preset threshold in the detector. The maximum number of errors allowed in the unique word detection is called *detection threshold*. The unique word detection occurs at the instant of reception of the last bit of the UW. It marks the receive frame timing if the unique word belongs to the primary burst, or it marks the receive traffic burst timing if the unique word belongs to traffic burst.

Unique Word Detection (False or Miss!)

The accurate detection of the unique word is of utmost importance in a TDMA system, and the entire traffic burst is lost if the UW of a traffic burst is missed. It is also possible to have false detection which generates the wrong receive frame timing, consequently incorrect transmit frame timing, resulting in overlap with other bursts at the satellite.

Miss-detection probability — If ε represents the maximum number of errors allowed in a UW of length *n* bits and *i* is the actual number of errors in the unique word as received, then when $i < \varepsilon$, the received sequence is declared to be the unique word. If $i > \varepsilon$, the detected sequence N is declared not to be the unique word.

Let *p* represent the average probability of errors in the link (BER), the probability of a miss-detection is when $i > \varepsilon$ and is given as:

$$P_{miss} = \sum_{i=\varepsilon+1}^{N} \frac{N!}{i!(N-1)!} p^i (1-p)^{N-i}$$

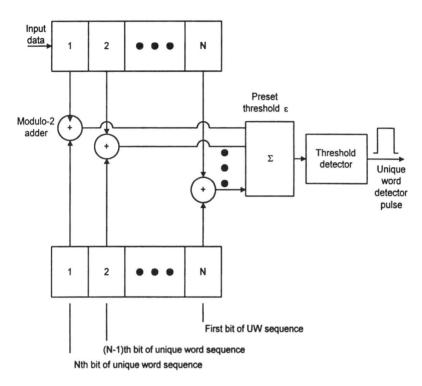

Figure 12.17 Unique word detector

This gives the average probability and it is not necessary to know the specific value of i.

UW false-detection probability — The false detection probability P_F is given by the probability that random data (bits 1 and 0 are assumed to be generated with equal probability) may accidentally match the stored UW pattern to the extent that the number of bits in disagreement does not exceed ε. For a UW of length n, there are 2^n combinations in which random data can occur, hence the probability of occurrence of one unique combination that corresponds to the stored UW is $1/2^n$, which is the situation of false detection probability when $\varepsilon = 0$. For a given value of ε, the total number of possible combinations in which ε or fewer errors can occur is

$$\sum_{i}^{\varepsilon} {}_N C_i$$

Thus, the probability that n random data bits will be detected as UW or false detection probability P_F is

$$P_F = \frac{1}{2^N} \sum_{i=0}^{\varepsilon} {}_N C_i$$

This is independent of link error probability.

In a typical link, the false detection probability is much higher than the probability of a miss. In practice, once frame synchronization has been achieved, a time window can be formed around the expected time of arrival for the UW, such that the correlation detector is in operation only for the window period, thus reducing the probability of miss-detection.

13 Internet/Mbone Audio

V. Hardman and O. Hodson

CONTENTS

0-8493-9594-1/00/$0.00+$.50
© 2000 by CRC Press LLC

Interest in Internet audio has grown with the popularity of the Internet. The technology presented in this chapter covers a range of Internet audio systems: Mbone audio tools, Internet telephony, and audio streaming applications. Mbone audio represents the super-set of the functionality of Internet audio applications and includes many aspects of the other two. This chapter therefore focuses on Mbone audio tools, and in particular discusses RAT (Robust Audio Tool). RAT is a second generation tool, incorporates an extremely rich set of functionality, and represents our own experience in Internet audio.

This chapter begins with an overview of technology that underpins Internet audio systems, audio compression algorithms and the Internet. It considers both speech and music algorithms, and the Internet, particularly with regard to the transport and application layer protocols. An overview of the components of an Mbone audio tool leads to an analysis of the current problems: gaps in the output audio and the effect on participants of an unnatural acoustic environment.

13.1 INTRODUCTION

Internet audio is a relatively new commercial application area that has grown with the popularity of the Internet and the availability of low cost audio hardware for multimedia PCs. For the purposes of this chapter, we consider Internet audio as covering three distinct areas: multimedia conferencing, broadcast applications such as music-on-demand, and Internet telephony. In comparison to existing technologies, such as the Public Switched Telecommunication Network (PSTN), Internet audio has the advantage that it can be integrated with other media, it can be mixed to provide multiway conferencing, and audio compression can be performed in software (and easily updated as algorithms improve).

The first packet audio system transmitted speech packets between a few sites in America over the ARPANET network. A follow-up project used a satellite connection to interconnect two sites outside the US (one of which was UCL, London) with the ARPANET network in the late seventies [Gold 73]. These early projects led to much subsequent research on both sides of the Atlantic (including the Universe, Admiral, and Unison projects [Clarke 90] in the UK). Research during the eighties developed individual media systems — an audio system and a video system — which used separate hosts and were integrated together using a Local Area Network (LAN). The recent proliferation of multimedia PCs, capable of compressing audio and video in real time, and the emergence of the global Internet have led to a rapid increase in the use of desktop collaborative multimedia systems running on a single host.

Despite many standardization activities in compression algorithms for the PSTN, most speech is still transmitted uncompressed. This situation is because of the need for efficient inter-working with other telecommunications operators and countries, and because telephone networks grow to cater for demand. In the

Internet, the situation is reversed, and there is often insufficient capacity to match demand. Internet audio competes with other traffic for a share of bandwidth, and compression is necessary to avoid congestion. Compression is also used to enable machines to be connected to the Internet through low capacity modem links, and to enable the heterogeneous Internet to carry all types of audio, such as toll quality speech and high quality music. Congestion in the Internet results in packet loss, because congested routers discard excess packets, and many schemes have been developed to overcome the effects of packet loss on audio. In the future, these schemes could be used to reduce costs, as services are likely to be charged for on a bandwidth-usage basis.

To transmit audio over the Internet, all a user needs is a multimedia PC, a headset, and network access. Internet audio tools are written as separate programs that a user starts from the UNIX command line or a windows interface. An audio tool smoothes the flow of arriving packets and converts them to a continuous stream of digital audio samples for playback to the audio device. The large processing capabilities of multimedia PCs mean that many techniques can be employed to improve the system at different stages in the pipeline: software compression algorithms can be swapped on an as-desired or as-needed basis to compress music as well as speech; error protection strategies can be matched to the quality requirements of the application and the current characteristics of the Internet; and digital audio techniques can be used to enhance the acoustic *feel* of the system.

Because of the global deployment of the Internet and the availability of Internet audio systems on low-cost general purpose PCs, the impact of Internet audio will be vast. An interactive aspect can easily be added to many existing applications, such as web-based distance learning, computer games, and mailing lists. Desktop multimedia conferencing applications which use multiway audio, video, and shared text will also evolve into more demanding applications, such as distance learning, video-on-demand, and telepresence. Current research in Internet audio is considering ways of improving the received quality of audio transmitted over a datagram network. It is also promoting congestion avoidance using adaptive applications that back-off transmission rate in the event of loss.

This chapter begins with a survey of the technology of Internet audio tools, looking first at compression algorithms and Internet protocols. Combining these two aspects into an audio tool is examined in terms of the structure of a typical Internet audio tool, with problem solutions for packet loss, and considering in detail options for enhancing the acoustic feel of the system. The chapter then discusses the impact Internet audio will have, and it concludes with an assessment of the current focus of the research community.

13.2 CHARACTERISTICS OF THE TECHNOLOGY

Networked-audio application development relies upon knowledge of both audio and networks. When the network is a packet network, and especially a best-effort one such as the Internet, this dual understanding is especially important because there is an interaction between network effects and the operation of the audio compression algorithm.

13.2.1 CHARACTERISTICS OF AUDIO

At the heart of any networked audio system is a compression algorithm which is used to reduce the bandwidth consumed. There are two distinct classes of audio compression schemes, voice and music, which address sound compression in complementary ways. Speech compression algorithms exploit the fact that the speech production process depends on anatomical movement (therefore changes relatively slowly) and has a limited frequency range. Music compression schemes are often based on a model of the human auditory system, which responds to a wider range of frequencies than encountered in speech but has decreasing sensitivity outside this part of the spectrum.

13.2.1.1 Characteristics of Speech and Human Conversation

Speech is limited in frequency range to less than 10 kHz, with most of the signal energy residing between 300 and 3400 Hz [Rabiner 78]. Higher frequencies ease listener identification but contribute little to comprehension. Speech is composed of a limited set of individual sounds, known as phonemes, which can be broadly split into two main groups: voiced and unvoiced. Voiced sounds are generated by air from the lungs being forced over the vocal cords, which vibrate in a quasiperiodic manner and produce a series of pulses of air. These air pulses pass along the vocal tract that allows certain frequencies to resonate, while attenuating others. For unvoiced sounds, the vocal tract produces a constriction, and air is forced through it, producing a turbulent flow. The two types of sounds have different types of frequency spectra. Voiced sounds, such as vowels, show a characteristic high average energy level, with distinct formant (resonant) frequencies. Unvoiced sounds show much smaller average energy, with a noise-like appearance in the higher parts of the frequency domain. Both groups of sounds are shaped by the dynamics of the vocal tract and interactions of airflow with the lips, teeth, tongue, and nasal cavity. Compression advantage in most modern speech compression algorithms is achieved by modeling the speech production process as the combination of a linear digital filter fed by an excitation signal. The digital filter models the vocal tract, and its parameters change quite slowly. The excitation signal is either a stream of regularly spaced pulses to represent voiced sounds, or a series of random pulses (a noise-like sequence) to imitate unvoiced sounds. For a greater understanding of the acoustic analysis of speech, see Kent and Read [Kent 92].

   Conversation between two or more people usually consists of one person talking with the others listening. Changes of speaker occur when the speaker stops speaking (mutual silence) or the original speaker is interrupted (double-talk) [Brady 69]. Double-talk happens infrequently and lasts only for short periods of time. Periods of mutual silence typically occupy up to 50% of conversation time, and it is reasonable to assume that a two party conversation generates solitary talk-spurts for the other 50% of the conversation time. Because of these gaps in conversation, speech systems can save bandwidth by transmitting only when speakers are active. This technique is known as silence suppression and is widely deployed in telephony networks and Internet audio tools.

Interactive conversation also needs low end-to-end delay over the channel, if conversation patterns are not to break down. In the local analog PSTN, the delay has to be much smaller than this because of the generation of echoes, perception of which increases with increasing delay. With an end-to-end digital network, these echoes do not arise, and the delay value can be set at the maximum for interactive conversation (less than 250 ms end-to-end). [Montgomery 83]

13.2.1.2 Speech Compression Algorithms and Standards

Speech compression algorithms were originally standardized for use over the PSTN. The frequency response of the analog transmission parts of the telephone network has a frequency range which covers the greater part of the spectrum of human speech (300Hz to 3.4 kHz). Speech coding algorithms covering this range are known as telephone or toll quality.

The representation of discretely sampled audio as binary codes is pulse code modulation (PCM). The simplest scheme is to linearly quantize samples over the range of amplitudes. For speech, it has been found that 11 bits of precision are necessary for linear quantization not to be noticeable [Rabiner 78], but most computer systems round this up to 16 bits/sample. Linear quantization results in a varying signal-to-noise ratio with the amplitude of the signal. A widely used alternative is logarithmic companding which scales the signal logarithmically. The International Telephony Union (ITU) standard G.711 defines two companding schemes (A-law and μ-law), which both represent samples by 8 bits. Both schemes have near constant signal-to-noise ratios across the signal amplitude range and are near transparent to the human listener.

The mechanics of speech production are such that there is a high degree of correlation between adjacent samples. Further reductions in the number of bits per sample can be achieved by using a predictor in conjunction with an adaptive quantizer. This approach is described as adaptive differential pulse-coded modulation (ADPCM). ITU standard G.726 represents speech samples with 2, 3, 4, and 5 bits per sample using such a scheme. Another standard, G.722, covers speech in the range 0-7kHz (wideband speech) using this technique on sub-bands within the frequency range.

An alternative way of representing speech signals is linear predictive coding (LPC). LPC coders model the speech production process as the linear sum of earlier samples (digital synthesis filter), which is fed by an excitation (residual) signal (see Figure 13.1). The encoder algorithm estimates the coefficients that will be used in the synthesis filter (usually 8-10 filter coefficients) and tries to identify the pitch period, which will be used to generate the pulses in the excitation or residual signal. Since the method of pitch prediction is not very accurate,* modern techniques are hybrid arrangements, where the synthesis filter is fed by an excitation signal that is a preserved version of the real vocal tract excitation signal, not an estimate of its properties averaged over a window of samples. Linear prediction-based compression algorithms are such as LD-CELP (G.728, which gives an output bit-rate of 16kbps), and standards developed for wire-less communication, such as RPE-LTP (GSM speech coder, 13 kbps bit-rate), CS-ACELP (G.729 standard, 8kbps output bit-rate).

---

* Some sounds, such as voiced fricatives, cannot be represented by this model.

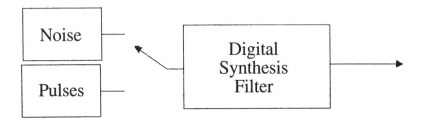

Figure 13.1 The linear predictive coding model of the speech production process

LPC-based codecs are named according to the representation of the residual; vector quantisation uses codes to represent the excitation signal, and the compression algorithm is called code-excited linear prediction (CELP). A comparison of algorithm against bit-rate can be found in Table 13.1.

For more information on speech coding see Spanias.

13.2.1.3   Characteristics of the Human Auditory System

The human auditory system is sensitive to frequencies from 20Hz up to 20kHz, and is able to sense sounds with a dynamic range of 130 dB. These properties are largely attributable to the anatomy of the ear, which consists of mechanisms to pass vibration to the inner ear oval window and on to the cochlea, a tapered tube wound into a helix and divided along its length by the basilar membrane. Sensitive hair cells along the length of the basilar membrane convey pressure variations to the aural nerves. The shape of the cochlea and the position of the hair cells along it determine the frequencies the hair cells are sensitive to and, if sensitive, the degree of sensitivity. Groups of hair cells act as overlapping band pass filters, and this behaviour is represented in the critical band model [Barnwell 96].

Popular music coders use the critical band model and also exploit the hearing phenomena of aural masking whereby a tone in a critical band is able to mask another tone in the same band; a musical signal in one band will mask noise in that

---

TABLE 13.1
A Comparison of Toll Quality Speech Coding Standards

Coding scheme	Sample/Frame based	Frame length (ms)	Bit-rate (kbps)
Linear 16	Sample	N/A	128
G.711	Sample	N/A	64
G.726	Sample	N/A	16/24/32/40
G.723.1	Frame	(4*7.5) 30	5.3/6.3
G.728	Frame	(8*2.5) 20	16
G.729/729A	Frame	10	8
GSM Full rate	Frame	20	13

band, but not noise in another band. Masking occurs because of the presence of higher level signals near the masked signal. It can occur in both the time (both forward and backward temporal masking occurs) and frequency domains. In music coders, spectral analysis of a small section of the incoming audio (along the lines of the critical band model) allows masked tones to be not represented in the output coded stream. Spectral analysis in this way also allows noise levels in bands where music tones exist to be higher than in bands where there are no music tones. In this way the compression algorithms produce transparent quality, with no perceived noise having been added.

13.2.1.4 Music Compression Algorithms

Two widely deployed music compression algorithms used on the Internet are MPEG-1 and AC-3. MPEG-1 exists as three variants: layers I, II, III, with each layer having a lower bit-rate and higher complexity for a given quality. A diagram of the principles of music compression can be seen in Figure 13.2.

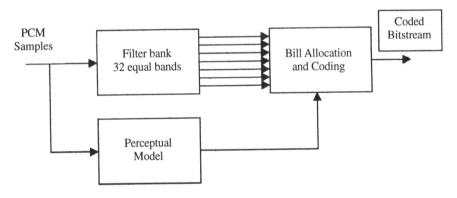

Figure 13.2 Principle of MPEG-1 Audio Encoder

MPEG-1 uses the perceptual model to determine how to allocate bits to the output of the filter bank. In layer I the perceptual model handles frequency masking. Layer II improves encoding efficiency by encoding across larger groups of samples, using improved coding techniques, and adding temporal masking to the perceptual model. Layer III improves the resolution of the filter bank by performing a modified discrete cosine transform (MDCT) on the output of the filter bank. It is also able to perform improved temporal resolution by changing the length of the sample blocks passed through the filter bank. Layer III also has a greatly improved bit-allocation scheme and uses entropy coding to provide additional bit-rate reduction. The bit-rates of the MPEG-1 layers can be seen in Table 13.2.

For a more detailed description of the principles of music coding in general, see Pan [Pan 93].

TABLE 13.2
Per Channel Bit Rates for Each of the MPEG1
Layers. If Stereo is Desired, Twice the Output
Bit rate is Needed

Layer no.	Per channel bit rate (kbps)
Layer I	192
Layer II	96
Layer III	64

13.2.2 CHARACTERISTICS OF THE INTERNET

The Internet transports information over a wide range of networking technologies. Central to the Internet philosophy is the concept of protocol layering and encapsulation, in which each successive protocol layer adds a more refined service to that offered by the layer below. The Internet Protocol (IP) conceals details of the underlying network technologies and offers a best-effort delivery service to the layer above. The basic unit of IP transmission is the datagram, which is carried independently from all other datagrams — the network is connectionless and there are no reception guarantees. For compressed continuous media, such as audio, the best-effort service of the Internet represents a serious problem for successful transmission. Audio compression algorithms are designed to operate on a stream of samples, not a series of unrelated packets; in order for the stream of samples to be successfully decoded, they must be presented to the decoder in the same order and essentially without loss.

This section provides an overview of Internet network characteristics, focusing specifically on those that relate to real-time audio transmission. The operation of the connectionless datagram switch (IP router) leads to effects that are especially important in real-time audio transmission: jitter and packet loss. Also, transport and application layer protocols build a set of tailor-made facilities for each application, and this section focuses on those relating to real-time media. In addition, multiway delivery mechanisms that are being used in commercial and research applications are considered. A better understanding of the Internet in general can be found in Tannenbaum and Peterson [Tannenbaum 96][Peterson 96].

13.2.2.1 Routers and Network Effects

The Internet consists of a large number of routers, which forward datagrams between hosts; forwarding is on the basis of the destination address in each packet. Routers make routing decisions on a per packet basis, which enables the network to be more robust, albeit at the expense of efficiency. Internet routers offer a best-effort service with no guarantees on timely delivery or even actual delivery.

Variations in the delay encountered by datagrams across the network occur because they are buffered in routers; buffering is used to absorb traffic burstiness. The amount of delay experienced at any router varies with the handling of routing updates, the commencement and termination of streams, the handling of one-off

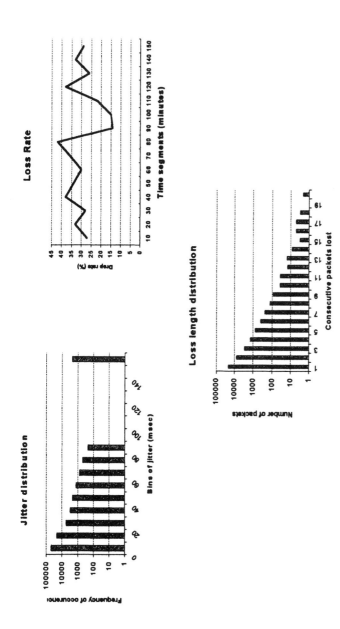

Figure 13.3 Graph showing sample loss and delay variation for real-time audio between two machines[OH1].

requests, etc. This means that the end-to-end delay between any two hosts is per-petually changing.

Routers suffer congestive losses in two ways. First, they might be over-whelmed by the volume of datagrams they receive and resort to discarding the excess. Second, the volume of incoming traffic might be such that traffic directed to an outgoing interface is greater than the bandwidth of the output link. The current mechanism for servicing packets buffered at a router is on a FIFO basis, and the tail of the buffer is dropped during congestion; consequently packets can be lost singularly or in bursts.

Real-time audio tools need to be able to adapt to both of these problems, and that issue is discussed in Section 13.3.2.2.

13.2.2.2 Internet Protocol Stack

The IP protocol provides a best effort service using a minimal set of facilities, including source and destination identification, packet fragmentation and re-assembly, header checksum, and type of service (widely ignored by routers) etc. Any further facilities required by an application are provided by transport and application layer protocols. The diagram below (Figure 13.4) shows how an application can build up a set of facilities by using different transport and application layer protocols.

Application Layer	HTTP	FTP	SMTP	RTP
Transport Layer	TCP			UDP
Network Layer	IP			

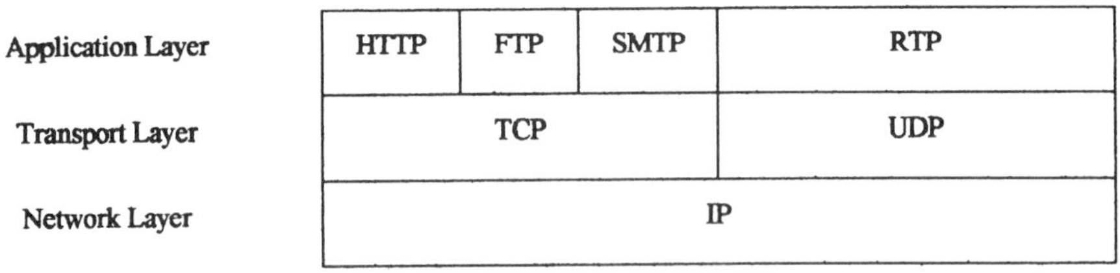

Figure 13.4 An Illustration of the IP protocol stack with a selection of application layer protocols.

13.2.2.2.1 Transport Level Protocols

Transport protocols, such as the Transport Control Protocol (TCP) and the User Datagram Protocol (UDP), essentially provide a multiplexing service to applica-tions. TCP also provides congestion control by scaling the rate of transmission, and reliability achieved via re-transmissions. TCP is inappropriate for real-time applications because the delay required for retransmission in the event of loss is too large for interactivity considerations. The arrival of a retransmission also means that applications cannot determine the amount of instantaneous available bandwidth. In contrast, UDP provides no delivery guarantees, or congestion control, and is consequently more amenable to real-time data. Providing band-width scalability and packet-loss robustness to real-time applications is an open research issue.

13.2.2.2.2 Multimedia Application Protocols

Application layer protocols provide services to applications by enhancing the service provided by the transport layer. Real-time media-specific application layer protocols can be further subdivided into those responsible for the streaming multimedia information and those responsible for negotiating and controlling the flows. Control protocols use TCP at the transport layer (because they need to receive all of the control data), and the streaming protocols use UDP (because they need timely delivery of data rather than guaranteed receipt).

A widely used multimedia streaming protocol is the RTP, Real-time Transport Protocol [RTP], which has an associated control protocol — RTCP, the Real-time Transport Control Protocol. RTP essentially provides applications with sequence number information for ordering purposes, media timestamping for calculating rendering times, a payload descriptor for codec identification, and a marker bit for application-defined purposes, such as talkspurt starts. RTCP conveys sender and receiver statistics and participant information. The RTCP bandwidth is limited to five percent of the total consumed by both RTP and RTCP streams. RTP is designed primarily for audio and video, and it specifies statically assigned payload types for popular compression algorithms (others can be defined through an external codec mapping mechanism). Some companies, such as Microsoft and Apple, are now working on generic coding schemes which can multiplex multiple inputs into a single stream and include not only codec-type information, but also the codec itself [ASF][RTP Generic].

A number of proprietary streaming protocols have found prominence in the marketplace with products such as RealNetwork's RealAudio and Xing Technology's Streamworks. These protocols are designed for streaming applications, which are tolerant to delay, and consequently include retransmission capabilities in the event of loss. RealNetwork's RTSP, Real-time Streaming Protocol [RTSP] has met with approval within the Internet Engineering Task Force (IETF) and has wide industry support. Despite its name, the primary purpose of RTSP is to control media streams, and it includes functionality, such as rewind/fastforward, and synchronization.

There are two competing standards for conference negotiation: the IETF's Session Invitation Protocol (SIP) [SIP] and the ITU's H.323 protocol [H.323]. In addition to call negotiation and control, H.323 also specifies which audio and video codecs may be used, transport mechanisms, and signaling. In contrast, SIP is a lightweight protocol that is not bound to specific audio and video compression algorithms.

13.2.2.3 Multiway Media Delivery

It is often necessary to deliver media data from one-to-many or from many-to-many. IP Multicast [Deering 89] was specifically designed for this purpose, but it is not yet deployed in all parts of the network. A widely used interim solution is one of packet reflectors, which forward and replicate packets between hosts. The reflector approach is used for commercial streaming products, such as RealAudio. However, it has a number of inherent drawbacks as a multiway delivery mechanism: the reflector is a single point of failure, reflector placement is ad-hoc, and the delivery tree is extremely likely to be suboptimal as a result. Reflector schemes might eventually be replaced as more Internet Service Providers (ISPs) support multicast. Several streaming products actually support multicast but currently use reflectors.

13.2.2.3.1   Multicast Backbone (Mbone)

The Mbone provides multiway communication facilities, by setting up a distribution tree from a sender to all interested receivers. In its simplest forms, there is a separate distribution tree for each sender. Receivers can join or leave a conference at will, and there is no explicit means of membership control. A receiver that wishes to join a conference sends an IGMP (Internet Group Management Protocol) join message to the local router. The local router then sends a graft message back towards the distribution tree. A receiver that wishes to leave a conference sends an IGMP leave message. If there are no other receivers attached to the router, the router will then send a prune message back towards the root of the tree. The early forms of multicast routing protocols were based on source-based distribution trees. More recently, newer multicast routing protocols set up shared distribution trees, which may be slightly suboptimal, but which provide other advantages. To use the multiway communication facilities provided by multicast, the sender merely has to use a class D address (in the range 239.225.225.225 to 225.0.0.0), assuming the local router is multicast capable. For a fuller discussion of multicast routing, see Deering [Deering 89] and Peterson [Peterson 96].

13.3   STRUCTURE AND TECHNOLOGY OF PACKET AUDIO SYSTEMS AND TOOLS

The availability of cheap multimedia capable PCs means systems that are able to handle real-time audio processing are widely deployed. Current processor performance means that quite highly complex compression may be performed in software, and, because many audio and video cards are now supporting compression with on-board DSPs, many more algorithms can be run in real-time. Networked audio tool programs can be developed with a minimal set of requirements: an Internet connection, headset or loudspeakers, and an audio card, all of which are standard components on many PCs. These factors have encouraged the rapid adoption of the new technology, and have led to a proliferation of different types of audio tool. There are three audio tool types that stem from different application requirements: Internet telephony (point-to-point, low delay), audio-on-demand (point-to-point, delay tolerant), and Mbone audio (multiway, low delay).

This section focuses on the technology of an Mbone audio tool and is based on our experience with the Robust Audio Tool (RAT) developed at UCL [Hardman 98], but other examples exist, such as Vat [Jacobson 92]. The same techniques are employed in point-to-point and audio-on-demand applications. An Mbone audio tool is a program that interfaces to the audio hardware, the Internet protocol stack, and the user interface. Writing a successful audio tool requires knowledge of network characteristics, audio card hardware, headsets, microphones, acoustics, and compression algorithms.

Internet/Mbone audio tools are used in many applications piloting projects and by users worldwide (research software source code and executables are often made freely available). This has allowed researchers to identify and solve some of the problems with current audio tools — those relating to transmission over the best-effort Internet and to manipulation of real-time audio on a general purpose host.

13.3.1 Low-Cost General Purpose Hosts Suitable for Multimedia Internet Access

Internet audio tools collect samples from the audio hardware and put them into packets. Most general purpose workstations and PCs have the IP stack implemented in the operating system (including selection of the transport layer protocol, UDP or TCP/IP, and unicast or multicast facilities). Multimedia-specific application layer protocols (e.g., RTP) are implemented in the user program, in-line with the recommendations of application layer framing [Clark 90]. In a multimedia PC or UNIX workstation, the usual method of communication with an audio card is the operating system device driver that collects samples from the hardware, and a chunk of memory containing perhaps 20 ms of linear PCM audio samples is given to the user program.

Our audio tool is supported on a range of general purpose workstation and PC operating systems (Solaris, SunOS, IRIX, FreeBSD, Linux, HP-UX, Windows 95/98, and Windows NT). The program code is written in C, with the user interface written in TCL/TK [Ousterhout 94], a portable code scripting language. Hardware-specific code that interfaces with the audio hardware is determined at linkage time and is hidden from the program by a standard audio interface.

On a UNIX workstation, reading audio from within a user program is via the *select* system call, which indicates that one or more buffers of audio are ready. The number available depends on how long it has been since audio was last read from the device, and the size of each buffer is fixed (perhaps 20 ms each — the value can be set). From within a program operating under Windows reading audio is different. When an application opens the audio device under Windows, it creates a multimedia task that generates messages every time a block of audio is available. The audio application uses a callback to handle these messages, and the audio application can sleep between messages. Both methods of reading audio mean that the audio tool does no processing while waiting for audio from the device and that the CPU can be used by other programs.

The preferred features of an audio card are full-duplex operation (standard on all workstations and available on PCs), 16-bit sample audio resolution, and multiple sampling frequency support. A number of legacy cards that deliver 8-bit samples are usable, but they result in lower quality, especially because they are often compressed by the audio card using μ-law PCM, which leaves artifacts in the audio. In order to cater to the full complement of audio codecs, the card should support a range of sampling frequencies (8, 16, 32, 44.1, 48 kHz). For more computer-intensive facilities, such as fully localised 3D audio, audio cards with on-board DSPs are emerging.

Some audio/video hardware supports on-card compression through high capacity digital signal processors, which can be utilized by an audio tool if an appropriate API is available. Most of the major processor families have single-instruction multiple data (SIMD) operations, such as Intel MMX and Sun VIS architectures, that enable general purpose processors to enjoy strong DSP performance. SIMD instructions allow many audio operations to be optimized, such as filtering and sample rate conversion.

Operating systems (such as Microsoft Windows) provide audio compression algorithms that can be accessed through a standard interface, the audio compression manager (ACM), that provides access to a range of codecs (G.711, G.721, GSM, MPEG-I layer-3). In addition the ACM has the ability to perform sample format conversion and sample rate conversion.

13.3.2   Technology of an Internet/Mbone Audio Tool

The basic structure of an audio tool can be seen in the block diagram of Figure 13.5. The user interface presents controls, such as volume, together with run-time statistics reporting mechanisms. A silence detection algorithm determines if the buffer is speech or silence, because Mbone audio tools conserve bandwidth by transmitting audio only when the user is speaking. Buffers determined to contain speech samples are compressed and formatted into packets with an RTP header. The packet is then passed to the operating system for standard UDP/IP transmission using a unicast or multicast address. The address is specified as a command-line option at program start.

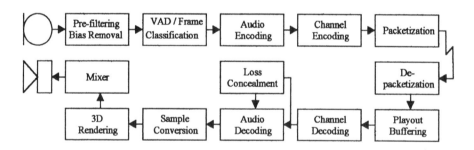

Figure 13.5 Block diagram of an audio tool

If the compression algorithm generates a continuous bit-rate, packets are generated at regular intervals. Because of the characteristics of the Internet, jitter will be added to datagrams as they traverse the network. At the receiver, a buffer is needed to smooth out the effects of jitter. After software decompression, mixing takes place before the audio is passed to the device driver for play.

13.3.2.1   User Interface

The user interface of our audio tool (see Figure 13.6) consists of a list of participants, volume controls, and activity bars. The names of active speakers are highlighted to help participants identify who is talking; when a participant stops talking, the highlighting slowly fades. Fading helps users identify who has spoken recently, which is especially useful in a multiway conference. An options page (launched by

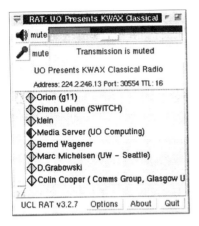

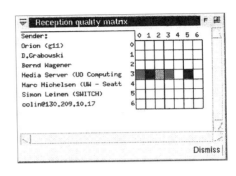

Figure 13.6 RAT user interface

clicking a button on the main window) is available to allow the more experienced user to change compression level/packet rate.

13.3.2.2 Buffering

Transmitting continuous media in a flow over a packet network incurs delay. That delay has the following components:

$$\text{Delay at receiver} = \text{packetisation delay} + \text{network propogation} \\ + \text{buffering at routers}$$

Delay is associated with packetisation because the transmitter must collect enough samples to fill a packet before it can send the packet onto the network. This delay is usually fairly stationary during the lifetime of a flow. Network propagation delay is often fixed or is more variable when retransmission at the data-link level occurs, such as over mobile links. Buffering at routers introduces a delay component that depends upon other traffic in the network and varies considerably.

Buffering is needed at the receiver to remove the effects of jitter from an incoming flow. The buffer adds delay to early datagrams, and correspondingly less to later datagrams, to increase the likelihood of many packets arriving in time for playback. Each incoming datagram of audio therefore has a play-out point associated with it. Average jitter levels are different for every flow, and they change with congestion on the network which means that an adaptive play-out point calculation is performed for each source. Because some packets might have spuriously high levels of jitter added to their arrival time or might not arrive at all, a budget of slightly less than 100% is commonly used to calculate the amount of buffering required. Recommendations for calculating the play-out point can be found in Schulzerine, et al. [RTP], Moon, Kurose, and Towsley [Moon 98], and Diot, Huitema, and Turletti [Diot 95].

13.3.2.3 Compression

The most computational intensive part of an audio tool is the compression algorithm. As a rule of thumb, compression algorithms increase logarithmically in complexity for a doubling of compression advantage. Because of the high level of complexity in many compression algorithms, the performance of an individual codec is optimized for a range of platforms. Within the audio tool, a compression algorithm will be called to compress a buffer, rather than individual samples of audio, from the audio hardware. Between calls to compress (and decompress at the receiver), the internal variables of the codec are stored for the next time the compression module is called.

Early research into the effects of loss in interactive packet audio shows that small packets should be used, preferably carrying no more than 16 ms of audio [Minoli 79]. This use of small packets balanced the impact of packet rate on the network, with the likely perception of loss. Within the Internet, the desire to maintain a reasonable packet header-to-payload ratio has meant that the nominal minimum size of a packet over the Internet is 160 bytes, which translates into 20 ms of toll-quality PCM. If the packet rate needs to be reduced to ease loss rates at the receiver(s), then ADPCM might be used to code 40 ms of audio into a datagram (or RPE-LPC to code 80 ms), while keeping the payload size the same.

13.3.2.4 Talking in a Multiway Environment and Silence Detection

Mbone audio sessions commonly do not use floor control mechanisms and are based on dynamic group membership (participants can join and leave a conference at will without any explicit reconfiguration of the audio tool). Chaos is largely avoided by reliance on the protocol of polite conversation to facilitate meaningful multiway discourse.

Silence detection is commonly used in multiway conferences to restrict traffic generation to roughly one half of a PSTN bearer circuit, regardless of the number of participants. Packets are generated only when someone talks, and this means that a series of packets will be generated, interspaced by gaps (the series of packets is called a talk spurt). The detection of speech in the presence of background noise is nontrivial, and silence detection algorithms must be adaptive, because background noise levels vary [Rabiner 78]. A rough energy measure is commonly used, and this can be increased by the addition of spectral analysis algorithms to identify low energy fricatives, which frequently fall below the threshold of the background noise. Silence detection algorithms should be disabled for broadcast type applications and for music, as these algorithms frequently lead to clipping effects.

13.3.3 AUDIO TOOL PROBLEMS

By far the most annoying problem with audio transmission is the gaps in the audio, and significant effort has been directed towards identifying the causes of this problem and eradicating them. Other audio tool problems and causes of frustration are associated with an unnatural auditory environment.

13.3.3.1   Cause Identification of Gaps in the Output Audio

Gaps in the output audio can be caused by any of the following situations, and the cause is not always obvious:

- incorrectly operating silence detection algorithm
- actual datagram loss over the Internet
- excessive jitter on the datagram arrival time
- packet(s) dumped by the audio tool (because of scheduling anomalies)
- faulty microphone or headsets

Some of the problems (such as faulty microphone or headsets) can be isolated by the activity bars provided by the user interface. The identification of other problems relies on detailed examination of loss statistics from the network (such as actual datagram loss over the Internet) and from the audio tool itself (packet dumped by the audio tool because of scheduling anomalies). Excessive jitter on packet arrival times is an infrequent event, catered to by the adaptive play-out point calculation, and it is displayed in the network statistics window. The design of successful silence detection algorithms is nontrivial and requires examination of the frequency spectrum to perform well. This extra facility is commonly not available within most audio tools because of the large amount of processing power taken to analyze frequency spectra.

Of the causes of gaps in the output audio, the loss of datagrams over the network and the occurrence of scheduling anomalies are by far the most common.

Solutions for datagram loss over the Internet — Packet loss is especially a problem for speech, where the use of relatively large packets means that the impact of loss is severe [Hardman 95]. Packet loss is also a problem for music, although delay tolerance means that a greater range of repair techniques are feasible than is the case for interactive communication [Hodson 98]. Possible approaches to alleviating the impact of packet loss are receiver-only repair, channel coding at transmitter, and combined source and channel coding at transmitter [Perkins 98].

Receiver-only solutions rely on the assumption that the audio characteristics have not changed over the duration of the packet and are usually taken to work for packet sizes of 16 ms. Different approaches to tackling this problem include repeating the audio from before the loss [Goodman 86], applying pattern-matching techniques on one or both sides of the loss [Waseem 88], interpolating over the gap [Sanneck 96], and applying repair in the transform domain (Figure 13.7). A number of codecs, such as the GSM, specify transform domain repair mechanisms, which suffer less from discontinuities at the boundaries of loss. Receiver-only techniques don't work very well for the larger packet sizes commonly used over the Internet because with an increasing packet size there is an increasing probability that the source characteristics will have changed (phoneme sizes are 5-100 ms, with the average being about 80 ms) [Hardman 95]. Interleaving components of audio across multiple datagrams can produce improvements because gaps are then significantly reduced in size, and receiver-only solutions are then used at the receiver. The disadvantage associated with this method, however, is an increase in delay, because multiple packets have to be collected at the transmitter before they can be sent.

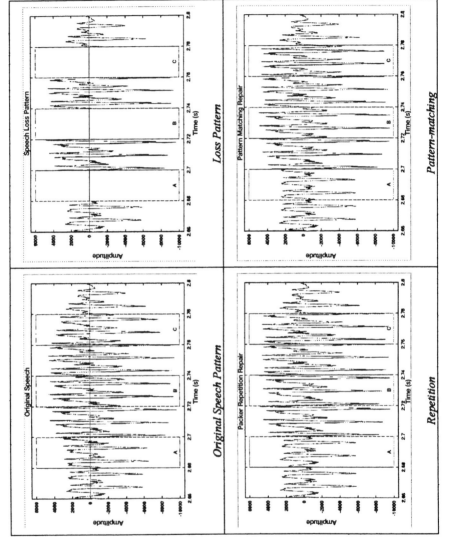

Figure 13.7 Examples of error concealment

Channel coding is the traditional approach to the problem of packet loss, where the source is assumed to produce a bit-stream that is preserved at the bit level. Common channel coding techniques are forward error correction (FEC) and interleaving and retransmission [Perkins 98]. FEC takes k blocks data and transforms it into n blocks of data, and, as long as the number of blocks lost is no more than $(n - k)/2$, all of the original k blocks can be determined. The inherent problem with these techniques is that preservation of speech at the bit level is not necessary, and these techniques tend to fail at a particular error level, rather than gracefully deteriorating. For natural conversation, the end-to-end delay needs to be less than 250 ms [Montgomery 83]. Interleaving, retransmission, and forward error correction increase the amount of buffering required by receivers and increase the end-to-end delay of the audio [Perkins 98]. Over local area networks, the additional delay required by retransmission means that it may be a viable repair method, but it is not suitable for interactive audio over wide area links.

Combined source and channel approaches to packet loss repair use analysis results from the compression algorithm to gauge the effect of loss and to transmit extra information to help repair at the receiver. In mobile communications, FEC is applied to parts of speech frames that are most sensitive to loss (e.g., GSM) [Vary 88]. The use of FEC in this manner is necessary because mobile communication suffers from large bursts of bit errors. If not corrected, bursts of errors introduce serious distortion into the output speech and cause the decoder to mis-track for a period after the loss. Using FEC for the important parts of speech or music frames is also a technique being employed in the Internet. Efforts by Mbone audio tool researchers have developed redundancy [Hardman 95], which piggy-backs redundant information onto subsequent datagrams in the stream in order to repair single losses (which are the most common), as shown in Figure 13.8.

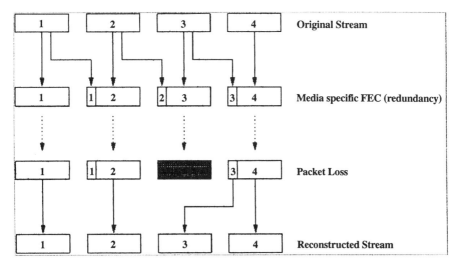

Figure 13.8 The use of redundancy to repair packet losses over the Internet

Redundancy can be extended to recover from bursts of losses, by lagging the redundancy from the original packet by different amounts [Kouvelas 97a]. The redundant information can be the output of the same codec (the most important consideration in reducing loss as a result of congestion is to reduce the packet rate), or can be the output of a different codec (that uses a high level of compression, or produces reduced quality speech). This scheme utilizes the fact that an exact fill-in to cover the loss is not necessary, and short degradations in sound quality are almost transparent. In addition, it incurs only extra delay at the receiver and none at the sender, because the piggybacked copy is from an earlier instant.

When interactive operation is not required, retransmission can be used. Retransmission is viable for streaming applications, because clients can ensure they extend the receiver buffering enough for every packet to cover the cost of sending the retransmission request and waiting for the data. Effort has recently been directed towards defining loss repair mechanisms for the Internet [Hardman 95], [Perkins 98].

Solution for lack of real-time scheduling support — Lack of real-time scheduling support means that an audio tool process may not be able to write enough audio to the device driver to cover the next period when the audio tool process is suspended. Similarly, it might be that the audio tool cannot read packets from the network interface socket before they have been overwritten. Both of these effects manifest themselves as packet loss.

Our RAT assumes that the clocking rate of the input audio from the audio card is in phase with the output (which is true if 1 crystal is used to provide input and output sampling frequencies). The audio tool reads whatever is available at the input and writes-out a corresponding amount. Using this mechanism it is possible to estimate the elapsed time since the last read and to estimate if the host machine is lightly or heavily loaded. If the host is heavily loaded, scheduling intervals might get large, and the output buffer in the device driver might run dry, leading to a gap in the output audio. To avoid this, RAT uses an adaptive algorithm that maintains a history of gaps and determines whether or not to increase the amount of samples in the device driver buffer to minimise gaps [Kouvelas 97b].

13.3.3.2   Cause Identification of an Unnatural Audio Environment

The acoustic feel of an Mbone audio tool is disrupted by many factors that can be categorized as two-way and multiway effects:

Two-way effects

- Difficulty in setting microphone and loudspeaker amplitudes correctly
- Silence suppression between sentences — high levels of background noise appear to cut in and out
- System appears dead because of a lack of reverberation
- Echoes because of the closed audio loop

Multiway effects

- No localisation abilities means it is difficult to separate voices from each other and to recognise individuals
- Multiple speakers have as many different background noise environments, and the change among them is disconcerting

13.3.3.2.1   Two-way Effects

Difficulty in setting microphone and loudspeaker amplitudes correctly — The acoustic part of the system suffers from problems when used with general purpose hosts. The need to support a wide range of headsets or microphones and loudspeakers means that impedance level mismatches result, and volume levels from participants are different. Input gain control is provided at the transmitter for this purpose, but participants cannot easily monitor their output volume at the far end. Because of this difference in volume between individual speakers, participants have to change the receiver volume gain on a regular basis as different people talk. This problem also causes input audio not to occupy the full dynamic range of the compression algorithm, which means that larger values of quantisation noise will be added to the signal than might be expected in the ideal case.

A solution to this problem is to use automatic gain control (AGC) at the transmitter. The AGC algorithm tracks the energy levels in the input talkspurt and multiplies every sample by a suitable scaling factor. Unfortunately, AGC suffers from problems, in particular, rapid gain changes during words reduce the intelligibility of what is being said. A system that is too slow to react will result in either greater quantisation noise being added by the compression algorithm. However, if the scaling factor is too slow to react, overloading results. Despite these problems, an AGC mechanism can be made to work reasonably well in an Mbone audio tool.

Silence suppression between sentences — Silence suppression is needed in multicast audio because it reduces the bandwidth consumed to manageable levels over the Mbone. It is also needed in order to sensibly mix multiple incoming flows from different speakers, because to mix the audio from hundreds of participants would mean that the few (usually one) speech streams would be masked out by many contributions of background noise.

Silence detection causes speech to be transmitted in talk spurts and cuts inter-talkspurt audio. This means that background noise cuts in and out, which has been shown to reduce the intelligibility of what is being said. In mobile point-to-point network connections, the traditional approach has been to transmit comfort noise, which is an average version of background noise at the transmitter, to the receiver [Hanzo 94]. Bandwidth is restricted because rough spectral information is sent, which is then used at the receiver to simulate frequency-correct noise for inter-talkspurt gaps.

System appears dead — In the natural world, reverberation is caused when sounds are delayed and reflected off the walls of the environment the listener is in. The speaker hears his or her own speech, plus added reverberation. The addition of reverberation to speech actually improves intelligibility [Olson 57]. It also gives the speaker some idea of the size of the room they are in and how loudly they need to speak to enable other people to hear them (compare the acoustic environment of a small room to that of a cathedral).

When using headsets in a multiway conference, the audio channel itself is perceived as dead; no reverberant information is present for the speaker to determine the acoustic map of the space into which they are speaking; the result is that people tend to shout. A cheap version of reverberation commonly used in telephone handsets is side-tone [Richards 73] — the speech input leaks through the handset to the earpiece, with essentially a very low delay. This technique is imitating the first reflection in the reverberation, which tends to be the most prominent anyway. Full reverberant audio can easily be generated using special audio cards, often designed to give the ability to localize individual speakers.

Previous work has also shown that a feedback loop is required between the AGC algorithm and the reverberation part of the system because it is possible to influence how loudly a person speaks by altering the level of reverberation [Sharifi 96].

Echoes — When an audio conference system is used with loudspeakers and microphones, there is often feedback through the system. This echo is extremely annoying for a remote participant, as their own voice is echoed back to them with often quite a large delay, and the perception of echoes increases with time after the original sound [Brady 71]. Frequently, feedback continuously circulates around the closed audio loop, and if the gain is greater than unity the system will howl-round. The use of more directional microphones eases the problem because then it is less likely that a microphone's pick-up pattern will overlap with the loudspeaker's directional output. A better solution is to use echo cancellation techniques, which adaptively estimate the delay path through the system and subtract the delayed version from the input audio. An alternative to echo cancellation is to use headsets, but these are neither comfortable for long conferences, nor do they provide a natural acoustic environment.

Some simple echo management functionality is available in audio tools to attenuate the microphone if the loudspeaker is active and vice versa, but the speech then suffers from clipping effects; parts of received speech are lost if the listening participant coughs, for example.

13.3.3.2.2   Multiway Effects

Lack of localisation ability — In a natural environment, a human's ability to localize sounds is used to focus attention and gaze on a speaker and to improve the audibility of a single conversation in a busy room (the cocktail party effect) [Cherry 53]. In general, this effect is associated with the urge to tip and turn the head to improve the localisation of the sound source. The visual image of the speaker is then matched with auditory input, and the combination of image and sound direction increases the intelligibility of what is being said [Begault 94].

The lack of auditory clues, when heard over a mono channel, makes individuals' speech sound as if it is coming from the same position and direction — the back of the listener's head, if the listener is using a headset. This is especially problematic in multiway conferences, in which the participants are suffering from visual overload and cannot focus on the list of speakers in the audio tool window to determine who is talking.

Different background noise environments for different speakers — Even if the rest of the causes of an unnatural acoustic environment are eliminated, a multiway conversation will never sound natural because multiple different acoustic environments are being switched among. What is really needed is elimination of a speaker's acoustic environment from the audio signal and addition of either an average environment at the receiver, or addition of the actual listeners' environment at the receiver. Such a solution is extremely difficult or impossible to produce.

13.4 IMPACT OF LOW-COST AUDIO COMMUNICATION

The impact of low-cost Internet audio communication will be vast. The deployment of Internet and Mbone audio is expected to increase rapidly with the availability of the global Internet and inexpensive personal computers in business and the home. Not only will the cost of existing applications (such as point-to-point telephony) be substantially reduced, but new applications will enhance interactive facilities well beyond audio-only communication.

Packet audio applications can be split into two broad groups — people-to-people and people-to-systems. Low delay and relatively small, but guaranteed, bandwidth requirements in both directions characterize people-to-people applications, such as Internet telephony and Mbone audio. In contrast, people-to-system applications (such as audio streaming applications) are much more delay tolerant and can function with substantially less bandwidth than is needed for real-time communication.

13.4.1 INTERNET TELEPHONY

Internet telephony servers and software are already commercially available, and their popularity is expected to increase [Minoli 98]. Such systems also need to be integrated with the existing PSTN networks, and most commercially available products provide a comprehensive solution, including end application software, and gateway cards and software. Mechanisms for remote service location are also being standardised for the Internet and enhanced to provide capabilities such as the remote location of a gateway based on cost [Rosenberg 98].

Point-to-point Internet telephones use unicast transmission facilities. Multiway audio tools, such as CuSeeMe and Netmeeting, use unicast transmission facilities and manually set-up reflectors to achieve multiway communication. When multicast becomes more widely available in the Internet, many more people will use Internet audio to add an interactive dimension to existing applications, for example mailing lists, news groups, and computer games (some games already use multipoint interactive audio communication over the Internet).

13.4.2 MBONE AUDIO AND MULTIMEDIA CONFERENCING

Many-to-many audio tools use multicast, a more efficient evolution of unicast and reflectors. Mbone audio is regularly used as part of multimedia conference sessions, and the Mbone is moving from a pilot to a service [Perkins 97][Buckett 95]. Multicast multimedia conferencing uses a combination of media to provide communication, often audio, video, and shared text. The availability of low-cost multimedia conferencing also means that distance learning applications are becoming popular, and telepresence applications are being researched. Not only will further deployment and development of the Mbone benefit multiway communication, but the increased integration of multimedia conferencing tools (such as the integration of audio with video to provide lip synchronisation [Kouvelas 96]) will substantially improve the attractiveness of this method of communication.

13.4.3 AUDIO AND VIDEO ON DEMAND

Initially delivered via file transfer, music is now streamed over the web using unicast transmission and reflectors (e.g., RealAudio). Similar applications might use AC-3 as a compression algorithm and error protection to successfully stream audio via point-to-point communication (at 44.1 or 48 kHz sampling frequency). Mbone audio tools are also widely used to multicast live radio stations. These tools use toll-quality speech compression algorithms (8 kHz sampling frequency), although support for multiple sampling rates up to 48 kHz is now available in some audio tools [Hodson 98].

Interactive systems also frequently need to access stored clips of material — in a business meeting, this could be a record of previous meetings; in distance learning, this could be access to other educational material. Multimedia jukeboxes are also being developed in the research community to provide stored clip access [Lambrinos 98]. Such streams would be controlled using RTSP, an Internet draft standard protocol that allows hosts to remotely control playback and recording of streamed audio and video [RTSP].

13.5 CONCLUSION AND FUTURE DIRECTIONS

This chapter presents an overview of Internet audio. A full understanding of Internet audio systems can be gained only from a comprehensive assessment of compression algorithms, acoustics, Internet protocols, and general purpose host implications. Analysis of these aspects leads to Internet audio tools and the problems associated with them.

Speech compression algorithms exploit the characteristics of speech production to gain a compression advantage and quite low bit-rates. Unfortunately, we can hear better than we can talk, and because music compression algorithms rely on the characteristics of hearing, the bit-rates produced are substantially larger than those for speech. The standards that relate to speech and music compression have been developed primarily for networks other than the Internet and tend to suffer substantially from packet loss over the Internet.

Real-time applications use UDP over IP because they do not need the reliable service offered by TCP. The IETF has standardised an application layer protocol, RTP, to provide facilities such as timestamps and sequence numbers for real-time applications. Of particular importance for audio is the ability to convert point-to-point applications into multiway ones. Multiway facilities are currently provided over the Internet by unicast transmission and reflectors. However, the Internet can support a more efficient method of multiway communication — multicast — which has distinct advantages over reflector-based systems. In the future, it is likely that multiway applications will use multicast for multiway multimedia delivery.

This chapter focused on a particular Mbone audio tool, rather than consider examples of each of the application classes, because Mbone audio represents a superset of the technology. Mbone audio tools are made up of a pipeline of components designed to reduce environmental problems, such as the effect of the best-effort Internet, the use of compression algorithms designed for guaranteed bit-rate transmission, and the use of general purpose workstations that provide no support for real-time applications. These problems manifest themselves primarily as gaps in the output audio, and this chapter identified their many causes, and suggested solutions. Another complex audio tool problem addressed here was an unnatural acoustic environment, which is associated with the transmission of one channel of audio to single and multiple participants.

Internet audio can be split into three types of applications: Internet telephony, music-on-demand, and Mbone audio. In the future, it is likely that many current Internet applications will have an interactive edge added to them because Internet audio has a low cost. The advantage of transmitting audio over the Internet is not just that it is low cost, or that multiway transmission is possible to hundreds of receivers, but that audio can easily be integrated with other media, such as video, graphics, and text. From this we can conclude that the impact of Internet audio will be vast.

13.5.1 FUTURE DIRECTIONS

The best-effort Internet will evolve to be able to support some guaranteed quality of service (QoS) levels. An initial attempt at providing QoS over the Internet produced the reservation protocol (RSVP), which allows receivers to specify that a guaranteed QoS be used over the links in that part of the distribution tree. Routers will provide QoS by employing queuing algorithms such as class-based queuing [Floyd 95], which might process delay-sensitive (or higher QoS) traffic with preference to other traffic.

In tandem with the drive to enable the Internet to support QoS levels is a desire to encourage all traffic to share evenly the available bandwidth and to back-off when congestion levels increase. Some applications are currently fair in their use of bandwidth (such as TCP/IP), but interactive applications do not behave so well. It is probable that current router buffering will change from a FIFO discipline to using techniques such as random early drop [Floyd 94], which might give advanced notification of congestion to receivers and penalize badly behaving sources. Audio applications might become adaptive to congestion [Kouvelas 98], although this might be restricted to applications that transmit higher quality audio. In such situations, the audio compression algorithm must be able to deliver multiple, different qualities and therefore bit-rates.

Multimedia conferencing and telepresence applications will evolve far better interaction capabilities than can currently be provided by independent media tools. Those tools will be integrated by a framework, which will support both remote configuration and control capabilities [McCanne 98] and increased media integration. Instead of monolithic tools, it is likely that software will be in the form of a series of reusable objects in a pipeline [DirectX].

ACKNOWLEDGMENTS

RAT is the work of a group, and we acknowledge Colin Perkins and Isidor Kouvelas, amongst others. The work is funded by EPSRC projects RAT (#GR/K72780) and MEDAL (#GR/L06614) and British Telecom plc (JAVIC project). In addition we thank Phil Lane for comments on the text.

REFERENCES

[ASF]	Online documentation http://www.microsoft.com/asf/
[Barnwell 96]	T. Barnwell III, K. Nayebi, C. Richardson, *Speech Coding: A computer laboratory Textbook*, Georgia Tech. Digital Signal Processing Series, 1996.
[Begault 94]	D.R. Begault, *3D Sound for Virtual Reality and Multimedia*, Academic Press, 1994.
[Brady 71]	P.T. Brady, "Effects of Transmission delay on Conversational Behaviour on Echo-Free Telephone Circuits," *Bell System Tech. J.*, Vol. 50, No. 1, 115–134, Jan 1971.
[Brady 69]	P.T. Brady, "A model for generating on-off speech patterns in two-way conversation" *Bell System Tech. J.*, Vol. 48, No. 9, 2445–2472, Sep 1969.
[Buckett 95]	J. Buckett, I. Campbell, T.J. Watson, M.A. Sasse, V.J. Hardman, A. Watson, "ReLaTe: Remote Language Teaching over SuperJANET" *Proc. UKERNA 95.*
[Gold 73]	B. Gold, "Digital Speech Networks," *Proc. IEEE*, Vol. 65, No. 12, 1636–58, 1973.
[Cherry 53]	C.E. Cherry, "Some Experiments on the Recognition of Speech, with One and Two Ears," *J. Acoust. Soc. Am.*, Vol. 25, 975–979, 1953.
[Clarke 90]	D. Clarke, "Final Report from the Unison Project," ALVEY Projects of the DTI, 1990.
[Clark 90]	D. Clark, D. Tennenhouse, "Architectural Considerations for a new Generation of Protocols," *Proc. ACM SIGCOMM*, 1990.
[Deering 89]	S. Deering, "Host Extensions for IP Multicasting," RFC1112, Internet Engineering Task Force, Aug 1989.
[Diot 95]	C. Diot, C. Huitema, T. Turletti, "Multimedia Applications should be adaptive," *Proc. High Performance Comp. Syst. '95*, 1995.
[DirectX]	http://www.microsoft.com/directx.
[Floyd 95]	S. Floyd, V. Jacobson, "Link sharing and resource management models for packet networks," *IEEE/ACM Trans. Networking*, Vol. 3, No. 4, Aug 1995.
[Floyd 93]	S. Floyd, V. Jacobson, "Random early detection for congestion avoidance," *IEEE/ACM Trans. Networking*, Vol. 1, No. 4, Aug 1993.

[Goodman 86]   D. Goodman, G. Lockhart, "Waveform Substitution Techniques for
                    Recovering Missing Speech Segments in Packet Voice Communica-
                    tions," *IEEE Trans. Acous., Speech and Signal Processing*, Vol.
                    ASSP-34, No. 6, 1440-1448, Dec 1986.

[H.323]   "Recommendation H.323 (02/98) Packet-based multimedia commu-
                    nications systems," http://www.itu.int/.

[Hanzo 94]   L. Hanzo, R. Steele, "The Pan-European Mobile Radio System Part
                    II" *Eur. Trans. Telecommun. and Relat. Tech.*, Vol. 5, No. 2, Mar-
                    Apr 1994.

[Hardman 95]   V.J. Hardman, M.A. Sasse, A. Watson, M. Handley, "Reliable Audio
                    for Use over the Internet," *Proc. INET95*, 1995.

[Hardman 98]   V. Hardman, M.A. Sasse, I. Kouvelas, "Successful Multi-party Audio
                    Communication over the Internet," *Communi. ACM*, May 1998.

[Hodson 98]   O. Hodson, S. Varakliotis, V. Hardman, "A software platform suitable
                    for multiway audio distribution over the Internet," in *Audio and
                    Music technology: the challenge of creative DSP*, IEE Colloquium,
                    London, U.K., 1998.

[Jacobson 92]   V. Jacobson, "VAT manual pages," Lawrence Berkeley Laboratory
                    (LBL), Also http://www-nrg.ee.lbl.gov/vat/, Feb 1992.

[Kent 92]   R.D. Kent, C. Read, "The Acoustic Analysis of Speech," Whurr
                    Publishers, 1992.

[Kouvelas 96]   I. Kouvelas, V. Hardman, A. Watson, "Lip Synchronisation for use
                    over the Internet: Analysis and Implementation," *Proc. Globecom96*,
                    893–898, 1996.

[Kouvelas 97a]   I. Kouvelas, O. Hodson, V. Hardman, J. Crowcroft, "Redundancy
                    Control in Real-Time Internet Audio Conferencing," International
                    workshop on Audio-Visual Services over Packet Networks, 1997.

[Kouvelas 97b]   I. Kouvelas, V.J. Hardman, "Overcoming Workstation Scheduling
                    Problems in a Real-Time Audio Tool," *Proc. USENIX Ann. Tech.
                    Conf.*, 1997.

[Kouvelas 98]   I. Kouvelas, V.J. Hardman, J. Crowcroft, "Network Adaptive Con-
                    tinuous-Media Applications Through Self Organised Transcoding,"
                    Proc. NOSDAV, 1998.

[Lambrinos 98]   L. Lambrinos, P.T. Kirstein, V.J. Hardman, "The Multicast Multime-
                    dia Conference Recorder," *7th Int. Conf. Communi. and Networks*,
                    IEEE, 1998.

[McCanne 97]   S. McCanne, et al., "Toward a Common Infrastructure for Multime-
                    dia-Networking Middleware," *Proc. 7th Intl. Workshop on Network
                    and Operating Systems Support for Digital Audio and Video* (NOSS-
                    DAV 97), 1997.

[Minoli 98]   D. Minoli, E. Minoli, *Delivering Voice over IP Networks*, Wiley
                    Computer Publishing, 1998.

[Minoli 79]   D. Minoli, "Optimal packet length for packet voice communication,"
                    IEEE Trans. Commun., Vol. COM-27, 607–611, Mar 1979.

[Moon 98]   S. B. Moon, J. Kurose, and D. Towsley, "Packet Audio Playout Delay
                    Adjustment: Performance Bounds and Algorithms," ACM / Springer
                    Multimedia Systems, Vol. 6, 17–28, Jan 1998.

[Montgomery 83]   W.A. Montgomery, "Techniques for packet voice synchronization,"
                    IEEE J. Sel. Areas Commun., SAC-1 (6), 1022–28, Dec 1983.

[Olson 57]   H.F. Olson, *Acoustical Engineering*, Van Nostrand, 1957.

[Ousterhout 94]   J.K. Ousterhout, *Tcl and the Tk Toolkit*, Addison-Wesley, 1994.

[Pan 93]   D.Y. Pan, "Digital audio compression," in *Digital Tech. J.* Vol. 5, No. 2, 1993.

[Perkins 98]   C.S. Perkins, O. Hodson, V. Hardman, "A Survey of Packet-Loss Recovery Techniques for Streaming Audio," *IEEE Net.*, Sept/Oct 1998.

[Perkins 97]   C. Perkins, J. Crowcroft, "Real-time Audio and Video Transmission of IEEE Globecom 96 over the Internet," *IEEE Commun. Mag.*, Apr 1997.

[Peterson 96]   L. Peterson, B. Davie, *Computer Networks: A systems Approach*, Morgan Kaufmann Publishers, Inc., 1996.

[Rabiner 78]   L.R. Rabiner, R.W. Schafer, *Digital Processing of Speech Signals*, Prentice-Hall, 1978.

[RealAudio]   'RealAudio' http://www.real.com/.

[Richards 73]   D.L. Richards, *Telecommunication by Speech*, Butterworth and Co. 1973.

[Rogerson 97]   D. Rogerson, *Inside COM*, Microsoft Press, 1997.

[Rosenberg 98]   J. Rosenberg, B. Suter, "Wide Area Service Location," IETF Internet Draft, draft-ietf-srvloc-wasrv-00.txt.

[RSVP]   L. Zhang, S. Deering, D. Estrin, S. Shenker, D. Zappala, "RSVP: A New Resource Reservation Protocol," *IEEE Net*, Vol. 7, No. 5, 8–18, Sept 1993.

[rtp]   "RTP: A Transport Protocol for Real-Time Applications," Audio-Video Transport WG, RFC 1889.

[RTP]   H. Schulzerine, S. Casner, R. Frederick, and V. Jacobson, "RTP: A Transport Protocol for Real-Time Applications," RFC1889, IETF, Jan 1996.

[RTP Generic]   A. Periyannan, D. Singer, M. Speer, "Delivery Media Generically over RTP," ftp://ftp.nordu.net/internet-drafts/draft-periyannan-generic-rtp-00.txt.

[RTSP]   http://www.real.com/library/fireprot/rtsp/ Also rfc2326.txt.

[Sanneck 96]   H. Sanneck, K. Stenger, B. Younes, B. Girod, "A new technique for audio packet loss concealment," *IEEE Global Internet*, 48–52, 1996.

[Sharifi 96]   H. Sharifi, "An investigation of Acoustic Enhancing Features for Use in Audio Tools," MSC Thesis Report, 1996.

[SIP]   M. Handley, H. Schulzerinne, E. Schooler, J. Rosenburg, "SIP: Session Invitation Protocol," http://www.cs.columbia.edu/~jdrosen/sip/drafts/draft-ietf-mmusic-sip-10.txt.

[Spanias 94]   A. Spanias, "Speech Coding: A tutorial review," *Proc. IEEE*, 82(10):1541–1582, 1994.

[Tannenbaum 96]   A. Tannenbaum *Computer Networks*, Prentice-Hall Int. Ed., 1996.

[Vary 88]   P. Vary, R. Hofmann, K. Hellwig, R.J. Sluyter, "A Regular-Pulse Excited Linear Predictive Codec," *Speech Communication* 7, 209–215, 1988.

[VIS]   http://www.sun.com./microelectronics/vis/.

[Waseem 88]   O.J. Wasem, D.J. Goodman, C.A. Dvorak, H.G. Page, "The effect of waveform substitution on the quality of PCM packet communications," *IEEE Trans. Acoust., Speech, and Signal Processing*, 36 (3), 342–8, 1988.

14 Broadcasting in the Internet Age

Emerging Business Models for Broadcasting

Richard V. Ducey

AS GREGOR SAMSA awoke one morning from uneasy dreams he found himself transformed in his bed into a gigantic insect . . .

Franz Kafka, *The Metamorphosis*

CONTENTS

0-8493-9594-1/00/$0.00+$.50
© 2000 by CRC Press LLC

14.1   INTRODUCTION

The radio and television broadcasting industries, like most other industries that have anything to do with communication and information, have embarked on a digital metamorphosis. Our dawning consciousness of digitally induced change forms a restless bridge between the "uneasy dreams" of society's wizards of technology and the determined entrepreneurs who turn what is possible into what is done. Broadcasters will be waking up in the next decade to find, like Gregor, that their corporate bodies have undergone massive and magical transformations. To paraphrase the car ad, "this is not the radio and television your parents grew up with."

What is changing and what new business models will result? It is a far simpler job to identify change drivers than it is to understand their ultimate impacts on industry and society. Clearly, a major change driver is the Internet and its relentless appetite for connectivity and interoperability among information and systems. Contemporary society places a preeminent value on intellectual capital as an intangible asset. This is in contrast, for example, to the Industrial Revolution when value was placed on physical assets. The whole notion of not just destroying old technology barriers that create gulfs between information systems but actually to *erect* new and empowering means of accessing, creating, and sharing all information in whatever form is incredibly revolutionary. This notion, at its core, is what the Internet is all about. And the Internet is the logical culmination of humans operating technology, for it embodies the most human of our behaviors – the desire to communicate. At this level, the Internet is as much a metaphor for human connectivity as it is any description of networking protocols.

The magnitude of the Internet's unfolding influence upon society may be framed by considering how much radio and television broadcasting has meant to the world. Beginning with the introduction of radio broadcasting in the twenties, the average person had the ability to receive a constant stream of electronic, real-time information, entertainment, and news – all professionally created, collected, edited, and presented – and all for no cost to the end consumer. Radio and television broadcasting keep people informed, involved, and entertained at a level of service and consumption no alternative has matched. Today, the average person in the U.S. watches almost three hours of television and listens to about three hours of radio *every day*. Certainly, radio and television consumption varies around the world, but the typical pattern is that the broadcast media form the bulk of the media diet.

Internet technologies provide extensions to what we can do with *data* and *information* (i.e., usefully structured data). The Internet is open and free to all. There are decreasing requirements for accessing Internet services. Service to the home is available at very low cost (for far less than basic cable) and the equipment needed to get to the Internet is getting less expensive all the time – about the price of television sets. Rather than dozens of radio, television, and cable options, the Internet opens up to every person on the planet literally tens of millions of destinations with hundreds of millions of web pages – regardless of a person's physical, spatial, or temporal circumstance. And the Internet is interactive and increasingly personalized via sophisticated navigational and user configurable profiles.

The magical thing about the Internet is that for some reason it has become the *lingua franca* of machine communications. Any device that can speak the TCP/IP (Telecommunications Control Protocol/Internet Protocol) Internet language can become part of the whole. That means personal computers, personal digital assistants, set-top box-equipped television sets — and in the near future, digital phones, pagers, and even household utilities — can all be accessed and controlled via Internet connections. The Internet is wired and wireless. It is becoming the ultimate in anytime, anything, anywhere, anyone communications.

Given this backdrop, conventional services such as radio and television must change – there simply is no choice. For broadcasters, this is at once a very frightening and very exciting moment. They are beginning to realize they compose their messages for presentation on speakers and screens, and these need not be confined to television and radio receivers. No longer must people gather themselves around television sets and radio receivers to receive these messages. No longer must they wait until the message is composed and delivered. Instead, as broadcasters leverage their business and information assets onto Internet-enabled platforms, they begin to view their assets as resources for their listeners and viewers to tap into, interact with, and shape their own interactive and personalized experience.

The Internet permits broadcasters to hitch the economic engine of their mass production efficiencies to the ability to individualize the navigation, production, and consumption of data and information. Broadcasters bring many inherent advantages to this business premise. Facing up to the challenges and opportunities will require a broad understanding of enabling technology, emerging market directions, and viable business models. For example, several major radio companies (Jacor, Westwood One, Capstar, and Chancellor) announced a deal valued at over $30 million with National Advertising to create an entity that "will link infomercials and e-commerce, operating across the Internet, television and radio."*

14.2 CONVERGENCE

Convergence is a wonderfully ambiguous and hopeful term of art. Like a palette of oil paints, convergence is a reservoir of conceptual opportunity. The moving parts of technology, the product and service ideas, and the consumer response must all come together for digital convergence to be successful. In practice, convergence may be best circumscribed by defining it to mean the erosion of barriers to entry driven by digital communications, or, in other words, digital technology drives new business opportunity. Those pursuing this opportunity must be responsive to existing consumer demand or successfully convert unmet needs into marketplace demand in order to win – hardly a trivial undertaking.

The computing, media, entertainment, and telecommunication industries have been rather distinguishable in the past. From the consumers' viewpoint this was easy to understand because they paid their bills to different companies and used different means for consuming the services. Today, with a device like Microsoft's WebTV Plus (www.webtv.com), a consumer can watch television, browse the web, get television

---

* "Radio Execs Sign Definitive Agreement with National Media," *Inside Radio*, August 13, 1998, p. 1.

listings and program descriptions, send and receive e-mail, read the newspaper (on a web site), and even make a telephone call (using IP telephony technology).

As the devices and service offerings become increasingly integrated, service providers and content owners will be driven to establish a compelling presence in each of these product spaces. In fact, newspapers have been some of the early adopters of interactive technology, and now the Internet, because they foresaw a strong challenge to their cash cow business of classified advertising. Now they can offer instantly updateable, personalized, and detailed multimedia classified ads via their Internet platforms. Of course, their abilities to provide news services, including streaming *audio* and *video*, is greatly facilitated by Internet technologies and rapid deployment in the consumer market. Broadcasters are certainly not ignoring this burgeoning marketplace.

14.2.1 BROADCASTERS ENTER THE CONVERGENCE MARKETPLACE

14.2.1.1 Three Revolutions of Television

Columbia University's Professor Eli Noam offers his vision of the television marketplace in terms of three revolutions.* The first revolution was the introduction of television itself. This era was characterized by the presence of relatively few broadcast television outlets in any given market. Noam calls this the *privileged television* stage. Only a few privileged operators were able to operate these services and competition in programming and advertising was constrained. The second revolution was the move toward *multichannel television* fostered largely by cable television and also by other services, such as microwave and direct-to-home satellite operations. An explosion in outlets, program choice and diversity, and new entry characterizes the second revolution. It also accustomed consumers to paying for what they once got for free – television.

The third revolution of television is the era we are just entering now. This is *cyber television.* Cyber television is driven by client-server and networking technologies such as video servers, switches, routers, high capacity telecommunications links, personal software agents, and integrated PC/TV types of client devices. These technologies transcend the *privileged* and *multichannel* television revolutions by mooting the whole point of counting channels. Rather than assessing marketplace progress by counting channels, the paradigm shifts to considering how much relevant data and information is connected to "me" – i.e., the *Me Channel.*

Consumers tend not to care whether they have 50, 100, or 500 channels. They care only that "what is on" is interesting to "me." That is the key value proposition, and this is the premise on which the Internet can deliver. Media companies that desire to be successful in the future need to be sure they are organized around this premise. This is the key insight Noam offers to those willing to listen.

The transition to digital television, now well underway with the first couple dozen stations coming online in Fall 1998, will infect traditional broadcasters with the convergence like never before. Current policy calls for the end of analog television in the year 2006. Once television broadcasters go digital, they will be deep into convergence and be driven to consider new business models offering incremental revenue streams.

---

* Noam, Eli M., "Towards the Third Revolution of Television," Symposium on Productive Regulation in the TV Market, Gütersloh, Germany, December 1, 1995.

14.2.2 Valuing Convergence

14.2.2.1 Convergence Icons versus Upstarts

For those in the media industry who are a little slow to read the changing landscape, the daily bible of traditional capitalism, *The Wall Street Journal*, ran a very interesting article.* The stock market exists as a form of legalized gaming where the participants are essentially betting their money on the future success of companies in which they purchase equity. While often criticized for being short-term rather than long-term in valuations which leads public company executives to manage for quarterly performance, Wall Street is at least an indicator of where things are going.

So where are things going in the media, entertainment, and communications businesses? Are you familiar with the two bookstores Barnes and Noble ($3 billion market capitalization) and Borders ($2.9 billion)? These are icons in the publishing business; they define much of the consumer experience in browsing and purchasing books. Yet a bookstore, Amazon.com, that exists *nowhere else but the Internet* has now achieved a market capitalization ($6.2 billion) that exceeds the market caps of Barnes and Noble and Borders *combined*. When Wall Street values a four-year old Internet-only bookstore company that turned in a $29 million *loss* (on $87 million of revenues) in its most recent year of operations, clearly the betting is that a change is in store for this market segment. Amazon.com went public on May 15, 1997 at $18/share and was trading in mid-August 1998 at over $121, yielding a *net loss* of $1.23/share. This company is definitely not trading on its current value.

Let's consider another American business icon in the media business. When we think of news and influence, surely *The New York Times* will rank high on most people's list. This $7.6 billion (market capitalization) company is a defining influence in American and international society. However, Yahoo!, an upstart company and the brainchild of two Stanford University graduate students, at $8.2 billion exceeds the market capitalization of the venerable *New York Times*.

Let's come a little closer to home for broadcasters. According to the *Wall Street Journal*, America Online, with a market cap of $26 billion, is worth more on Wall Street than ABC, CBS, and NBC *combined*. America Online has begun to reposition itself away from simply just another online service and more as an information and entertainment company. In fact, America Online likes to cite that its evening usage rate puts it on a par with the ratings of some cable networks!

14.2.2.2 Inflection Points

Mapping the media landscape leads one to look for the inflection points that Noam so thoughtfully considers. The seeds of *cyber television* are being sown as stations install video servers, LANs, and Internet connectivity and move into digital television. By *inflection point* I mean the ramp-up point in the classical innovation adoption S-curve that has been observed so often in the social science and marketing

---

* *Wall Street Journal*, July 14, 1998, page A18.

literatures.* It is the point when a product or service moves from the innovator/early adopter category to the mass market.

In consumer electronics, the first million units sold is a relevant benchmark for assessing a make-it-or-break-it product introduction. Others suggest that a ten percent market penetration rate is relevant.** In any case, the inflection, or ramp-up, point usually comes quite a bit after the product introduction as many more elements than the technology must come together to create success. Even the Internet was around for decades before it became an *overnight success.*

Clearly, this is arbitrary territory and the key is what type of consumers are adopting new products, and ultimately the penetration rates and impacts on society.

There is a clear demarcation between the privileged television and multichannel television eras in terms of the supply side (i.e., when the services were launched). However, the launch of new service offerings by cable, satellite, and microwave multichannel services did not lead to immediate acceptance by consumers. Indeed, marketplace impacts in terms of consumer penetration lagged these service introductions by a fair amount. Often the inflection point does not occur until the right mix of price, product, and perception come together in the consumer psychology.

For example, cable television was introduced in the fifties but did not become much of a factor until the seventies and the roll-out of satellite-delivered television and the roll-out of Home Box Office (HBO). Aided by changes in policy that facilitated urban penetration, cable take-up rose relatively quickly in the next decade to break the 50% penetration level to settle where it is now with about two of every three U.S. households subscribing. Satellite-delivered television and particularly pay-TV in the form of HBO *made a difference to consumers,* so they responded in the marketplace by subscribing to cable in ever greater numbers.

Home VCRs offer the benefit of *time-shifting* to consumers (fair-use taping, legally endorsed by the Supreme Court, of broadcasts or cable networks for subsequent private viewing).*** VCR penetration took off after the Supreme Court decision, fueled particularly by the video rental business. Hollywood unsuccessfully fought VCRs, but, despite that defeat, they ultimately won because the video rental and sell-through businesses now generate more revenues than the theater ticket office. The ability to record and especially to play movies inexpensively drove VCR penetration and use.

Another inflection point was probably IBM's entry into the personal computer market in the early eighties. IBM, in a sense, legitimized this product class with its status as a blue chip company and paved the way for business and home adoption of the technology. The Internet itself faced a major inflection point in the early nineties with the introduction of the HTTP technology that created the World-Wide Web,**** developed by the Swiss nuclear research facility, CERN. Of course, the University of Illinois development team led by Marc Andreessen, now of Netscape

---

* Sachs, William S. and George Benson, *Product Planning and Management*, Tulsa, OK: Pennwell Books, pp. 335-338.

** Sachs and Benson, p. 337.

*** Sony Corp. v. Universal City Studios, Inc., 464 U.S. 417 (1984), 464 U.S. 417.

fame, brought this to a new level with the Mosaic graphical browser. The Internet has been growing at phenomenal rates ever since.

Now we are at another watershed period. TVs and PCs are developing new capabilities to the point where the boundaries separating them are beginning to blur. Early attempts at integrating these devices have not gone particularly well, at least not in terms of consumer response. These mostly high-end devices are abandoned in favor of lower cost, standalone TVs and PCs. However, the U.S. introduction of digital television by a couple dozen stations in the top markets is fostering some potentially late change on this topic.

14.3   THE CHANGING BROADCAST MARKETPLACE

Both radio and television are going digital. They are proceeding on different courses and at different paces. In each case, a much higher quality traditional service will be offered, but it will require new equipment. Additionally, the transition to digital will support new business models for broadcasters.

14.3.1   RADIO

In the broadcast radio business, there are three vendors offering so-called in-band/on-channel (IBOC) solutions for existing broadcasters: Digital Radio Express, Lucent Technologies, and USA Digital Radio. This set of IBOC solutions allows current broadcasters to continue providing their analog service while converting unused portions of the spectrum allocated to the radio broadcast service for digital operations. All three companies are proposing market roll outs by late 1999, and at least two of them, Lucent and USADR, have proposed a full digital transition (i.e., moving from IBOC plus analog to a fully digitized signal and the elimination of analog service). The radio data broadcasting business will expand with the full use of the radio broadcast service's digital capacities. There is no fixed timeframe for radio's rollout of digital services in the U.S.

Even with today's analog radio, particularly with FM stations, a fairly healthy data business (using FM subcarriers, pagers, background music, reading for the vision impaired, etc.) has emerged. A version of the European *Radio Data System* (RDS) technology for FM stations is now deployed with some success in the U.S. Approximately 600 FM stations are using this technology for data, broadcasting relatively simple information for display on radio receivers. This information includes, for example, station call letters, telephone numbers, traffic and weather information, and commercials. The display in early receivers is fairly limited in its abilities, although European work is underway to transmit graphic files and HTML pages. In addition to the RDS initiatives, there has been some other work with developing FM high speed subcarrier services to further expand radio's data-

---

**** The hypertext transport protocol (HTTP) and the associated hypertext markup language (HTML) forming the basis of the World Wide Web was proposed by Tim Berners-Lee in 1989 at the European Center for Nuclear Research (CERN) in Geneva. Originally, it was meant as a means for nuclear scientists to conveniently exchange information, even if they were using different computer environments and were located around the globe.

carrying capacities. It appears that further aggressive development is not in the cards at the moment.

On the other hand the television industry, while proving to be somewhat a moving target, has a firm timetable for conversion to digital operations. The remainder of this chapter regards broadcast television case as moving to digital more quickly than is radio. However, radio shares some of the same capabilities and market competitiveness of digital television broadcasting, so much of what is said for television extends to radio.

14.3.2 Television

14.3.2.1 Traditional Business Lines

Broadcast television is a *public good* in that it is free (nonexclusionary) and there is no incremental cost to the broadcaster for adding incremental viewers. Broadcasters produce and air programs for the purpose of collecting audiences at specific points in time (for specific broadcast programs). Audience estimates provided by the Nielsen company are used, along with other market value indicators, to produce advertising-generated revenues paid to the broadcasters who collect the audiences for their broadcast programs. Broadcasters provide different types of news, information, and entertainment programming with local, regional, and national appeal. That is the traditional broadcast business in a nutshell. It is a good business and generates about $40 billion per year in advertising revenues, which is essentially the only revenue source for commercial broadcasters.

14.3.2.2 Nontraditional Business Lines

Both commercial and public television broadcasters have been exploring nontraditional business lines. Using the natural advantages of wireless data communications capacity, and ubiquitous distribution in local, regional, and national markets with the added benefit of being able to cover operating costs with traditional business lines, television broadcasters could quite competitively enter the data broadcasting, bandwidth, or Internet market spaces.

14.3.2.3 Data Broadcasting

Convergence between the television and Internet worlds is still being built in terms of the underlying standards. While the digital television standard has been set in the U.S., the data broadcasting portion of the standard is still in a degree of flux. This is true even for the analog television data broadcasting service.

The Advanced Television Systems Committee (www.atsc.org) has efforts underway to standardize data broadcasting in the analog and digital television domains. There are several approaches to such a standard and no final decisions have been made. Several groups are working on standards solutions: the Internet Engineering Task Force (www.ietf.org) working on IP over VBI, the ATSC T3/S17 Data Broadcast Protocols subcommittee (http://toocan.philabs.research.philips.com/misc/atsc/t3s13/), and the ATSC Digital TV Application Software Environments (DASE) subcommittee (http://toocan.philabs.research.philips.com/ misc/atsc/dase). There now is another relatively new group, the Advanced TV Enhancement Forum (www.atvef.com). This work is at least loosely coordinated and will likely result in some very useful solutions to advance the television data broadcast business.

There are a number of prominent examples of television broadcasters entering the high speed mobile data broadcast market. In each case, the otherwise unused vertical blanking interval capacity of the television signal is used for data broadcasting. The vertical blanking interval (i.e., the equivalent of 10 horizontal lines out of 525 lines transmitted in NTSC television) yields over150 kbps of data capacity.*

As a point of reference, this unidirectional data capacity exceeds that of ISDN basic rate interface (128 kbps) which is the telephone industry's answer to providing high speed data bandwidths to the consumer and business marketplace. Therefore, this is not trivial data capacity, but an economic opportunity for television broadcasters to exploit. It is also an incredibly cost-effective means for providing data access to millions of homes, schools, offices, cars, and other places people find themselves – even walking down the street. Television signals are among the country's most ubiquitous data service available.

Commercial ventures are testing the market for television data broadcasting. For example, the Public Broadcast System (www.pbs.org) has teamed with WavePhore (www.wavephore.com) to create the for-profit subsidiary, PBS National Datacast. Other examples include the Intercast consortium (www.intercast.org), Datacast, LLC, WinkTV (www.wink.com), and Norpak, the Canadian firm responsible for much of the data broadcasting technology installed in stations. There are also some international efforts, pioneered by companies such as IO Research (www.iores.com.au) and the Japanese work by NHK (www.strl.nhk.or.jp) on *Integrated Services Digital Broadcasting* (ISDB).

14.3.2.4 Program-Related and Nonprogram-Related Data

Data broadcasting services fall into two broad categories – program-related and nonprogram-related. The data broadcasting services that are related to on-air programming are typically broadcast in real-time – i.e., enhanced content relevant to the program is broadcast in the VBI simultaneously with the program itself. The display device for the data broadcast service would typically be the television receiver connected to a set-top box of some sort and destined for a target group of *viewers/users* (or as Gary Arlen, of Arlen Communications prefers, *viewers*). That would put the traditional television service and the enhanced broadcast service in the same user environment, on the same platform – a consumer plus!

For program-related data broadcasting there are many possibilities. So-called *enhanced broadcasting* is a way to add value to the traditional television viewing experience. One option is to use crossover links that send the URL (Uniform Resource Locator) of a web site via the VBI to a properly equipped device (TV, PC,

---

* There are 21 vertical blanking interval lines. Lines 10-20 are used for data. Line 21 is reserved for closed-captioning and TV program rating codes. Each VBI line has a data rate of about 15,000 bits per second, or 150,000 bits per second (150 kbps) for all ten lines.

etc.). This device then connects a viewer to the web site.* To signal the viewer that such crosslinks are available, some kind of icon or onscreen clue would be provided. Broadcasters and advertisers can broadcast their crossover links to enhance the experience of watching a program or an advertisement.

Producers of programs such as *Baywatch*, and companies such as Disney and N2K are experimenting with the use of crossover links in the production process. Advertisers can also use this technology for viewers to crossover between an on-air commercial directly to a web site for additional information – or to make a purchase. During sports broadcasts, for example, an "I" icon might appear and offer the viewer an opportunity to travel to a web site where in-depth statistical information is available. Or a viewer of a car ad could link to a dealer's web site, find what is in inventory, see sticker prices, and even make an appointment with a sales representative.

Nonprogram-related data broadcasting refers to data not necessarily dependent (temporally or contextually) on main channel television program content. Typically, the PC is used as the display device for these services. Microsoft has been working with a number of broadcast and cable groups, including companies such as Cox Broadcasting, Capitol Broadcasting, Sinclair Broadcast Group, Media General Broadcast Group, and some public stations for proof-of-technology and market tests. PBS has again partnered with WavePhore, Inc. to rollout the new service WaveTop (www.wavetop.com), a value-added, content aggregation service delivering branded content, advertising, and software updates. It requires a PC card that is available for approximately $100, and the PC software capability is bundled into Windows98. This service is initially available in the top 100 television markets.

14.3.2.5 Bandwidth

The raw data-carrying capacity of analog television is at least 150 kbps, in addition to traditional television service, yet few television stations in the U.S. make commercial use of this capacity. In digital television, the data-carrying capacity leaps to a minimum of 1 Mbps and can extend to nearly 20 Mbps. In a marketplace where data communications bandwidth is an increasingly valued service, particularly for mobile applications, this carrying capacity gives broadcasters a serious competitive advantage. Even if broadcasters want nothing to do with data broadcasting, they could at least rent this capacity to someone who does!

Broadcasters have killer advantages in the bandwidth marketplace. They have little incremental expense to incur in developing a *bandwidth for hire* business line and 100% market penetration, with 24 × 7 availability (often with standby redundancy), T-1 or better data rates, and mobile access. Moreover, the network is already built and operating now – there is no deployment wait time. This situation makes data broadcasting very attractive to anyone considering ISDN, T-1, cable, microwave, or a satellite platform alternative.

---

* The Consumer Electronics Manufacturers Association (CEMA) along with several major companies has proposed a standard for crossover links, EIA 746. This standard uses VBI line 21 for transporting Uniform Resource Locators (URLs).

Broadcasting is, of course, a unidirectional service. This works great for things such as software updates, catalog downloads, database updates, and e-mail distribution. However, if a return path is required, typically it is asymmetrical in its requirements. In other words, the downstream data path needs to be broadband to include large files such as graphics and streaming media. However, the return or upstream path is usually fairly limited in bandwidth requirements.

It takes just a few bits for a user to *click* on a URL, but that click might trigger a tsunami of streaming and multimedia data back downstream. Marrying television data bandwidth capacity to return links like wireless telephony (cellular phones, PCS, satellite phones), unregulated radio frequency applications (such as Metricom's Richochet service, www.ricochet.net), or wired alternatives (like the public, switched telephone network) can easily and affordably extend the commercial value and business utility of this service.

The bandwidth marketplace is headed toward integrated services and commodity pricing. Indeed, Sprint's (www.sprint.com) recent announcement of its Integrated On-Demand Network (ION) foretells the day when users simply order whatever bandwidth they need for a particular application when they need it, rather than maintaining high overhead, fixed-cost data pipes. The Sprint ION solution integrates various bandwidth services now, whether frame relay, IP, voice, data, etc., into a single platform. A market extension of this could be to marry wireless platforms such as TV VBI or DTV bandwidth to enable huge mobile downstream data pipes at a highly attractive price point.

14.3.2.6 Internet

What is the Internet marketplace worth? The numbers are incredible. Companies such as MindSpring (www.mindspring.com) have increased in value over 1000% in a year's time. Microsoft and Compaq each invested over $212 million for 10% stakes in the RoadRunner cable modem service, which values that business at over $2 billion. With about only 100,000 subscribers, this valuation is *well over $20,000 per-subscriber basis.* For broadcasters considering entry into the Internet marketplace, these numbers should be very encouraging.

What role can broadcasters play in the Internet services market? IBM's William Beckmann joins others in cautioning about the Internet infrastructure, particularly overburdening the backbone. He argues for staged infrastructure deployment to facilitate risk avoidance. In his view, the key strategic market is *bandwidth to the end user.* In this regard, he argues that *broadcasters* have the lead for at least 18 to 24 months. At this point, telephone deployment of digital services networks will start to reach critical mass.*

The threat to the Internet backbone is already evident and will get worse with the rollout of more streaming media relying on unicasting protocols. There is work underway by the IP Multicast Initiative (www.ipmulticast.com) to get the industry off its reliance on unicast streaming applications, but this will likely take time.

---

* Beckmann, William H., "How to Offer New Interactive Broadcast Services," Vice President, IBM Video Enabled Solutions, IBM Corporation, USA, *BroadcastAsia98 International Conference,* Singapore, June 1–4, 1998.

Beckmann and others argue that local backbone caching to avoid international or national backbone traffic is a wonderfully appealing solution. The move toward IP multicasting enhances this practice all the more.

When one goes looking for a *local* ISP capable of caching and downloading large data files, streaming media sessions, etc. to fixed and mobile users, the synergy television broadcasting offers is very compelling. The broadcast infrastructure, through standards development work to enable both analog and digital television to push IP packets, can easily become part of the Internet local caching solution.

14.4 CONCLUSION

Like Gregor, broadcasters are waking up to the fact that they have more legs to stand on and will start to look quite a bit different in the next age of broadcasting – the age of Internet broadcasting. As commercial analog television passes its 50th birthday, the path to the next decade of digital television is emblazoned with new opportunities for revenues, services, and advancing the convergence paradigm. The way things are adding up, broadcasters and their *viewsers* have a compelling and enriching next decade. Let the millenium roll in!

15 High Definition Television (HDTV)

Douglas A. Ferguson

CONTENTS

15.1 INTRODUCTION

For many years, the television industry has awaited the arrival of high-definition television (HDTV). An updated format is long overdue: regular-definition television picture has been in service in the U.S. since the late 1930s, based on technology from pre-World War II era.

The original screen in the U.S. (with 525 lines of vertical resolution) was designed to mimic the screen format (aspect ratio) of the motion pictures shown in movie theaters. But television became so successful that Hollywood widened its screen format in the fifties and later developed surround-sound and digital effect technologies.

Conventional television set makers and broadcasters were able to copy the stereo sound but yearned to duplicate the high-quality projected image. The

Japanese succeeded in perfecting HDTV, but the amount of video information filled the space (spectrum) of six television channels, making the system impractical in the U.S. where spectrum is scarce. HDTV seemed like an impractical dream only a few years ago.

Video compression of a digitized HDTV signal in the nineties was the key development that made feasible high-definition video transmission. But, unlike color television, the new system was incompatible with existing TV sets. Congress and the Federal Communications Commission (FCC) devised a plan to allow broadcast stations and their viewers to make a slow transition from standard-definition video to HDTV.

The benefits of HDTV are clear. First, it fosters interference-free transmission. Second, it has the capacity to send additional data transmissions during transitory still images. Of interest to advertisers, the HDTV viewing audience can attain greater involvement because the shorter relative viewing distance makes a substantial difference in the qualitative feel of watching TV. Instead of the stereotypic passive experience, HDTV is like looking at reality out the window.

15.1.1 Some Definitions

HDTV is not the same as DTV (digital TV). The high-definition was initially achieved by an analog signal, but now is standardized as a digital format. DTV, on the other hand, is always digital but not necessarily high-definition. Initially HDTV was IDTV (the I was for *improved*), but the 1000-line threshold of vertical resolution, as high as 1280 lines, produced twice the resolution.

Even today, many so-called HDTV monitors are not true HDTV because they operate around 800 lines. HDTV standards put the screen width at 16:9 ratio to emulate the wider motion picture screens adopted by Hollywood in the fifties. Compressed digital HDTV requires the full bandwidth of a television channel (6 MHz) to deliver a maximum pixel (picture element) count of 2,073,600.

Compressed DTV (MPEG-2) is also the format used on DVD (digital versatile disk), digitized video files (MPEG) on computers and web pages, and tape storage formats such as D1, D2, etc. DTV is typically a compressed format, also allowing satellite distribution of high-quality images to small receiving dishes.

DTV can also refer to the video receiver itself. There is little to differentiate a digital TV from a computer. An antenna substitutes for the modem, and the hard drive is unnecessary, but otherwise the microprocessor, memory chips, and binary data function as in a computer. Just as with a computer, the bit stream flows in megabits-per-second (Mbps), usually at very high speeds, depending on circuitry and operating systems. Microsoft and Thomson are developing an operating system for DTV receivers that will likely surpass another system being created by Sony, if Bill Gates gets his way. The distinction between using a home computer and watching TV will likely continue to blur.

15.1.2 SDTV versus HDTV

SDTV (standard definition TV) is roughly the digital equivalent of old-style (analog) video screens in the 4:3 aspect ratio. Because of its lower resolution (480 lines) it occupies much less bandwidth than HDTV, one-fourth to one-sixth, allowing the delivery of four to six separate SDTV signals on a single 6-MHz bandwidth. SDTV channels for DVD movies and DSS satellite channels typically operate at 2.5 Mbps, depending on the amount of rapid movement in the picture.

HDTV is the maximum resolution and the highest quality of video and sound. The huge bit rate (at least 108 Mbps) prevents HDTV transmission unless signals are squeezed into a smaller space. In the nineties, advances in digital compression allowed the huge bit stream necessary for HDTV to fit into the 19 Mbps range possible within the confines of a 6-MHz channel.

15.1.3 WHY BOTHER?

Although there are huge conversion costs involved, HDTV and its underlying DTV technology offer more channel capacity, significantly better picture quality, larger screens, more programming, interactivity, new digital services (e.g., high-speed Internet access, pay-per-view, one-to-one marketing). In his book *Defining Vision*, Joel Brinkley speculates that viewers will eventually have the ability to turn sex or violence up and down in a manner similar to volume control.

The primary goal in switching to HDTV is enhanced video quality. Just as digital formats have enhanced audio, digital formats such as D1 (component) and D2 (composite) have been used in video studios for years because there is no loss in quality from original to successive copies (generations) and because special effects/editing are possible.

The benefits of interactivity and offering several SDTV signals over one channel are less clear. Wireless cable operators (also called MMDS) failed to get multichannel service off the ground. Broadcasters may have the same trouble emulating the kind of distribution offered by cable and direct satellite.

Benefits of HDTV are offset by some mighty drawbacks, primarily the cost of conversion but also video artifacts (picture glitches), which are also associated with digital satellite delivery. Additionally, many critics have expressed doubt that the public will want to buy the more expensive digital receivers.

The plan, however, is to force the new standard, just as the recording industry did when it discontinued vinyl records. Every non-DTV analog receiver is expected to be effectively obsolete by 2007, requiring some kind of converter box to down-convert the new digital standard to the old analog signal. This plan means the new HDTV standard will take effect less than a decade after its introduction.

It is a monumental undertaking to switch standards in less than 10 years. The situation is analogous to the introduction of color, according to research by Mark Schubin. Color TV in 1954 cost $1295, while a 1954 Ford cost $1695. The high cost of HDTV is not unprecedented.

The projected length of time for HDTV to be widely adopted, however, is remarkably short. It took 15 years for color television to catch on, and another 10 years for it to dominate black-and-white (B&W) receivers. However, color was compatible with B&W; DTV is not compatible with NTSC. Another favorable aspect for color TV was its backing by RCA, which made the sets, and its programming subsidiary, NBC, which made the shows. It is not likely (with the possible exception of Sony) that HDTV set-makers will influence the spread of HDTV programming.

15.2 HISTORY

To understand where we are it might be useful to look at how we got here. Television in the U.S. began with the 525-line system delivered on 6-MHz channels, first available shortly before World War II but delayed until after the war. The format came to be known as NTSC (National Television Standards Committee) as it evolved a color standard in 1952. NTSC requires that RGB (red/green/blue) signals from the color camera, in order to be compatible with black-and-white receivers, be transformed, by linear combination, into three equivalent component signals: luminance (Y), where Y=0.587 G + 0.299 R + 0.114 B); blue chrominance (C_b), where C_b= 0.564 (B-Y); and red chrominance (C_r), where C_r= 0.713 (R-Y).

DTV kept the component signals (Y, B-Y, R-Y) but used digital sampling to reduce the information on an analog wave to discrete (binary) numbers. For example, the 4:2:2 component standard takes 4 digital samples of the Y signal and 2 samples of both the C_b and C_r signals. Videotape systems such as D1 and D2 offered the first DTV to video professionals who needed the quality or editing capabilities. DTV transmission began mid-1994 with DirecTV satellite delivery.

But HDTV was originally an analog system like NTSC, not because digital TV was impossible, but because the compression necessary to make it practical was (or seemed to be) impossible. In the eighties, the Japanese (MUSE standard) and then the Europeans developed analog HDTV standards. IDTV (Improved Definition) had 750 lines and HDTV ranged between 1125 and 1250 lines.

Being first, however, as Sony had discovered with its Betamax VCRs, is not always best. The analog standard that looked a clear winner in 1987 was not so 10 years later. Indeed, the MUSE-only TV sets in Japan will become obsolete in 2003.

The following timeline, inspired by Sara Brown in 1998, sets forth the two-decade progress of HDTV.

February 1981	first American demonstration of analog HDTV
February 1982	first demonstration to FCC
March 1985	ATSC adopts analog HDTV standard (1125 lines, 16:9)
October 1985	CCIR adopts same as international standard
Spring 1986	CBC shoots 13-hour miniseries in HDTV
January 1987	HDTV is first broadcast over standard TV channels
January 1988	ATSC unanimously adopts analog standard
September 1988	FCC adopts compatibility standard for HDTV
January 1989	Defense Department issues $30M RFP
July 1989	$50M pledged to fund Defense research for HDTV
June 1990	General Instruments develops digital HDTV
March 1992	GI and MIT demonstrate digital HDTV
September 1992	FCC gives broadcasters six years to begin HDTV
February 1993	NHK's analog standard withdraws from competition

May 1993	Grand Alliance formed by GI, Zenith, AT&T, and ATRC
October 1993	Grand Alliance adopts MPEG-2 compression
February 1994	Zenith wins the race to design the transmission standard
April 1995	First broadcast of digital TV
July 1996	WRAL-HD Raleigh begins HDTV transmission on Ch. 32
August 1997	ABC and Sinclair plan to multicast DTV channels
January 1998	First HDTV receivers shown at CES; KTLA covers Rose Bowl parade in HDTV
February 1998	FCC releases final allotment table
November 1998	Digital broadcasting begins
November 1998	FCC sets 5% fee for multicast pay-per-view
January 2007	End of NTSC broadcasts

15.3 THE HDTV TECHNOLOGY

The electronic HDTV camera transduces the real image with a lens and pick-up devices, generating an analog HDTV signal (video and audio). Bits are the ones and zeros that digitize the analog video and audio signals.

As described earlier, the component signals of the video source are sampled and reduced to a bitstream. Subsequent devices (e.g., editors, special-effects processors, recorders, and display screens) process the data or translate the bitstream into a viewable wide-screen image.

The Serial Digital Interface (SDI, also known as SMPTE 259M) is the standard used for the transmission of uncompressed digital video (with embedded audio) through a video facility, assuming a device separation no greater than 300 meters. Originally an 8-bit signal, the CCIR-601 specification on which SDI is based had been upgraded to a 10-bit component or composite format. There are two incompatible variations: Sony's SDDI (Serial Digital Data Interface) and Panasonic's CSDI (Compressed Serial Digital Interface). A possible open-standard resolution to incompatibility, called SDTI (Serial Digital Transport Interface), is being studied at this writing. Such a standard would greatly reduce the costs of building a digital video facility.

15.3.1 TRANSMISSION

A band of frequencies (measured in cycles per second, called hertz) is designated for each television channel assigned by the FCC. Each channel is the same width but its exact location is different for each channel number. Until the projected eventual switchover at the beginning of 2007, each television broadcaster has two channels: the original analog channel plus an additional digital channel. Digital TV broadcasts have a 6-MHz pipeline (bandwidth) through which only 19.4 million bits per second (Mbps) can be transmitted.

True HDTV signals have around 200 Mbps, which is beyond the capability of the 6-MHz pipeline. MPEG-2 data compression allows redundant information (from frame to frame) to be eliminated from the bitstream. Compression, also known as Bit Rate Reduction (BRR), was originally intended for digital transmission, but also applies to digital video in computers and digital video recorders. Depending on the desired picture quality, compression can be at either a set rate or a variable rate.

Two kinds of video compression schemes (algorithms) have been developed: lossless, which reduces the size of digital data files by eliminating redundant information, and lossy, which eliminates noncritical data. In lossless schemes, pictures are analyzed for redundancy and repetition and unnecessary data is discarded. Lossy compression systems digitize the picture quality according different criteria: what the human eye will notice (perceptual coding), which picture elements remain the same from frame to frame (predictive coding), and what movement can be inferred from inertia (motion compensation).

The device used by local stations to compress their digital signal into the 19.3 Mbps bit stream is called an encoder. Encoding for HDTV was first available via a $500,000 device from Mitsubishi. SDTV encoding was first available with Divicom's $65,000 MPEG-2 encoder.

15.3.2 RECORDING

SDTV signals are recorded on digital tape recorders. D1 recorders use a component digital video recording format that uses data conforming to the ITU-R BT.601-2 (CCIR-601) standard. D2 recorders use a composite digital video recording format that uses data conforming to SMPTE 244M. Both D1 and D2 record onto 19 mm magnetic tape. Other corresponding formats (D5 and D3) record onto half-inch tape. There are two other component formats for straight DTV: D9 (Digital-S) uses half-inch tape at 50 Mbps and D7 (DVCPRO) uses quarter-inch tape at 25 Mbps.

The digital HDTV format is D6, which uses D1 tape to record HDTV at 1200 Mbps, or 1.2 gigabits per second (Gbps). There is no D4 or D8 format, nor will there be.

The digital-video interface standard for HDTV is IEEE 1394 (FireWire). The standard (not specifically intended for video, but widely adopted for that purpose) handles digital information in both small and large packets. FireWire supports up to 63 devices on a single bus, with bridged buses allowing thousands of devices. It can handle at least 20 Mbps of continuous data, with speeds of 100 Mbps expected in the new millennium (allowing the transfer of fully uncompressed video). Even at slower speeds modest compression can yield outstanding quality.

The clear advantage to recording digital video is that multiple generations (layers) do not result in any loss of quality, allowing much more complex effects. Eventually consumer-grade video recorders (priced near $1,000) will permit up to 49 hours per cassette at extended-play mode, or 7 hours for standard mode, once equipment manufacturers and copyright holders hammer out a workable protection scheme.

15.4　HDTV AND SDTV FORMATS

Silent film was shot and projected at rates of 18 frames per second (fps) to take advantage of the human eye's persistence of vision; sometimes such slow rates produced a flicker which gave early films the nickname *flicks*. The coming of sound in the late twenties required faster speeds for adequate audio fidelity, so the film standard was raised to 24 fps, where it remains today.

As mentioned earlier, television in the U.S. is a 525-line system with 60 interlaced fields per second generating 30 fps (interlace scanning). Why 30 fps, and not 24 fps? Because the electric current in the U.S. is 60 cycles per second (60 Hz), so synchronization is easier with 60 cycles than with 24. Alternating electric current in Europe is 50 Hz, which explains why 25 fps video is in use on that side of the ocean.

The interlaced fields were an elegant solution to early technological shortcomings. The NTSC system scans the odd-numbered lines in a picture first, then scans the even-numbered lines to create two fields that comprise a single frame every 30th of a second. Because of the vertical blanking interval between frames, NTSC has only 480 visible lines of vertical resolution (330 lines horizontal).

By the time computer monitors were designed, television signals were capable of being created from lines scanned in sequential order (progressive scanning). Different types of video look better (or worse) in progressive scanning. For example, interlaced looks better with large-screen projection, but progressive is better for flat-TV screens. As for content, programs shot on film look better on progressive, but true video (especially live-action sports) looks better on interlaced.

With the advent of digital television, the old NTSC standard will be replaced by ATSC, which has 18 formats. Only four of these formats, perhaps five, will be used in the U.S.:

SDTV	480P	4:3	640x480	(307,200 pixels)	148Mbps	@30fps	b.c.
SDTV	480P	16:9	704x480	(337,920 pixels)	162Mbps	@30fps	b.c.
HDTV	720P	16:9	1280x720	(921,600 pixels)	884Mbps	@60fps	b.c.
HDTV	1080I	16:9	1920x1080	(2,073,600 pix)	995Mbps	@30fps	b.c.

The fifth format would be 1080P with the same resolution.

The true pixel difference between 720P and 1080I is deceptive because a true comparison counts pixels per 60th of a second. 1080I/30 displays 1,036,800 pixels versus the 720P/60 which displays 921,600 pixels. Those who want the 1080I standard argue that true HDTV is at least one million pixels.

By 1998, CBS and NBC had adopted 1080I and ABC had chosen 720P. As of this writing, Fox is predicted also to go with 720P, although it would prefer 480P for multicasting. PBS will likely adopt a new version of 1080 to complement its datacasting plans.

DirecTV and DSS satellite transmissions were expected to begin in 1999, with variants of the 1080I format delivered to double-sized dish receivers. There is no shortage of program content, because Hollywood movies from the last 40 years are already high-resolution wide-screen format. Unfortunately, most programs filmed (or taped) for TV were shot with the narrow format.

The new standards specify digital surround-sound (Dolby 5.1). HDTV receivers in the home, especially with very large screens, will thus rival the projection of movies in theaters with regard to video and audio quality (leading one to wonder whether theater owners will change their in-theater experience yet again to compete with the in-home experience).

15.4.1 TRANSMISSION

Once the signal is encoded and ready for broadcast, local stations need an 8-VSB (eight-level digital vestigial-sideband scheme) transmitter. In many cases, broadcasters seek to improve existing NTSC programming, by using upconverters or line-doublers, available in 1998 from Snell & Wilcox or Faroudja Labs.

The STL (studio transmitter link) is how the signal gets from the station to the transmitting antenna. One method uses 7-GHz digital microwave. Fiber-optic cable is another means, using the DS-3 standard (45 Mbps).

PSIP (Program and System Information Protocol) provides analog and digital channel information merely by entering the original channel assignment (e.g., Ch. 2 analog might be Ch. 51 digital, but PSIP shows the analog signal as 2-0, while 2-1 and 2-2 are the same station's digital channels for HDTV and SDTV); another scheme lets analog Ch. 2 keep that designation with the high-definition digital version a Ch. 2D and SDTV channels as 2D.1, 2D.2, etc.

Transmitted data is not limited to video and audio. Asynchronous Transfer Mode (ATM) is a very high bandwidth transmission, switching, and multiplexing technology that can move very large streams of data very quickly. ATM utilizes cells to move data from one point to another. It does not care which bits are which: audio, video, graphics, data, or voice. Everything is sent down a single pipeline.

ATM takes different types of data and converts them into fixed length cells that are 53 bytes long. Within the cell, five bytes contain control information and the other 48 bytes contain data. The cells are moved through very large data pipes at a high rate of speed and are then reassembled into their original form at the receiving end of the connection. Since the information is sent in small packages, ATM offers a highly reliable way of transmitting data from one point to another. If a cell does not get through, it can easily be resent (see Silbergleid).

On November 1, 1998, 42 stations began broadcasting digital television signals. DirecTV and DSS were scheduled to begin in 1999 and cable television delivery was slated for 2002 when converter box issues were likely to be resolved.

15.4.2 DISPLAY

Receivers made by CEMA (Consumer Electronics Manufacturers Association) have been designed to receive any of the 18 formats in the A/53 DTV standard but in 1998 only displayed some of the standards by converting them to 1080I, 960I, or 480P. For most first-generation sets, 1080I was the native format.

Even old TV displays (NTSC standard) were expected to look better with the new digital signal converted by a digital-TV tuner/decoder because signal defects are automatically corrected. At best, most conventional TV sets (as of this writing) offered S-video inputs which yielded DVD (digital video disk) quality.

Smaller-screen displays using a CRT (cathode ray tube) require masks or grilles with tiny holes that permit the phosphor dots to glow when struck by the electron beam; higher-resolution means higher cost of manufacturing very tiny perforations in the mask. Sony's small-screen set cost as much or more than rear-projection sets with much larger screens.

A possible advantage of DTV is the cliff effect of antenna reception: viewers see a perfect picture, regardless of distance from the source, or nothing at all. Indoor antennas are much more susceptible to multipath interference. One unanswered question is whether consumers will put up with adjusting an indoor antenna, if each adjustment requires that they have to back away from the antenna.

Certainly, directionality and antenna height are key variables in considering multipath interference. Rotating rooftop mounts and/or co-location of DTV transmitting antennas in a TV market were expected to hold the answer. In 1999, it remained to be seen if devices in the receivers that compensate for multipath interference, adaptive equalizers, would overcome the kind of problems uncovered in early test transmissions before the official U.S. launch of digital HDTV in November 1998.

15.5 MULTICASTING

Packet-switched statistical multiplexing (stat-mux) allows unused bits to be assigned to other tasks, such as computer data. Broadcasters can transmit up to six SDTV channels (fewer with high-action sports), as long as one channel is free. Pay TV is more difficult to establish because, as of this writing, no standard for conditional access has been decided.

Different types of programming require varying bit stream rates. The contrasting speeds needed to transmit video data are shown below:

Entertainment and news	3.8 Mbps
Sports	8.0 Mbps
Data	1.2 Mbps

15.6 IMPLICATIONS AND ISSUES

Standardization has been a very bumpy road for HDTV since at least the analog designs of the 1980s. Adoption of open-architecture technology, with which digital video equipment manufacturers can create compatible systems, was seen in 1998 as critical to the future of HDTV.

An initial lack of encryption standards also slowed the development of consumer-level digital video recorders. Motion picture and television program copyright holders feared piracy of their products, not unlike the problem faced by computer software distributors and the recorded music industry. Digital reproduction makes a perfect copy.

At the station level, choosing an equipment standard has presented a critical challenge. Chief engineers at broadcast facilities had to decide among options that included 4:2:2 digital component, 480P, 720P, and 1080I.

Nevertheless, the deployment of HDTV in the early years of the new century is expected to present manifold influences on media economies and media audiences. Beyond the crystal-clear images and sound and the general lack of signal interference, all sorts of digital information (web pages and e-mail) and data could coexist with video and audio signals.

Broadcast networks have not fared well in the multichannel environment. Advertising has provided an important but solitary revenue stream to over-the-air commercial television. Old-line companies in the late nineties embraced the opportunity to attract new audiences and revenue streams but feared the threat of new costs with uncertain rewards. Broadcast networks own and operate many of their largest affiliated stations. Costs to those stations included a new transmitter, more power consumption, and a DTV encoder. Because not all antenna towers can withstand the weight of an additional sidemounted DTV antenna, some stations needed a new tower and transmission line, assuming that FAA regulations and local objections could be surmounted. On average, the cost was $4–5 million per station to upgrade completely to HDTV.

Multicasting SDTV channels added to the expense of digital conversion. Broadcasters realized that they might need to subsidize the cost of converter boxes, similar to satellite multichannel television providers such as DirecTV and Echostar (DISH).

15.7   FCC POLICY

In addition to technical and economic concerns, important policy considerations have played an influential role in the transition to digital television. The old analog channels in the U.S. are slated to go off the air at the end of 2006 (provided that at least 85 percent of the viewers in a city own digital television receivers). The old NTSC 6-MHz frequencies, a huge portion of usable spectrum space, is scheduled to be auctioned to other wireless services (e.g., land mobile telephones). Auction proceeds for the U.S. government were expected to reach $6 billion. Budget-balancing plans and revenue projections in the mid-nineties were predicated on this revenue for the Federal Treasury.

The transition to HDTV receivers assumed that television audiences would want to replace their receivers to get better quality pictures and sound, which in turn presumed sufficient HDTV programming. But many broadcasters wanted to offer limited HDTV service until the critical mass of receivers were in place. By multicasting SDTV channels, they hoped to recoup their huge investment in digital conversion. By 1997, however, it became clear that Congress wanted HDTV, not SDTV. House Telecommunications Subcommittee Chairman Billy Tauzin (R-La.) and Senate Commerce Committee Chairman John McCain (R-Ariz.) pressed hard to hold broadcasters to their promise of HDTV instead of SDTV multicasting. This chicken-or-the-egg dilemma was an important policy matter at the FCC in the time leading up to digital deployment.

On another front, the White House also wanted free political time in exchange for additional spectrum set aside for digital broadcasting. Vice President Al Gore led the effort to extract additional public-interest commitments from digital broadcasters.

The FCC announced plans in 1998 requiring broadcasters to pay the government five percent of any pay-subscription revenue via DTV channels. Despite objections from public interest groups, stations would not have to share revenue from home shopping, infomercials, and other direct marketing enhanced by additional SDTV capability.

15.8 CABLE

In the nineties, multiple system operators (MSOs) for cable television spent huge sums to upgrade their set-top converter boxes, hoping to depreciate the cost over a period of years. Waiting for HDTV to decide on standards was not an option for many of them, and, once committed, these cable operators were reluctant to reinvest in new converters that cost hundreds of dollars apiece.

More important was the issue of shelf space. Cable operators had no vacant channels on which to locate the new digital channels for local signals. The FCC *must-carry* rules required cable operators to provide local channels on their systems, but cable systems announced that they were unable to find space for *two* channels for each local signal.

Two-thirds of all television homes received video reception via cable service in 1998. The alternative was rooftop antenna, which was a major inconvenience for some and an impossibility for others. Cable providers at the turn of this century find themselves in the unenviable position of not being able to provide, for technical and economic reasons, the HDTV signals readily available over-the-air and via direct satellite.

At this writing, it is unclear how this issue will be resolved. The FCC will, of course, play a major but reluctant role. Members of Congress are influenced by competing industries: broadcast (through the NAB) and cable (through the NCTA).

15.9 COSTS TO CONSUMERS

Buying a new, expensive television receiver is an important issue to consumers. The consumer electronics industry welcomed the opportunity to sell new receivers to 100 million television homes. But the decision was not simple for the average person. Some observers predict that consumers will not spend huge sums for television sets. Joel Brinkley in 1998, however, reports that tens of thousands of consumers had already bought high-end NTSC receivers that sold for $5000 to $50,000, so it is reasonable to predict that many viewers would continue their intense attraction to the video medium.

The set manufacturers decided to offer two formats: integrated and HDTV-ready. The latter requires a separate decoder, analogous to the way computer monitors are sold separate from the computer itself.

Panasonic offered the first DTV receiver, capable of reproducing only 800 lines (enhanced definition) because true 1080I resolution monitors cost $40,000 in 1998. Initial enhanced definition receivers, first sold to consumers in August 1998 and widely available by early 1999, were overbuilt and cost $6000 to $10,000. A retailer in San Diego sold 30 display units the first day, discount-priced at $5500 each, and took orders for another 18 the second day.

Except for the early adopters who would pay almost anything for the latest technology, ordinary consumers will decide how long to wait for the prices to drop. Analog equipment, such as VCRs, took a decade to drop to half of the original cost, but digital components are expected to get less expensive much more quickly, according to Moore's Law. After all, digital receivers are more like computers than television sets. In late 1998, HDTV was projected to cost $3500 by Christmas 2000 and $3000 by 2002. The price was not expected to reach below $1000 until 2004.

15.10 ANALYSIS AND PREDICTIONS

First, we can be suspicious about exaggerated cost estimates for HDTV transmission and reception. When costs are phrased as being "as much as" and "up to" inflated amounts, we can assume the translation is "a lot less than." Second, we have learned that competition is sufficient to foment an innovation like HDTV, but that cooperation (e.g., the Grand Alliance) is necessary to crystalize it, especially when government agencies are reluctant to set a standard.

My own view is the transition will take three to four years longer than the 2007 target year. But the transition will happen, as surely as telephones, radio, recorded music, and nearly everything else gone digital. As I write this, the digital signal already reaches 37 percent of the nation in its first month. It is inevitable, but how it will all unfold is not at all clear. What is clear is that HDTV is long overdue.

REFERENCES

Blake, Jonathan D., "The Origins of DTV Revisited," *Broadcasting & Cable*, October 26, 1998, 39.

Booth, Stephen A., "Digital TV Turns On," *Popular Science*, November 1998, 76–82.

Brinkley, Joel, *Defining Vision: The Battle for the Future of Television*, New York: Harcourt Brace, 1997.

Brinkley, Joel, "HDTV: High in Definition, High in Price," *The New York Times*, August 20, 1998, D1, D8.

Brown, Sara, "HDTV Highlights," *Broadcasting & Cable*, March 9, 1998, 38–46.

Dickson, Glen, "Counting Down to DTV," *Broadcasting & Cable*, July 20, 1998, 22–40.

Ferguson, Douglas and Klopfenstein, Bruce, "How Media Managers Deal with Change: The Case of HDTV." Panel paper at Assoc. for Ed. in Journalism and Mass Commun. (AEJMC), Baltimore, August 1998.

Higgins, John M., "Making Sense of Multicasting," *Broadcasting & Cable*, September 8, 1997, 14–17.

McClellan, Steve, "Ready and Not, Here Comes DTV," *Broadcasting & Cable*, March 9, 1998, 28–46.

McGann, Mike, "HDTV is Here (Finally)," *Home Theater*, November 1998, 42–53.

Schubin, Mark, "Metamorphosis," *Videography*, October 1998, 38–44.

Silbergleid, Michael, *The Guide to Digital Television*, New York: Miller Freeman, 1998.

West, Don, "The Dawn of Digital Television," supplement to *Broadcasting & Cable*, November 16, 1998, S1–S56.

16 The New Technologies of Radio

Terrestrial Digital Audio Broadcasting (DAB) and Satellite Digital Audio Radio Service (DARS)

Ted Carlin

CONTENTS

16.1 THE FOUR-STEP PROCESS OF DIGITAL RADIO CONVERSION

Like broadcast television and home video, the radio industry is contemplating a future dependent upon digital conversion. Many radio professionals and industry observers believe this transition to be necessary, if not unavoidable. According to Suren Pai, President of Lucent Digital Radio, "Digital itself is inevitable. You see that in every aspect of life, and radio is no exception. The world is going digital. There is no going back" (Merli, 46).

These thoughts are echoed by Feldman in his seminal work on digital media: "The idea of digital revolution is implicitly an image of humankind stepping through a doorway into an unknown and fundamentally changed future. And it is a one-way journey, a doorway through which we can never step back to return to the comfortable media certainties of the past."

For radio, this digital doorway has been open for over 15 years, but primarily in the areas of producing and recording audio. Compact disc, digital audio tape, computer hard drive, and MIDI technologies have been available to radio production staffs since the eighties, consistently replacing analog recording technologies such as LP records, reel-to-reel, and cartridge tape systems. In the Preface to *The Art of Digital Audio*, Watkinson describes the state of this digital conversion a decade ago in 1987:

> Digital audio is still developing, but it has reached a point where there is something solid to discuss. There are products in the marketplace which are dependable work-horses rather than laboratory curiosities. People use them to make a living, recording music with breathtaking clarity. Standards have been agreed for many common areas, and controversy over basic theory has largely ceased.

Digital audio production and recording are still developing. A recent, informal survey, conducted by the author, of local radio stations in Southern Pennsylvania and Northern Maryland found that all 20 medium and small market radio stations, covering various programming formats, had at least partially converted their audio production facilities to digital. Every station broadcast music from a digital platform (compact disc or computer hard drive) and digitally mastered all commercials and promotional announcements (on a computer hard drive, digital audio tape, or mini-disc). Three of these stations no longer used any form of analog tape for their broadcast material.

The decisions to convert analog production equipment to digital versions were made relatively easily. Each station tested various digital equipment models and adopted the equipment that best suited its internal needs. Basically, station personnel evaluated the equipment on price, ease-of-use, and effectiveness, with little help or pressure from outside interests.

Yet, as radio stations continue into the next century to convert their production facilities to digital, transmission to the listener, which is the primary focus of this chapter, is in the midst of receiving a digital makeover. For radio broadcasters, this conversion will be a much more difficult task than was the digital conversion of production facilities. As Sedman points out, this digital transmission conversion will be a four-step process which affects many people:

> Major changes in radio service are very difficult to institute. To be successful, a new service generally requires four levels of adoption: (1) approval by a governing body (such as the Federal Communications Commission (FCC) in the United States); (2) acceptance by broadcast stations; (3) consent from the consumer electronics industry to design and market the new technology; and, (4) adoption by the public.

This process applies not only to radio services, but to all communication technologies, including television. The television industry is currently implementing its second major transmission conversion; the first was the conversion from black and white to NTSC color in the mid-fifties. A brief examination of the current conversion from NTSC color to advanced television (ATV) illustrates how this four-step process works:

16.1.1   STEP ONE: APPROVAL BY A GOVERNING BODY

The FCC begins a series of policy initiatives for ATV, including the establishment of the Advisory Committee on Advanced Television Service to investigate the policies, standards, and regulations that would facilitate the introduction of digital ATV in the U.S. (FCC, 1987). Following recommendations from the Advisory Committee, the FCC outlines a simulcast strategy for the transition to an ATV standard (FCC, 1990) over a 15 year period (FCC, 1992). The FCC updates this simulcast strategy in 1997 (FCC, 1997a) and issues ATV broadcast licenses in 1998 (FCC, 1998). However, the FCC is allowing the marketplace to decide on a transmission standard.

16.1.2   STEP TWO: ACCEPTANCE BY BROADCAST STATIONS

Reluctantly, broadcast stations agree to the ATV conversion. The reluctance comes from the FCC's decision to phase out the current NTSC color system and replace it with a totally new ATV system. This move means that all existing NTSC color TV production, transmission, and reception hardware will have to be replaced with new equipment capable of processing the ATV signal. All of the ABC, CBS, FOX, and NBC affiliates in the top 30 markets have committed to broadcasting an ATV signal by November 1999.

Because the television industry, like the radio industry, must make profits to survive, this ATV conversion, estimated to cost $12 million per station (Dupagne and Seel), forces TV stations to spend large sums of money (solely from existing revenue streams) that drastically erode, or erase, their profit margins. Stations see the long-term potential for an ATV system, but must face the short-term realities of implementing an expensive overhaul of their fully functioning, existing NTSC color system (Fedele).

16.1.3   STEP THREE: CONSENT FROM THE CONSUMER ELECTRONICS   MANUFACTURERS

Without an ATV transmission standard (the Advanced Television Committee approved 18 variations of ATV), electronics manufacturers are cautiously producing ATV sets that cost thousands of dollars, hoping to capture affluent early adopters of ATV. According to Yang, the TV manufacturers "think digital TV is a gold mine. The new sets, costing $5500 and up, carry cushy premiums, and may also spur sales of DVD players, VCRs, and audio gear. But technical glitches could foil the launch. And confused shoppers may decide to wait — slowing sales of regular TVs as well" (Yang, 146).

The primary reason that an ATV transmission standard has not been selected is the late arrival of computer manufacturers to the debate (Tedesco). Computer monitors do not display digital information in the same method as television sets; computers use the progressive scanning technique, while TV sets use interlace scanning. With the increased popularity of computer use for a variety of video applications, computer manufacturers want the ability to transmit ATV signals via their equipment as well. By making their computers a *TV set* in addition to its current functions, computer manufacturers are hoping to increase their dominance and longevity in the video marketplace (Yang).

16.1.4   STEP FOUR: ADOPTION BY THE PUBLIC

At some point, consumers will have to purchase ATV sets to replace their current NTSC color sets. To smooth the transition, the FCC has provided an ATV phase-in process that allows all parties to adapt. In addition, consumer electronic manufacturers have promised to provide ATV converters that will extend the usable life of current NTSC color sets. The major stumbling blocks will be consumer awareness and comprehension of the conversion to ATV, the cost of ATV sets, and the ability of cable TV providers to successfully distribute ATV signals over their systems (Dupagne and Seel).

16.2   TERRESTRIAL DAB

The radio industry is not as far along as television in its transition to digital transmission. In fact, there has been little impetus for change until recently. Unlike the proliferation of new and bigger TV sets, accompanied by computer and satellite-delivered video over the past decade, the reception of radio has remained virtually the same since the inception of FM broadcasting. Except for the addition of LCD tuners, radio transmission is still primarily designed for car and clock radios, portable units, and home stereos:

> Consumers are not crying out for new and better radios. Consumers are happy with radio — listenership is growing. It's free, it's portable. What else do they need? Outside of the addition of the FM band, radios themselves have changed very little since the medium's birth in the 1920s. Radios come in all shapes and sizes but essentially remain the same (Miles, 17).

In 1996, two enhancements to radio listening did finally occur: the expansion of the AM frequency band and the proliferation of radio data systems (RDS). The expansion of the AM band from 1605 MHz to 1705 MHz provided more AM channels for listeners to receive. The commitment to RDS by more than 300 radio stations provided listeners with an LCD display of information on their radio (such as song information, weather alerts, and vendor coupons). Both enhancements required consumers to purchase new radios to receive the new services. However, neither enhancement was interesting enough to entice consumers to buy enough new radio receivers to make the services profitable (Sedman). Little has been said about either service in the last few years.

At about the same time that these enhancements were making their way to the consumer, the radio industry, like television, was attempting to embrace society's discovery and acceptance of digital technology. The radio industry wanted to ensure that their product was going to remain attractive to consumers in the future, so research was begun into the use of digital audio broadcasting (DAB) by existing terrestrial AM and FM radio stations. Three types of terrestrial DAB were to be considered: (1) in-band adjacent-channel [IBAC]; (2) in-band on-channel [IBOC]; and (3) out-of-band (Jurgen).

IBAC systems would allow FM stations to keep their current frequency assignment while broadcasting digitally in their sidebands. No IBAC system has been developed for digital AM broadcasting, and there is concern about potential interference from the DAB signals to existing adjacent FM stations if an IBAC system were implemented in the FM band (Spangler). IBAC has, therefore, been virtually eliminated from consideration in the U.S. as a DAB option.

IBOC systems would also allow AM and FM stations to keep their existing frequency assignments. Stations would be able to simulcast an analog and a DAB broadcast on the same frequency. Analog radio receivers would not become obsolete, but consumers would also have the option of purchasing a DAB receiver to be able to receive the digital version of the broadcast. A successful IBOC system has yet to be successfully tested under real-world conditions, but it is strongly endorsed by existing radio stations and their lobbying group, the National Association of Broadcasters (NAB) (Meadows).

Out-of-band DAB service would require existing radio stations to use a frequency spectrum other than AM or FM to transmit its programming. This dual frequency approach is the one being used by television stations to deliver the ATV programming described above. Stations would continue to transmit their programming on their current frequency while simulcasting a digital version on the new, expanded frequency assignment. At some point, the FCC would phase out the simulcasting and take control of the original AM or FM frequency for reuse by another technology, which would then make existing analog receivers obsolete.

This out-of-band approach is being followed by the radio industry in Europe, Canada, Japan, Mexico, South Africa, and Australia. The only major country that has not adopted this out-of-band transmission method for DAB is the U.S., mainly because of successful lobbying efforts by the NAB to protect existing AM and FM stations. The United Kingdom, for example, is using the Eureka 147 DAB transmission standard to broadcast DAB in the L frequency band (217.5 to 230 MHz).

Eureka 147 allows one radio station to transmit through multiplexing up to six stereo radio services on one frequency. For example, one frequency assignment would be able to deliver six types of country music programming: traditional country, young country, country love songs, women of country, country groups, and top-40 country hits. Because of frequency and interference limitations, this multiple delivery is not possible with current AM and FM frequency assignments. FM stations do have the opportunity, however, to practice a limited version of multiplexing by broadcasting Muzak and data over their subchannels. More frequency space is needed to provide true stereo multiplexing, hence the move to a new frequency band would be required for this type of DAB.

Using the four-step adoption process, a closer examination of the status of these terrestrial DAB options can be provided.

16.2.1 STEP ONE: APPROVAL BY A GOVERNING BODY

There have been no IBAC proposals submitted to the FCC to date.

In October, 1998 the first IBOC proposal was delivered to the FCC by USA Digital Radio (Stimson, 1998b). Until this time, research and testing of IBOC systems were being conducted by the National Radio Systems Committee (NRSC). The NAB and Consumer Electronics Manufacturers Association (CEMA) decided to combine forces to establish the NRSC in 1994 as an independent group to evaluate new radio technology. Two subcommittees were established: one to investigate the possible implementation of RDS, and the other to analyze the feasibility of IBOC DAB. Such a joint effort was designed to speed the process of new radio product development by keeping the program transmitters (NAB radio stations) and program receiver manufacturers (CEMA members) working together through real-world development and testing.

The RDS subcommittee has been working steadily since 1994 on the development of RDS *smart* radios. These RDS systems are in operation and are actively being tested by approximately 60 radio stations.

After failing to successfully test an IBOC DAB system for three years, however, the IBOC DAB subcommittee was disbanded in 1997. Original attempts by the USA Digital Radio (USADR) partnership (CBS, Westinghouse, and Gannett) and AT&T's Lucent Digital Radio to test IBOC systems were simultaneously withdrawn (Spangler).

One year later, following the 1998 NAB Radio Show, the IBOC subcommittee was reconvened after news of significant progress by USADR, Lucent Digital Radio, and a third IBOC proponent, Digital Radio Express (DRE) (Stimson, 1998a). However, citing the need to protect intellectual property, USADR bypassed the NRSC and directly submitted its own petition to the FCC for a rulemaking designed to establish IBOC DAB service in the U.S. Lucent and DRE are still working with the NRSC (Stimson, 1998b). Future IBOC petitions may be forthcoming from the NRSC, Lucent, or DRE.

In his speech to the NAB Radio Convention, FCC Chairman William Kennard described the FCC's role now that they have an IBOC petition to consider:

But let me be very clear. Here is what we will not do. We will not undermine the technical integrity of the FM band. Our job is to be the guardian of the spectrum, not to degrade it. And we will not do anything to prevent the conversion to digital. Just last week, Michael Jordan of CBS presented me with USA Digital Radio's (USADR) petition to establish an in-band, on-channel digital broadcasting service. While we're considering this petition, we'll also continue to follow the testing and development of in-band digital systems by the National Radio Systems Committee (NRSC) set up by the NAB and CEMA. This is a great start, and I will do my part to make sure that local radio is not left on the sidelines of the digital revolution (Kennard, 3).

The FCC's next step is to seek public comment on USADR's petition, and eventually issue a rulemaking. This includes comment, analysis, and IBOC test results performed by all interested parties.

No petitions for out-of-band DAB service have been submitted to the FCC to date. The FCC would probably not authorize out-of-band terrestrial DAB because of the lack of available frequencies within the L band. This frequency spectrum is currently being used by the U.S. government for mobile aeronautical telemetry systems (Spangler). It would be a risky proposition for an entrepreneur to propose a new radio service without first having available frequency spectrum, or worse yet, to propose using one that is already being used by the government.

16.2.2 STEP TWO: ACCEPTANCE BY BROADCAST STATIONS

Existing AM and FM stations, through the NAB, have definitively backed the IBOC system approach (Stimson, 1998b) which allows stations to keep their frequency assignments and simulcast a DAB signal with the current analog signal on it. This arrangement would alleviate consumer confusion concerning the location of the DAB channel because all transmissions would be located on the existing frequency. Unlike the upcoming digital television conversion process, the IBOC digital conversion of radio would be fairly seamless from the broadcaster's perspective:

IBOC technology provides a unique opportunity for broadcasters and consumers to convert from analog to digital radio without new frequencies or service disruption. Broadcasters will use their current frequency allocations to transmit simultaneous analog and digital audio, in addition to new mobile data services. Consumers will receive familiar radio stations with superior CD-quality sound along with broadcasted in-vehicle data information (USA Digital Radio, 1490).

The cost of the digital transmission equipment will be the main concern for existing stations. As conglomeration continues within the radio industry, these equipment costs may be more easily absorbed by the resulting large station groups, but burdensome for smaller groups and independents. To convert to IBOC, broadcasters will need to make investments in the following equipment: (1) a transmitter (most FM and about half of all AM stations); (2) studio-to-transmitter link (every AM and FM station); (3) an exciter (every AM and FM station); and, (4) an antenna diplexer (about 100 FM stations) (Merli).

16.2.3 STEP THREE: CONSENT FROM THE CONSUMER ELECTRONICS MANUFACTURERS

CEMA, through its participation in the NRSC, has decided to explore the merits of an IBOC system with radio broadcasters. However, in 1997, after broadcasters had failed to produce a working IBOC system, CEMA was endorsing the only existent working DAB system at that time, the out-of-band Eureka 147 system. Logically, CEMA wants to ensure that equipment manufacturers have a viable product to produce and will support any digital transmission system that makes it to market. Because Eureka 147 has been proven to function in the real-world, equipment manufacturers would understandably feel more at ease supporting it instead of an unproven IBOC system. Such has been the case with the manufacturing of other communication technology equipment, including ATV, AM stereo, and DBS systems:

> There's a lot riding on the outcome of IBOC DAB development. But we've been burned before with technology that wasn't ready for prime time. Remember when the FCC picked Magnavox AM stereo as the US standard? AM stereo never recovered and never had a chance thereafter. If [IBOC] works and proves to be everything US radio needs to take it into the next millennium, I'll be one happy radio guy. The rest of the world may then see the advantages [of IBOC vs. Eureka 147] and climb aboard ... if Eureka doesn't grab a strong foothold first. There's a great case to be made for a system that can use existing radios: There are almost more radios in the developed world than there are people (Wire, 2).

16.2.4   STEP FOUR: ADOPTION BY THE PUBLIC

Similar to ATV, DAB listeners will have to purchase a new radio receiver to receive DAB programming. Like every other new communication product introduction, DAB will likely be adopted earliest by those who can afford to purchase the expensive DAB receivers (Klopfenstein). It is highly unlikely that equipment manufacturers will deviate from past practices of pricing new communication technology high. This pricing strategy continues until a critical mass of adopters is reached, or competition forces prices down. The risk associated with the potential failure of the new technology is the primary factor the new product is priced so highly.

By supporting the IBOC system, broadcasters and equipment manufacturers have decided to lower the risk by keeping radio listeners tuning into the same frequencies. It will be up to the consumer to decide if the quality of the DAB signal is superior enough to the existing analog signal to warrant the purchase of a DAB receiver. According to Klopfenstein (188), "Only when new media provide potential adopters with a service that fills a need at a reasonable cost will they have a chance to be successful."

Because most radio listening occurs in vehicles, automobile and truck manufacturers could be deciding factors in the consumer adoption of DAB. If vehicle manufacturers decide to install DAB receivers in their new vehicles, consumers will have a much greater chance of accessing DAB transmissions than if they are forced to buy a DAB receiver to adapt to their existing vehicle radio (similar to adapting a portable CD player to a car's installed cassette player). It is likely that vehicle manufacturers would first install DAB receivers on the most expensive vehicles, where the costs can be most easily absorbed by the manufacturer and the luxury vehicle buyer.

16.3   SATELLITE DARS

In 1990, Satellite CD Radio, Inc. (CD Radio) filed a petition with the FCC to allocate spectrum for a new satellite-delivered radio service — satellite digital audio radio service (DARS) — and also applied to provide the service to American consumers. CD Radio wanted to create a totally new audio delivery system that would enable consumers to receive audio programming via satellite anywhere in the continental U.S. In essence, a consumer would purchase a small (size of a silver dollar) antenna and digital radio then receive CD-quality audio that would not fade out anywhere across the country. Consumers would then have two choices for broadcast audio: terrestrial AM and FM stations that are subject to quality and interference limitations, or satellite DARS services that eliminate these difficulties. The NAB, and terrestrial radio stations, were obviously not happy with the request:

> The current number of operating FM and AM stations serving the United States public represents the highest level of audio program diversity available in the world ... any continued policy of simply adding more and more stations to the commercial radio environment will ultimately disserve the public interest. (Flint, 29).

A closer examination of satellite DARS using the four-step adoption process will provide a useful analysis of this digital audio option.

16.3.1 Step One: Approval by a Governing Body

In the five years that followed CD Radio's 1990 petition, the FCC was urged to examine the impact of this new digital radio service on existing AM and FM stations. In November 1992, the FCC established a proceeding to allocate satellite DARS spectrum domestically and announced a December 15, 1992 cut-off date for satellite DARS license applications to be considered with CD Radio's. In January 1995, the Commission allocated the 2310-2360 MHz band for satellite DARS on a primary basis.

In June 1995, the FCC requested detailed information on satellite DARS' potential economic impact on terrestrial broadcasters. The Notice asked about the most appropriate service design and regulatory classification, about what public interest obligations to impose, and whether providers should be permitted to offer ancillary services. The Notice proposed three possible licensing options and rules to allow expeditious licensing after an option was chosen. After the Notice was released, Congress directed the Commission to reallocate spectrum at 2305-2320 MHz and 2345-2360 MHz for satellite DARS to be consistent with international allocations, and to award licenses in that portion of the band using competitive bidding (i.e., auction).

In a regulatory move that has angered the NAB and caused concern for most U.S. radio stations, the FCC authorized satellite DARS as a competitor to terrestrial AM and FM stations (FCC, 1997b). Seven years of petitions from the NAB argued that satellite DARS would present "a potential danger to the U.S.' universal, free, local radio service and thus to the public interest it serves. The erosion of audiences and advertising revenues caused by satellite radio would inevitably destroy the ability of many community stations to offer these services" (Flint 29). Despite the petitions' argument, the FCC approved this new digital radio service on March 3, 1997. The FCC's final response to these claims, after analyzing information from both sides of the issues, came out in favor of the development of satellite DARS:

Given the distinguishing features of satellite DARS — it is a national service, it will require new and relatively costly equipment, and it may be offered via paid subscription — we find that the effect of satellite DARS on terrestrial radio is likely to be significantly smaller than the effect of additional terrestrial radio stations (FCC, 6).

The FCC was even more specific when analyzing the impact of satellite DARS on advertising revenues of existing radio stations:

While we recognize that satellite DARS has significant competitive advantages in offering advertising to a national audience with satellite DARS receivers, several factors may limit the possible significance to terrestrial radio of such additional competition. First, at this time, only one out of the four satellite DARS applicants has indicated an intention to implement its system on a non-subscription, advertiser-supported basis. Second, a large share of the national radio audience is not likely to have satellite DARS receivers, at least for a significant period of time. Third, national advertising revenue amounts to only 18% of terrestrial radio advertising revenue and is on average less important for small-market stations than for large-market stations. Local advertising revenue is much more important than national advertising revenue for terrestrial radio's viability and prevalence, and, at this time, we have no evidence that satellite DARS would be able to compete for local advertising revenue (FCC, 1997b, 7).

In April 1997 the FCC successfully auctioned two segments of the S frequency band, 2320–2332.5 MHz and 2332.5–2345 MHz, among four applicants: CD Radio, American Mobile Radio Corporation (AMRC), Digital Satellite Broadcasting Corporation, and Primosphere Limited Partnership. The two winners were CD Radio ($83.3 million) and AMRC ($89.8 million) (Holland).

CD Radio and AMRC, which has since changed its name to XM Satellite Radio Inc., are now in the process of implementing their services, with CD Radio hoping to begin service to consumers in 1999 and XM a year later. Because of the long lead time necessary for satellite construction, the FCC proposed that these satellite DARS licensees begin construction of their space stations within one year of the auction, launch and begin operating their first satellite within four years, and begin operating their entire system within six years. The FCC also proposed that licensees file annual reports on the status of their systems, and, because the Communications Act limits broadcast license terms to eight years, the FCC determined that satellite DARS license terms should be eight years. The license term will commence when each service is put into operation, and be subject to renewal or termination after the initial eight-year period (FCC, 1997b).

16.3.2 STEP TWO: ACCEPTANCE BY BROADCAST STATIONS

Because satellite DARS will not be implemented by existing broadcast stations, this adoption step is not necessary for satellite DARS to succeed. However, it would be logical to expect continued resistance from existing stations in the form of further FCC petitions, NAB lobbying efforts, and consumer marketing similar to the cable industry's approach to DBS systems.

The main concerns of radio broadcasters, the potential loss of audience ratings and advertising revenues, will certainly be scrutinized as the satellite DARS providers begin operations. A key factor yet to be addressed by those involved is the audience measurement methodology to be used with satellite DARS. Because broadcast stations are the largest clients of the audience measurement services, including Arbitron, will stations ask them to include satellite DARS in their measurements? This information will give stations, and advertisers, a clearer picture of satellite DARS usage, but will stations want this type of view? Broadcast TV stations have been very slow to support the inclusion of accurate cable television ratings with television ratings (Eastman and Ferguson).

16.3.3   STEP THREE: CONSENT FROM THE CONSUMER ELECTRONICS MANUFACTURERS

Because of their close ties with the NAB, CEMA has not been a vocal participant in the satellite DARS debate. Both CD Radio and XM have publicly stated that they are close to signing manufacturing agreements with individual equipment manufacturers (Stimson, 1998c). Unlike IBOC DAB, the risk factor associated with satellite DARS is lower because it has been authorized by the FCC.

As with IBOC DAB systems, satellite DARS equipment will come in two stages: new and aftermarket. There were approximately 8 million aftermarket car radios sold in the U.S. in 1997, and XM's CEO Hugh Panero says this shows "a significant turnover of people who don't like their current (car) radio and want to upgrade" (Stimson, 1998c, 14). It is likely that both satellite DARS companies will also produce converters that utilize existing vehicle cassette players, although only CD Radio has specifically stated an intention to do so.

16.3.4   STEP FOUR: ADOPTION BY THE PUBLIC

CD Radio expects to break even at 1 million consumers in its first two years of operation, while XM hopes to be in the black by 2003 with 2 million users (Curran). Services will probably not be launched before the fourth quarter of 2000 or early 2001.

CD Radio has announced contracts with Space Systems/Loral Incorporated to build its fleet of four satellites, and with Lucent Technologies Microelectronics Group to develop and supply the digital transmission technology. XM has announced contracts with Hughes Space and Communications to build its three satellites, and with German-based Fraunhofer to develop its digital technology (Curran).

Both companies are planning to offer 100 channels of CD-quality audio by subscriptions for around $10 a month. In addition, consumers will have to purchase the radio receiver or converter and the receiver antenna. Original estimates by both companies priced a new radio/antenna combination between $400 and $600 (Curran). The $10-a-month fee will provide subscribers with 50 channels of commercial-free music and 50 channels of commercial-supported news/talk/sports. Up-to-date descriptions of each service can be obtained from their WWW sites: http://www.cdradio.com (CD Radio) and http://www.amrc.com (XM Satellite Radio).

As Klopfenstein points out in Chapter 6, any new radio service will have to fulfill consumers' needs at a reasonable price to attract subscribers away from existing services. This is the key to success for any entertainment provider. Successful AM and FM radio programmers have known this for decades: "Ultimately, it's going to be a question of compelling programming. If [competition is] not compelling, they'll siphon only a little off. There's nothing you can do to stave off technology. It's going to be a question of programming — if you're programming a better product than they are" ("Contemplating digital," 117).

And this compelling programming might not be the CD-quality, commercial-free music, according to some FM radio programmers: "If satellite radio is targeting music lovers, it won't get enough audience to survive. Some of those people will probably do it, but not the masses. Call up Sony and ask them how their MiniDisc is going" ("Contemplating digital," 118). Many programmers see personality as the key to every successful radio service, including satellite radio:

> I would try to find the most compelling, unique personalities on the planet, pay them way too much money, and lock them in so they could not be available any other way. That's how you do it. We saw what Rush Limbaugh and his followers have done for AM. There's no reason why a breakthrough personality, or several breakthrough personalities, can't do that for satellite radio. That's the key, as opposed to 200 channels of Montavani ("Contemplating digital," 117).

It would be reasonable to expect that satellite DARS, as well as IBOC DAB, will follow the consumer adoption paths of previous new communication technologies: high-tech affluent early adopters followed by dissatisfied AM and FM users once the price of the equipment falls to a reasonable level. This reasonable level will be determined by the consumer's perception of value for the new service.

REFERENCES

1. Contemplating digital radio's threat. (1998, September 5). *Billboard,* 110, 117-118.
2. Curran, L. (1998, August). Satellite radio: Along for the ride. *Electronic Business,* 24, 26.
3. Dupagne, M. and Seel, P. (1998). Advanced television. In A.E. Grant and J.H. Meadows (Eds.) *Communication Technology Update* 6th Ed., Boston: Focal Press 64-78.
4. Eastman, S. and Ferguson, D. (1997). *Broadcast & Cable Programming* 3rd Ed. Belmont, CA: Wadsworth.
5. Fedele, J. (1997, September 25). DTV schedule breeds apprehension. *TV Technology,* 16.
6. Federal Communications Commission. (1987). Formation of Advisory Committee on Advances Television Service and announcement of first meeting. 52 *Fed. Reg.* 38523.
7. Federal Communications Commission. (1990). Advanced television systems and their impact on the existing television broadcast service. First Report and Order. 5 FCC Rcd. 5627.

8. Federal Communications Commission. (1992). Advanced television systems and their impact on the existing television broadcast service. Second Report and Order/Further Notice of Proposed Rulemaking. 7 FCC Rcd. 3340.

9. Federal Communications Commission. (1997a). Advanced television systems and their impact on the existing television broadcast service. Sixth Report and Order. 12 FCC Rcd. 14588.

10. Federal Communications Commission. (1997b). Establishment of rules and policies for the digital audio radio satellite service in the 2310-2360 MHz frequency band. Report and Order/Further Notice of Proposed Rulemaking. FCC Rcd. 8610. Washington, D.C.: FCC.

11. Federal Communications Commission. (1998). Advanced television systems and their impact on the existing television broadcast service. Memorandum Opinion and Order on Reconsideration of the Sixth Report and Order. MM Docket No. 87-268.

12. Feldman, T. (1997). *An introduction to digital media.* London: Routledge.

13. Flint, J. (1993, February 8). FCC told DAB threatens local radio service. *Broadcasting, 123,* 29.

14. Holland, B. (1997, April 19). FCC auctions off 1st 2 digital channels; CD Radio, American Mobile Radio win licenses. *Broadcasting & Cable,* 109, 94.

15. Jurgen, R. (1996, March). Broadcasting with digital audio. *IEEE Spectrum,* 52-59.

16. Kennard, W. (1998). Speech to the 1998 NAB Radio Show. On-line: http://www.fcc.gov/Speeches/Kennard/spwek832.html].

17. Klopfenstein, B. (1998). The digital revolution in home video. In A.E. Grant and J.H. Meadows (Eds.), *Communication Technology Update* 6th Ed., Boston: Focal Press. 174-189.

18. Meadows, L. (1998, September 16). IBOC DAB process moves ahead. *Radio World,* 1.

19. Merli, J. (1997, October 27). Local digital radio gets closer to reality. *Broadcasting & Cable,* 127, 46-48.

20. Miles, L. (1993, December 13). *MediaWeek,* 3, 17.

21. Sedman, D. (1998). Radio broadcasting. In A.E. Grant and J.H. Meadows (Eds.), *Communication Technology Update* 6th Ed. Boston: Focal Press. 79-86

22. Spangler, M. (1997, March 5). CEMA maneuvers on DAR, takes on DAB. *Radio World,* 1.

23. Stimson, L. (1998, June 10). In-band, on-channel: Now what? *Radio World,* 1,19.

24. Stimson, L. (1998, October 14). USADR to file 'historic' petition. *Radio World,* 1,12.

25. Stimson, L. (1998, October 28). The new XM satellite radio. *Radio World,* 34.

26. Tedesco, R. (1998, March 9). Intel broke ranks but computer coalition holds. *Broadcasting & Cable,* 46.

27. USA Digital Radio unveils the future. (1998, October 14). *PR Newswire,* 1490.

28. Watkinson, J. (1987). *The art of digital audio.* Boston: Focal Press.

29. Wire, G. (1998). A little DAB will do ya.....or will it? *Radio World* On-line: www.rwonline.com/newsroom/archives/gw398.html.

30. Yang, C. (1998, October 26). Digital D-Day. *Business Week,* 144-158.

17 Key Concepts in Internet Commerce

Bruce Klopfenstein

CONTENTS

This chapter briefly introduces the key concepts surrounding Internet commerce. This fluid topic is tremendously broad, the implications are vast, and the landscape is rapidly changing. Internet commerce includes complex issues related to technology, policy, privacy, and security, each of which has both national and global implications. Definitions of terms surrounding Internet commerce are still being formulated (Smedinghoff 1998). Media scholars are loath to talk about revolutions because new media tend to evolve on the established media infrastructure. Internet commerce is, perhaps, as close to a *bona fide* revolution (which implies the displacement of an established way of doing things with a new one) in business as anything witnessed in this generation. It would be difficult, for example, for most businesses to reasonably argue against having some presence on the World-Wide Web (WWW). As this chapter discusses, the remaining barriers to rapid adoption and diffusion of Internet commerce are more social and psychological than they are technical.

17.1 BACKGROUND

What is Internet commerce? Theoretically, Internet commerce is a subset of electronic commerce. It and electronic commerce are perceived to be virtually synonymous in 1999. Finding a standard definition for electronic commerce, however, is more difficult than one would expect. In fact, one recent definition of electronic commerce includes the Internet:

> Electronic Commerce is the buying and selling of goods and services or the transfer of money over the Internet or an Intranet. This can involve stores or banking activities. Standards have established to make the process easier and more secure. (Electronic Commerce, 1998)

On the other hand, this definition could be far too narrow because of its focus on the use of currency. A broader definition includes customer service over the Internet, electronic responses to requests for proposals, or simply the automation of any business-to-business or consumer-to-business relationship (Jordan, 1998; see also Minoli and Minoli, 1998, 8). It must also be noted that the parties for Internet commerce can be business-to-business, business-to-consumer, and/or intraorganizational (Mougayar, 1998). For the purposes of this chapter, no distinctions are made between electronic commerce and Internet commerce. Most of the focus here is also on business-to-consumer Internet commerce.

Research indicates that the market for electronic commerce is burgeoning. According to figures cited by Howell (1998), nearly 50 million U.S. users will be purchasing goods and services online by the year 2000. As of 1998, the average online purchase was $350 and the electronic commerce market was in the neighborhood of $16 billion. While 40% of U.S. businesses were involved in electronic commerce in 1997, that figure grew to nearly two-thirds in 1998. Worth noting is that the number of online shoppers in 1998 was much higher than the number of online buyers, but this is simply a harbinger of things to come. Chrysler, for example, expects 25% of its sales to come from the Internet by 2001. As reported by Jordan (1998), citing Forrester Research Inc., total commerce over the Internet may already have been $17 billion in 1998 and will grow to $350 billion by 2002. The 1998 Christmas season may be remembered as the turning point for shopping online as far more consumers took the plunge.

The U.S. Commerce Department estimates that by 2002, the $300 billion-plus in annual revenue from goods and services sold over the web will represent more revenue than the annual sales of General Motors Corp. and General Electric Co. combined. Channel companies are selling nearly $9 billion a year in products and services related to building e-commerce sites for end-user businesses, according to Channel Information Services (CIS), a business unit of the Channel Group of CMP Media Inc. Sales of web servers and e-commerce products are growing faster than sales of any other software products (Jordan, 1998).

A number of factors have coalesced to create a highly favorable environment for electronic commerce. At the start of 1999, approximately half of all U.S. households had a personal computer. The number of people with access to the Internet is further buttressed via connections at school, work, and public places such as libraries. Internet access appears to be well on its way to having the same ubiquity as the telephone. (If that prediction sounds overly optimistic, consider the fact that nearly as many American households now have VCRs as have telephones.) Access via other online technologies such as WebTV actually may bring the number of online households above the number of computing households by 2002 (NFO Interactive, 1998).

Another factor that is serving as a catalyst for electronic commerce is the real and perceived time constraints under which people are living. It can be less time-consuming (if not faster) to do business online (note that e-commerce is not limited to electronic shopping) than physically traveling to a business place. Book buyers, for example, can either drive to a bookstore or order books online via an online bookseller such as Amazon.com. The tradeoffs, which include waiting for books to be shipped and paying an additional shipping charge, should be compared to the positives of a tremendous selection of titles (many of which can be delivered more quickly than if special ordered via a traditional bookstore) at prices often discounted from the suggested retail price.

A third catalyst for electronic commerce is the underlying computer hardware and software infrastructure. Electronic databases have made the process of shopping online efficient and simple. One of the keys to technology is making the interface transparent to the user; in other words, we simply want things to work without having to worry about how they are being done. This is happening.

People's own experiences with the web at home and work are making its use a normal part of their living environment. The number of people who have not used the web continues to decline. Meanwhile, households with web users are generally better educated with higher incomes than those who do not. This group often includes consumers attractive to a variety of businesses, but it also includes enterprise decision-makers who have the influence to move their own organizations to integrate the web into their normal operating procedures.

Finally, the price of consumer Internet access decreased in the 1990s. Many observers had believed that metered Internet pricing was on the way, but this has yet to happen. Indeed, the real price of consumer Internet access has actually declined (especially when accounting for inflation) despite contrary predictions by a number of Internet economists (Klopfenstein, 1997a; 1997b). As Internet access is bundled with other telecommunications services, such as regular telephone or cable television service, prices may decline further. Advertiser-supported services will also serve to lower the direct cost to consumers. This premise adds to the likelihood that Internet access is becoming as ubiquitous as telephone access.

17.2 THE INTERNET VERSUS OTHER NETWORKS

E-commerce began in the 1970s in large corporations with Electronic Data Interchange (EDI), a set of standards that allows companies to send invoices to and order from other companies, all electronically by means of data networks, via value-added networks (VANs). EDI also implies closed, private networks. VANs are developed for a specific application (Cook, 1998), but they are disappearing. The Internet is driving today's burst of e-commerce. The Internet is not as robust a medium as EDI via VAN, but it is less expensive by a factor of 100. Expensive VANs still make sense in one-to-many distribution, such as discount retail chain Wal-Mart (Jordan, 1998). Interestingly though, companies are moving from EDI to the Internet, a network designed to be open.

17.2.1 INTERNET SECURITY AND ENCRYPTION

Although transparent to the user, a secure environment for conducting electronic commerce is essential for both pragmatic and psychological reasons. Tremendous legal liabilities are also present for anyone who wishes to offer products and services over the Internet in return for electronic payments such as those made possible by credit cards (Crocker and Stevenson, 1998). Zimits and Montaño (1998) provide an excellent overview of technical issues related to Internet commerce.

With its roots firmly in military applications, encryption is the conversion of data into a form, called a cipher, that cannot be easily read by unauthorized people. Decryption is the process of converting encrypted data back into its original form. Simple ciphers include the substitution of letters for numbers, the rotation of letters in the alphabet, and the scrambling of voice signals. More complex ciphers work according to sophisticated computer algorithms that rearrange the data bits in digital signals.

In order to easily recover the contents of an encrypted signal, the correct decryption key is required. The key is an algorithm that undoes the work of the encryption algorithm. Alternatively, a computer can be used in an attempt to "break" the cipher. The more complex the encryption algorithm, the more difficult it becomes to eavesdrop on the communications without access to the key. Encryption is critical in electronic commerce applications such as a credit-card purchase online, or the discussion of a company secret between different departments in the organization. The stronger the cipher — in general, that is, the harder it is for unauthorized people to break it — the better. However, as the strength of encryption/decryption increases, so does the cost (Thing, 1998).

17.2.2 PUBLIC KEY INFRASTRUCTURE

Public key infrastructure, PKI (see Table 17.1 below), includes dual public and private encryption keys, digital certificates, digital signatures, key-management protocols, and certificate authorities. By many measures, the PKI provides mechanisms for establishing trust and binding commitments that are superior to accepted business practices. Over time, electronic commerce tools based on public key technology will substitute for and eventually replace established commerce archetypes such as paper contracts, personal signatures, and currency (Zimits and Montaño, 1998).

Public-key encryption is a method of encrypting and decrypting data using a pair of keys: a public key available to everyone and a private key owned by an individual and available to only that individual (Technology Overview, 1998).

Table 17.1 lists important concepts related to encryption. The public and private keys in a key pair are related in that: data encrypted with a public key can be decrypted only with the corresponding private key. For example, to send someone confidential information, you can encrypt data with a person's public key. The encrypted data can be decrypted only by the private key, which should be accessible to only the intended recipient. Data encrypted by a private key can be decrypted only with the corresponding public key. For example, to prove that you are the person sending the data, you can encrypt data with your private key. Anyone using

TABLE 17.1
Public Key Infrastructure Building Blocks

Encryption Algorithms	The basic mathematical algorithms used to scramble information. Symmetric encryption uses the same keys to encrypt and decrypt, whereas asymmetric encryption uses separate keys to encrypt and decrypt information.
Private and Public Keys	A secret, private key and a mathematically related public key are generated for each party in a transmission. Given the public key, it is nearly impossible to determine the private key.
Digital Signatures	An electronic signature that is irrefutable, unique, and virtually impossible to copy or transfer.
Digital Certificates	An electronic document comprising a public key, digital signature, owner identity, serial number, issuer, and expiration date.
Certificate Authorities	Issuers of digital certificates acting as a trusted third party in electronic transactions.

Source: Zimits and Montaño, 1998

your public key can decrypt the data, which verifies that the data was encrypted (and sent) by you.

The public key infrastructure is a system centered on the use of public key cryptography that provides each of the following critical elements:

- Information confidentiality
- Information integrity
- Authentication
- Difficult repudiation

The last point involves disallowing easy claims by a purchaser that he or she did not actually initiate an exchange.

How can one be certain that the name on a given public key truly represents the person with whom a transaction is desired? Digital certificates use a public key and owner information that together have been digitally signed (certified) by a trusted third party organization (Schnell, 1996). It gives Internet commerce customers the assurance that a web site, for example, is legitimate and not that of an impostor. It also provides a legal basis for transactions on the Internet. However, a problem with digital certificates is that if one person uses another's computer he can also thereby use the owner's certificate (Howell, 1998).

Use of digital certificates should, however, reduce fraudulent transactions (i.e., the classic ordering of a pizza delivery for an unsuspecting recipient). Certificates contain information about the certification authority, the owners of the certificate, a public key, the period for which the certificate is valid, and the host to which the certificate was issued. The token is designed in such a way that none of its details can be changed without invalidating the digital signature. Eventually, digital certificates could be built into web browsers and *virtual wallets*, which are stored on a person's computer hard drive and contain encrypted payment and billing information for ordering online. Eventually, wallets could contain checks, coins, and credit cards (VNU E Commerce Glossary, 1998).

Howell (1998) gives an example of how a secure transaction on the WWW can take place. The person initiating the transaction on a client PC requests an item from a web server. The server returns its digital signature saying, in effect, it is what the client thinks it is. The client next passes ths digital signature on to a trusted third party, the digital signature registry, which then confirms (or refutes) the server's identity. Once this is successfully completed, the client completes the transaction. In each instance, all data sent uses dual key encryption.

17.3   INTERNET COMMERCE PAYMENT SYSTEMS

Crocker and Stevenson (1998) note that there are three basic architectures for Internet commerce payment systems: wallets, cash registers, and gateways. As noted above, a wallet is software that runs on a consumer's PC (a term originally coined by the firm CyberCash). A second network computer operates as the merchant's Internet *cash register*. The gateway server is operated by the system operator, a bank, or a transaction processor. Each provides cryptographic capacity for secure transactions. The greatest security threats come from scam artists who can create fake online businesses with the sole purpose of getting consumer credit card information, and from honest merchants who record credit card information on insecure computer systems.

Other Internet commerce payment systems are in various stages of development. Internet check transactions are possible whereby a consumer's actual checking account could be linked to an electronic wallet. Consumers who are accustomed to check floating (taking advantage of the time it takes from the point at which a check is written to the point at which money is actually removed from the checking account) may be dissuaded from using Internet checking systems. Also in development are account-based and token-based *digital cash* systems that create cash-like payment systems allowing immediate and even anonymous payments. Smart cards are an example of token-based payment, and they may require a card reader at the consumer's PC (Crocker and Stevenson, 1998). A smart card is similar to a credit card with embedded electronics and/or a microchip that stores cash in encrypted form to be used with PCs, telephones, ATMs, and other devices with built-in card readers (VNU E Commerce Glossary, 1998).

17.4   OTHER INTERNET COMMERCE TECHNICAL
CONCERNS

Internet pioneer and enthusiast Vinton Cerf (1998) worries about utopian views of the encroaching virtual world of the Internet. The opportunities for fraud and deception are there, and anonymous digital cash systems allow for possibilities such as money laundering. Law enforcement agencies are acutely aware of this potential, which has created tension between business and government (see portions of U.S. Department of Commerce, 1998; Canadian Minister of National Revenues, 1998). Minoli and Minoli (1998) is an excellent and thorough text that reviews many key threats to successful Internet commerce systems.

Encryption is a fairly obvious requirement for Internet commerce transactions, but there are other equally significant concerns important to one or both parties in an electronic commerce transaction. Some of these issues are listed in Table 17.2. Access control is especially of concern to would-be providers of Internet commerce; Surveys continue to show that it is one of the barriers to commercial web site implementation by existing businesses.

TABLE 17.2
Fundamental Internet Commerce Security Requirements

Access Control	Determines who may have access to information within a system
Authentication	Verifies the identity of communicating parties
Privacy	Protects sensitive information from being viewed indiscriminately
Integrity	Guarantees that information is not tampered with or altered
Non-Repudiation	Provides inability to disavow a transaction

Source: Zimits and Montaño, 1998

It is worth noting that authentication may be on the verge of moving beyond digitally encrypted signatures. Retinal scans, palm prints, and voice verification are three examples of authentication technologies that might be close to market introduction. Merkow (1998) reviews the state of these technologies in 1998.

The topic of Internet privacy, even if limited to Internet commerce, is one worthy of doctoral dissertations, scholarly books and other works, Congressional hearings, and commercial writers. Beyond the ethics of privacy is the pragmatic concern many consumers have about their online privacy. Research repeatedly demonstrates that concerns about privacy are among the most important barriers to consumer online shopping. The Electronic Frontier Foundation (www.eff.org), one of the most well known and respected Internet advocacy groups, has a repository of privacy-related information online. The Internet Privacy Coalition (www.privacy.org/ipc), an organization devoted to enlightening people about issues of privacy, and the Internet Privacy Information Center (www.epic.org) also have similar information. Readers are encouraged to visit these and other sites for current information and debates related to privacy. Responsible companies know that protecting consumer privacy is in their own best interests; to do otherwise can harm their efforts to conduct business online.

17.4.1   SET: Secure Electronic Transaction

On February 1, 1996, MasterCard International and Visa International jointly announced the development of a single technical standard for safeguarding payment card purchases made over open networks such as the Internet. This trademarked standard, the SET Secure Electronic Transaction™ specification, also known as the SET™ specification, includes digital certificates and will provide financial institutions, merchants, and vendors with a safe way of getting the most from the emerging electronic commerce marketplace (see http://www.setco.org/).

A fairly thorough review of what SET is trying to accomplish can be found in Minoli and Minoli (1998, Chapter 6). SET features include:

- Confidentiality of information
- Integrity of data
- Cardholder account authentication
- Merchant authentication
- Interoperability

Interoperability means that for SET to work, it must be hardware and software independent. SET is not the only avenue available for Internet commerce, but its importance is clear given its support by MasterCard and Visa.

17.4.2   SSL (Secure Sockets Layer)

SSL is a program layer created by Netscape for managing the security of message transmissions in a network. Netscape's idea is that the programming for keeping messages confidential ought to be contained in a program layer between an application (such as a web browser or HTTP) and the Internet's TCP/IP layers. The sockets part of the term refers to the sockets method of passing between a client and a server program in a network or between program layers in the same computer. Netscape's SSL uses the public-and-private key encryption system from RSA, which also includes the use of a digital certificate.

Netscape includes the client part of SSL in the Netscape web browser. If a web site is on a Netscape server, SSL can be enabled and specific web pages can be identified as requiring SSL access. Other servers can be enabled by using Netscape's SSLRef program library which can be downloaded for noncommercial use and licensed for commercial use. Netscape has offered SSL as a proposed protocol to the World Wide Web Consortium (W3C) and the Internet Engineering Task Force (IETF) as a standard security approach for web browsers and servers. In order for SSL to work, both the client and the server must be SSL-enabled (Minoli and Minoli, 1998). For better or worse, Microsoft will have something to say about the adoption of SSL in the world of Internet commerce.

17.5 SUMMARY AND CONCLUSIONS

Electronic commerce on the Internet implies the integration of various sophisticated technologies. The primary barriers to the growth of electronic commerce are most likely not of a technical nature. Instead, consumers and businesses both must develop confidence in electronic commerce systems. Both worry about privacy and security, and credit card companies (among others) must be very concerned with issues of liability.

Electronic commerce is literally a new way of doing business. As was the case with innovations that preceded it, some people have rapidly adopted electronic commerce while others are taking a more conservative approach, allowing the technology to prove itself before they jump in. The rapidity with which this might happen should not be underestimated. Many believed that consumers would not accept automatic teller machines for many legitimate reasons, such as a lack of confidence in electronic bookkeeping and concern for their as physical safety while retrieving cash in open and public places. However, the advance of electronic commerce is as inevitable as the drive for lower costs of doing business and better service. This drive is what will push electronic commerce, as an acceptable way of completing transactions, into the next millennium.

REFERENCES

Canadian Minister of National Revenues (1998). *Report of the Committee on Electronic Commerce.* North York, Ont.: CCH Canadian.

Cerf, V. G. (1998). Stranger than truth or fiction: Fraud, deception, and the Internet. In Tapscott, D., Lowy, A. and Ticoll. D. (eds.), *Blueprint to the digital economy: Creating wealth in the era of e-business* (pp. 371-383). New York : McGraw-Hill.

Cook Report Internet Glossary by Subject (1998). [Online http://www.cookreport.com/cook/glossary.html, as of 1998, November 27].

Crocker, S.D. and Stevenson, R. B. (1998). Paying up: Payment systems for digital commerce. In Leebaert, D. (Ed.), *The Future of the Electronic Marketplace.* Cambridge, Massachusetts: The MIT Press.

Electronic Commerce (1998). *WDVL: The Illustrated Encyclopedia of Web Technology* [Online http://wdvl.com/Internet/Commerce/index.html, as of 1998, November 24].

Howell, G. (1998, November 8). *Electronic Commerce Rev. Proc. WebNet 98 World Confer. of the Assoc. for the Adv. of Comput. in Educ.,* Orlando, Florida. Available on CD-ROM.

Insights into Online Advertising: Highlights from "The Online Consumer Survey." (1998, August). Northwood, Ohio: NFO Interactive and Jupiter Communications.

Jordan, P. (1998, November 16). E-Business Report Part 1: E-Business Click On Profit — Electronic commerce reaches beyond simple transactions. It's a whole new way of doing business. *VarBusiness, 88.*

Klopfenstein, B. C. (1997a). Internet economics: An annotated bibliography. *J. Media Econ.,* 11(1), 33-48.

Klopfenstein, B. C. (1997b). Internet economics: Pricing Internet access. *Convergence: J. Res. New Media Technol.,* 3(4), 10-20.

Merkow, M. (1998, June 17). Your Body IS Your PIN! [Online http://www.webreference.com/ecommerce/mm/column3, as of 29 November 1998 Westport, CT: Mecklermedia.

Minoli, D. and Minoli, E. (1998). *Web commerce technology handbook*. New York: McGraw-Hill.

Mougayar, W. (1998). *Opening digital markets: Battle plans and business strategies for Internet commerce* 2nd Ed. New York: McGraw-Hill.

Schnell, S. (1996, November). Codes, Commerce, and National Security: A 10 year Perspective on Cryptography [Online http://www.rsa.com/oracle4/, as of 28 November 1998] Bedford, Massachusetts: RSA Data Security, Inc.

Smedinghoff, T. J. (1998, November 25). Summary of Electronic Commerce and Digital Signature Legislation [Online http://www.mbc.com/ds_sum.html, as of 28 November 1998] Chicago: McBride, Baker & Coles [law firm].

Tapscott, D., Lowy, A. and Ticoll. D., (eds.) (1998). *Blueprint to the digital economy: Creating wealth in the era of e-business*. New York: McGraw-Hill.

Technology Overview (1998). Netscape Security Features Evaluation Guide [Online http://www.netscape.com/products/security/resources/evalguide/tech.html, as of 28 November 1998].

Thing, L. (1998). Encryption and Decryption [Online http://whatis.com/encrypti.htm, as of 28 November 1998] Kingston, New York: whatis®.

United States Department of Commerce (1998, April). *The Emerging Digital Economy*. Washington, D.C.: National Technical Information Service.

VNU E Commerce Glossary (1998). [Online http://www2.vnu.co.uk/e_com/e_07_01.htm, as of 29 November 1998] London: VNU Business Publications.

Zimits, E.C. and Montaño, C. (1998, April). Public Key Infrastructure: Unlocking the Internet's Economic Potential. iWord, 3(2). [Online http://www.iword.com/iword32/istory32.html, as of 28 November 1998].

18 Emerging Security Testing, Evaluation, and Validation

The Key to Enhancing Consumer Trust in Security-Enhanced Products

Paul J. Brusil, L. Arnold Johnson, and Edwin F. Steeble

CONTENTS

0-8493-9594-1/00/$0.00+$.50
© 2000 by CRC Press LLC

18.1   OVERVIEW

Networked information technologies are changing the way the world interacts and the way industry, government, and other sectors do business. In the emerging electronic and global society, new electronic business (e-business) models are replacing traditional models propped up by trust built through personal interaction. With the proliferation of e-business powered by evolving network and information technology (IT) products, industry must earn consumer confidence by demonstrating that it has taken effective measures to protect the information being handled electronically. A key method of illustrating its commitment to safe transmission, processing, and storage of information is through validated, impartial, standards-based evaluations. Conducting such evaluations increases the confidence, or trust, that security features of network and information technology (IT) products are correctly and completely implemented and that these products behave as promised. Given the implications of vulnerabilities for the national economy and national security, the government has recognized the importance of safeguarding networks, particularly of critical national infrastructures. To promote an international marketplace for trusted, security-enhanced network and IT products, and in so doing protect its national interests, the U.S. Government formed the National Information Assurance Partnership (NIAP), a collaboration between the National Institute of Standards and Technology (NIST) and the National Security Agency (NSA). The NIAP program has seeded and is furthering the growth of a robust, state-of-the-art, commercial, security testing and evaluation industry. The NIAP is fielding a flexible national scheme to accredit private-sector security testing and evaluation laboratories and to oversee laboratory activities to ensure that security tests and evaluations are conducted in accordance with new, internationally recognized standards.

18.2   A NEW WAY OF DOING BUSINESS

Trade journals have documented that electronic businesses are transforming traditional business models in all market sectors. Business partners, suppliers, regulators, customers, and users worldwide are increasingly sharing critical information 7 days a week, 24 hours a day. Financial reports, business plans, inventory data, supply-chain management data, customer order data, health care claims and referrals, and other sensitive information are being communicated. Ready availability of such information is dramatically improving old business relationships and introducing new ones, to the detriment of competitors who do not make information easily accessible.

According to the trade press, new e-business models, based on ready and widespread availability of business data, are having profound effects. These new models are globalizing business operations by erasing traditional boundaries of time and space. They are improving business efficiency by more tightly integrating business processes and units, decreasing time to market, and reinventing and improving personalized customer services. They are attracting new business opportunities and stimulating new business relationships in ever more complex supply chains. They are improving customer satisfaction, decreasing delays, and lowering the cost of doing business by orders of magnitude in diverse commercial market sectors. Businesses and service providers that fail to offer online e-business services to consumers and trading partners are losing out.

The following are just a few examples of emerging business changes and their impact:

- In the banking world, costs to process an electronic transaction are now under a penny compared to a dollar or more when customers deal directly with tellers.
- In retailing, online ordering and processing costs are being comparably reduced relative to telephone-based orders or in-store purchases, and online sales activities are being more highly targeted.
- In the transportation industry, the National Transportation Exchange (NTE) has introduced a new e-business model by providing an electronic trading floor to balance supply and demand by matching buyers and sellers of trucking space.
- In the automobile industry, the Automotive Exchange Network (ANX) provides unprecedented communication among competitors by effectively lowering the cost structure of the entire auto supply chain, thus allowing everyone in the chain to benefit.

18.3   ALONG WITH E-BUSINESS COMES A NEED TO PROTECT INFORMATION

The more business is conducted electronically, the greater the need to protect the information being handled. Security has become a necessary enabling technology for e-business. New business models depend on trust of the network and IT infrastructure used to conduct e-business among consumers and suppliers. Security has become a necessary revenue generation enabler, a necessary precursor to consummate new business ventures, and a significant cost-avoidance factor. One recent survey* showed that cyber attacks in the U.S. are rising and the direct costs attributed to such attacks are significant and growing. In 1998 such direct financial losses amounted to about $150 million. But direct costs may not be as significant as other indirect costs. If sensitive information, such as corporate strategies, customer profiles, and trade secrets, is not adequately protected, businesses can be destabilized, competitive advantage can be lost, and tenuous buyer/seller allegiances can be altered. In worst case scenarios, the consequences of compromised information can be dire and even fatal, such as in the medical sector where subversion of critical information concerning diagnoses, histories, or treatments during electronic referrals can lead to inappropriate and life-threatening medical decisions.

---

* "Issues and Trends: 1998 Computer Crime and Security Survey" by the Computer Security Institute and the Federal Bureau of Investigation, San Francisco, http://www.gocsi.com/prelea11.htm.

As the information age engulfs society, there is unprecedented demand in the commercial sector for security-enhanced network and interconnected IT products that can be trusted. Indeed, trust in the products they buy and use is what compels the telecommunications service providers to offer security service-level agreements that contractually guarantee the delivery of networking and IT security services. Trust evidenced through recognized security testing and evaluation is proving to be a powerful legal case builder to demonstrate "prudent business practice and due diligence" when seeking to reduce financial cost in liability suits associated with e-business.

In addition to the commercial world, society at large is beginning to depend on trusted network and IT products. In 1998, a Presidential Decision Directive[1] recognized that the viability of the U.S. (and accordingly the world's strongest military and largest national economy) relies upon critical national infrastructures. These infrastructures exist in both the public and private sectors: telecommunications, energy, banking and finance, transportation, water systems, and emergency services (including law enforcement, public health, and disaster recovery services). The directive recognizes that the network and IT systems that effectively link these infrastructures are key to ensuring minimal orderly functioning of the economy and government. The directive acknowledged that future enemies might seek to launch nontraditional offensives against the U.S. by attacking the critical infrastructures and the networking and IT upon which they rely. The directive requires "that the United States take all necessary measures to swiftly eliminate any significant vulnerability to both physical and cyber attacks on our critical infrastructures, including especially our cyber systems."

The U.S. is not alone in its resolve to protect itself from cyber attack. Other countries agree that determining the "... trustworthiness of [network and IT-based] products for national security systems has become a necessary objective of governments and businesses around the world."[2]

18.4 COMPETITION HEATS UP FOR TRUSTED PRODUCTS

System vulnerability concerns are yet one more problem being faced by modern enterprises already grappling with downsizing, market share, profits, and time-to-market questions. In today's e-business world, organizations have access to a growing number of security-enhanced network and IT products with various claimed security capabilities and with various less well-known limitations. Indeed, the size of the security product market is growing at a rate of over 75% per year and should reach $5 billion by the year 2000.* Customers must make important decisions about which of the products in such a large market provide an appropriate degree of protection for their assets. More and more, organizations

---

* 1997 Network Security Market Growth Analysis by The Computer Security Institute, as reported in Solutions Partner Presentation, RedCreek Communications, Inc., 3900 Newpark Mall Road, Newark, CA 94560, May 13, 1998.

find themselves needing to place increased trust in the security-enhanced products that they acquire.

In light of the several other business problems needing attention, organizations are looking for help to assist them in confirming, or validating, the level of trust they can place in products. Reliance on valid information about the degree of trust that can be placed in a particular product will become ever more critical as

- networking and IT continue to evolve rapidly,
- new networking and information technologies emerge,
- networking and IT products and systems become increasingly complex,
- business develops a more critical dependence on such products and systems, and
- businesses' technical support staffs get even less time to keep up with all such changes.

18.5 SECURITY TESTING AND EVALUATION IS KEY TO ENHANCING TRUST

But how can trust be developed in a security-enhanced network or IT product? Assessments of the security soundness of products can provide trust that the products are reliable and perform as expected. Such assessments are especially compelling when they are made according to well-known, well-engineered, and well-understood security testing and evaluation practices.*

Such testing and evaluation benefits all organizations within the chain of designing, building, marketing, procuring, and using products that are intended to be trusted. Designers and builders need effective product testing and test methods before shipping products. Vendors rely on testing to demonstrate compliance with consumer requirements regarding trust. Vendors also rely on testing and evaluation to increase the value and marketability of their products to would-be consumers. Consumers rely on testing and evaluation as a way of developing trust by ensuring product conformance to their security requirements. They also see testing and evaluation as providing a *mark of quality* and a way to differentiate between competing products. Users are beginning to rely on testing and evaluation to help establish due diligence in legal disputes.

Formal testing of network and IT products has traditionally been used to ensure product conformance to functional, performance, reliability, or interoperability standards. But testing the implementations of security is different. Implemented security services are intended to protect the functionality within, the performance achievable by, or the reliability expected of a component. When testing and evalu-

---

* To those readers familiar with security and the testing and evaluation of security-enhanced products, the term "evaluation" is typically used to mean both the testing of a product as well as the evaluation and analysis of its architecture, design, documentation, code, etc. In this chapter, the terms "testing" and "evaluation" are used in their traditional, colloquial senses. Herein the word testing implies the stimulation of an implementation and the observation of responses from the implementation; and, the word evaluation implies just the analysis of a product's architecture, design, code, etc.

ating a product that claims to provide security services, the confidence demands are greater than those associated with testing the functionality, performance, or reliability alone. Testing and evaluation requirements for security become more complex and difficult. For example, in addition to testing that certain behaviors exist in accordance with the specifications, security testing and evaluation must also help ensure that unwanted behaviors do not happen.

What further makes security assessment difficult is that the amount and type of analyses performed as part of security evaluation efforts will need to be increased as the degree of trust desired to be established is increased. Such analyses may include scrutiny of a product's architecture, design, and source code, depending on how much trust is desired. Furthermore, trust in a product can be enhanced when the testing and evaluations are performed and validated by competent, independent (ideally separate), third parties, i.e., one party for testing and evaluation, and another party for validating the testing and evaluation results.

18.6 NUMEROUS APPROACHES EXIST FOR SECURITY TESTING AND EVALUATION

Until recently, there have only been a few organizations able to perform competent, impartial security testing and evaluation — and even fewer effective methods for conducting such security assessments. Current and traditional approaches to security testing and evaluation include the following.

18.6.1 HACKING

De facto assurance of the underlying security in a product can arise from aggressive students as well as professional technicians and users who actively probe new products for security flaws. Hacking does not necessarily follow a consistent or comprehensive approach to evaluating the quality of the security functions and services that are implemented. Hence, the assurance achieved is to some uncertain, typically very modest, level of trust.

18.6.2 INITIAL COMMERCIAL APPROACHES

Initial commercial approaches arose typically to support trade press surveys or to provide surface-level testing results for vendor brochures. These approaches are often based on simple, one-size-fits-all (often-called *low-hanging-fruit*) testing that provides minimal, cursory checks of some of the implemented security functionality. No evaluation is made of the confidence (assurance) that can be associated with the soundness (or lack thereof) of the security implementation.

18.6.3 GOVERNMENT-INTERNAL APPROACHES

In order to introduce consistency in describing the security features and levels of trust of a limited set of security-enhanced products, and in order to facilitate comprehensive testing and evaluation of such products, the U.S. Department of Defense

(DOD) developed the Trusted Computer System Evaluation Criteria (TCSEC).[3] The TCSEC — or more colloquially, the "Orange Book" — defined a small set of six classes of increasing security functionality and increasing assurance (from a so-called C1 class to an A1 class) that applied to operating systems. The TCSEC was extended to networking devices[4] and database management systems.[5] Government in-house evaluations were offered first, followed by comparable, government-sponsored commercial evaluation services. Use of TCSEC concepts extended beyond the DOD to other U.S. government agencies and to foreign governments as well.

18.6.4   VENDOR SELF-DECLARATIONS

Another initial approach was based on vendor self-declarations that a specific product meets the needs of their customers in terms of the appropriate amount of confidence to be placed in its implemented security features. In part, such confidence was implicitly tied to the reputation of, or past experience in dealing with, a specific vendor to do a good and adequate job of implementing security.

18.6.5   GOOD SOFTWARE ENGINEERING APPROACH

Another approach is based on the notion of providing trust through use of sound engineering practices during product architecting, design, and implementation, rather than through post-implementation testing and evaluation. One way a software developer can demonstrate competence in building products is through recognized, so-called *capability maturity* assessments of the developer and the developer's software engineering processes. Security-enhanced products built by organizations with demonstrated expertise and maturity can be viewed to merit greater trust than products built by organizations that do not demonstrate mature, competent, software design and engineering capabilities.

18.6.6   CONSUMER EVALUATIONS

Consumers can develop the requisite substantial technical expertise in-house to test and evaluate specific security-enhanced products directly. Alternatively, consumers can contract a private evaluator or evaluation organization to do such testing and evaluation.

18.7   SHORTCOMINGS OF OLD APPROACHES PROVIDE REQUIREMENTS FOR A NEW APPROACH

All the previous security testing and evaluation approaches added value to the commercial networking and IT marketplace. The degree to which different approaches helped tended to depend on the situation to which they were applied. The lessons learned from these early approaches have been analyzed, used, and integrated to fuel the development of a new, best-of-all-previous-breeds approach (see "The National Information Assurance Partnership Embraces the New Approach to Security, Testing, and Evaluation," Section 18.11).

The types of shortcomings discovered in the old approaches had greater or lesser degrees of significance dependent on the approach and situation in which it was used. Not all shortcomings accompanied all approaches. Often, the old approaches were not as simple, effective, or inexpensive as they were thought to be. In any event, these shortcomings provided valuable insights, experience, and expertise that have benefited the development of the new, emerging approach.

The categories of the shortcomings that surfaced, and that have since been leveraged to develop the new approach, include a lack of

- a flexible common approach to specify security requirements,
- flexibility of the security testing and evaluation methodologies,
- a common security testing and evaluation methodology,
- security testing and evaluation expertise,
- an independent third party to conduct impartial testing and evaluations,
- an impartial third party to validate the quality of such independent testing and evaluations.

The ramifications of these shortcomings were many, as summarized in the following sections. But, in general, security testing tended to add cost to security-enhanced goods and services. It tended to be insensitive to vendors' time-to-market pressures, sometimes adding delays to product rollout schedules (in the worst cases to beyond the full lifecycle of the product undergoing testing). Sometimes delays arose because of the specific testing or evaluation methodology used. For example, some methodologies relied on a heavily iterated and sequential approach that cycled from testing, to patching and fixing, to retesting and regression testing, and then on to other new test scenarios. In the case of the TCSEC approach, delays sometimes seemed to arise because vendors were not prepared in terms of making detailed documentation and code available, or they sometimes seemed to lack the incentive to expedite testing and evaluation efforts that were subsidized by the government.

Specific ramifications of the shortcomings of early approaches surfaced.

18.7.1 THE LACK OF SECURITY REQUIREMENTS SPECIFICATION STANDARDS

There has been no common and flexible approach that could be used to specify, or to identify, just the right amount of security functionality and assurance for differing products or classes of products. The impact of a lack of a commonly understood, standard, specification language is particularly important in today's fast changing world of networking and IT. Accordingly, users have tended to be confused or unsure as to what security features are really being claimed for a product. Conversely, vendors have tended to find it difficult to articulate their security claims in ways that users can understand and in ways that distinguished one product from another. Time and money were wasted in conveying knowledge about security features or security implementations.

18.7.2 THE LACK OF FLEXIBILITY OF SECURITY TESTING AND EVALUATION METHODOLOGIES

Many testing and evaluation approaches were inflexible. They were not always able to provide desired trust in products because they could often not be tightly tied to products' security claims. Some approaches were only able to provide coarse, surface-level testing. The lesson of such approaches is that they did not always prove to be cost-effective or meaningful.

18.7.3 THE LACK OF COMMON SECURITY TESTING AND EVALUATION STANDARDS

For users, it has been typically difficult, if not impossible, to compare security test and evaluation results associated with differing products. This is especially so if the tests and evaluations were performed by different testing and evaluation facilities or if the tests and evaluations were associated with different testing and evaluation methodologies. Accordingly, users could not readily compare various products to understand their relative security capabilities and limitations.

From the vendor's perspective, lack of a common, internationally accepted testing and evaluation methodology reduced international competitiveness for product sales. This resulted because tested products still often needed to undergo further user-specific, country-specific, or region-specific, retesting efforts. It was not uncommon for some vendors, especially those supporting large enterprise customers, to conduct a plethora of multiple, often redundant, tests and evaluations on a single product. The costs of retesting prevented vendors from being able to realize many economies of scale with respect to security testing and evaluation.

Accordingly, many leading-edge vendors developed a business case demand to institute a standard set of tests and evaluations that could cover the majority of their testing and evaluation needs across most customers (domestic or foreign). Such standards could be used to eliminate or minimize duplicate testing, could be used for cost avoidance associated with retesting, and could therefore positively impact product price and profit margins.

18.7.4 THE LACK OF SECURITY TESTING AND EVALUATION EXPERTISE

Only a handful of organizations have had sufficient testing and evaluation expertise to be able to conduct a single, competent, independent security testing and evaluation campaign, let alone many concurrent security testing and evaluation campaigns. Fewer still were security testing and evaluation approaches that were both effective and efficient. In the highly competitive, fast-paced network and IT world, testing and evaluation approaches that added extensive costs or delays proved not to be relevant.

18.7.5 THE LACK OF THIRD PARTY TESTING AND EVALUATION

There have been few third party testing laboratories available to conduct impartial security assessments to increase the perceived level of confidence in products; many vendors and consumers feel it necessary to use a competent, recognized, impartial, third party security testing and evaluation laboratory to conduct security assessments. From customers' perspectives, such third party testing and evaluation dramatically lowers any sense of impropriety.

18.7.6 LACK OF ACCREDITATION, VERIFICATION, AND VALIDATION OF INDEPENDENT SECURITY TESTING AND EVALUATION

There has been a lack of third parties to accredit the worthiness of independent testing and evaluation laboratories, to verify that tests and evaluations have been properly performed, and to verify that testing and evaluation results are appropriate and valid. This has lead to uncertainty and diminished consumer trust regarding the consistency and robustness of existing testing and evaluation approaches. It has also led to difficulty in comparing the results of testing and evaluation performed by different organizations. Vendors and consumers feel it valuable to have a third party to independently oversee, review, and/or otherwise validate the security testing and evaluation and to make sure the testing and evaluation approaches are used consistently across different product security assessments.

18.8 WHAT ELSE IS NEEDED

As networking and IT continue to change rapidly, as products become increasingly more complex, and as our dependence on them becomes more pronounced, effective and economic testing and evaluation of the security implemented in such products becomes even more critical. Security products and security-enhanced network and IT products must change to stay ahead of evolving threats. The tests and the test and evaluation methods and metrics used to evaluate such rapidly proliferating and changing product offerings must also be able to evolve quickly.

Users want to be able to create tailored specifications for the security-enhanced products they need to solve their specific problems. They do not want the completion of such specifications to depend on a lengthy standardization process. Instead, users now want to have control over their own specifications, rather than a long-term involvement in some standardization process. They want to be able to take full advantage of the COTS (commercial-off-the-shelf) marketplace and be able to stipulate their specific requirements in a language that both the user and vendor understand. They also do not want to be at risk that the language they use will become obsolete.

18.9 SETTING THE STAGE

Many other events and circumstances further set the stage for a new approach to security testing and evaluation of network and IT products. While the earlier described initial approaches for security testing and evaluation filled early needs, a changing marketplace demanded a new approach based on international standards. The marketplace for which the initial approaches were developed had been changing rapidly in the mid to late eighties both in terms of vendor production philosophies and erosion of U.S. centricity. Acquisition and development of unique, consumer-specific, network and IT products and systems were phased out in favor of an approach based on (a) acquisition of COTS network and IT products, coupled with (b) consumer integration of these COTS products into network and IT systems. Furthermore, the shift to networked computers and applications and use of new network and IT services brought forth a plethora of new security-enhanced network and IT products. These products began addressing security functional and assurance requirements far more diverse than could be handled by approaches such as those within DOD's "Orange Book."

Also, different countries and organizations began developing other security criteria divergent with U.S.-based approaches. In Europe, several countries jointly developed the Information Technology Security Evaluation Criteria (ITSEC) in 1991. The Canadian Trusted Computer Product Evaluation Criteria (CTCPEC) was developed in 1993 as a combination of ITSEC and DOD's "Orange Book" approaches. In the U.S., the Federal Criteria were also drafted in 1993 as another attempt to combine North American and European security evaluation criteria approaches.

These marketplace changes weakened the effectiveness of initial security testing and evaluation approaches. The shift of consumer reliance to COTS products with rapid vendor turnover cycles began rendering the slower reacting, government-internal, security testing and evaluation model less effective. Because of the various different security product evaluation criteria and because of concomitant marketplace confusion about the many criteria, many countries and organizations chose not to embrace government-oriented security testing and evaluation approaches.

18.10 STANDARDIZATION BEGINS

With the advent of such marketplace dynamics, a new approach for security testing and evaluation became even more evident. New international standards for specifying security requirements and features and for defining accompanying security testing and evaluation methods were needed. In 1993 the development of such standards began under the auspices of a multi-national Common Criteria (CC) project. One of the goals of the standardization efforts within the CC project was to develop new, state-of-the-art concepts by leveraging experiences gathered and lessons learned via all the earlier security specification, testing, and evaluation efforts.

The impacts of these new standards (see "Relying on Standards" under Section 18.14) eventually became significant. Their emerging ability to support specification and assessment of virtually unlimited, different combinations of security functionality and assurance for any security-enhanced network and IT product, or class of products, began reducing the utility of the more limited, initial approaches.

18.11 THE NATIONAL INFORMATION ASSURANCE PARTNERSHIP EMBRACES THE NEW APPROACH TO SECURITY, TESTING, AND EVALUATION

In light of the needs, problems, and factors presented above, the National Institute of Standards and Technology (NIST) and the National Security Agency (NSA) formed a collaborative venture to pursue the emerging new CC-based international approach to security assessment. This collaboration was dubbed the National Information Assurance Partnership or NIAP.*

The partnership combined the extensive IT security testing and evaluation experience in both agencies. The NIAP was initiated to help ensure the availability and quality of security-enhanced network and IT products and systems by means of new, commercial testing and evaluation services that are government-accredited, government-validated, and internationally recognized.

18.12 WHY THE U.S. GOVERNMENT GOT INVOLVED

The U.S. government got involved with this new approach in order to transfer its vast security assessment knowledge base and to represent the U.S. interests in the development of these international standards and agreements.

With the shift to new international standards, there were compelling reasons for the U.S. government to remain involved, but with a shifted focus, with the newly emerging CC-based security testing and evaluation paradigm. The U.S. government had significant security and security assessment knowledge and expertise it could contribute to the standardization efforts. The government also had a role in representing U.S. interests in developing the international CC standards and associated testing, evaluation, and implementation guides. The government was also needed to help develop and approve international agreements for recognizing the results of security tests and evaluations performed in other countries in accordance with these standards. In addition, the U.S. government could provide oversight to maintain quality and consistency of independent testing and evaluation facilities. It could foster research pertaining to new approaches for security testing and evaluation. Also, it had some ability to stimulate user-pull (demand) and vendor-push (investment) for the security-enhanced network and IT products that could arise from testing and evaluations done in accordance with the new CC-based security assessment approach.

NIST and NSA were the most appropriate elements of the U.S. government to bring together to form the NIAP. Each party was able to bring a distinct but complementary mission to this partnership. NIST is responsible for standards and guidance for the unclassified but sensitive systems within the U.S. government. Also, NIST has a statutory responsibility to assist the private sector when requested. NIST could leverage its traditional role in research, standards, accreditation of private laboratories, and metrology to address security testing, evaluation, and assurance needs. NSA is responsible for the security of classified

---

* The NIAP web site can be found at http://niap.nist.gov.

government systems. With an increasing shift to COTS products and a desire to shift evaluation of commercial products to the private sector, the NSA had a compelling need for its customers to use COTS products with a reasonable degree of assurance. The NSA brought many years of experience from working on the security of very sensitive, critical systems.

With the above factors as a backdrop, NIST and NSA signed a letter of partnership in August 1997 thus forming the NIAP.

18.13   NIAP'S GOALS

The strategic goals of NIAP are

- to improve the trust of citizens, private sector organizations, and government in the security and reliability of networking and IT that we rely on so heavily for conducting business,
- to help ensure the development, manufacture, and use of security-enhanced networking and IT products, with such trust, through the creation and maintenance of a standards-based, commercial, security testing and evaluation industry whose goal is to be cost effective, and
- to establish a program to ensure quality and validity of testing and evaluation and to foster an international marketplace for tested and evaluated products.

18.14   THE NIAP PROGRAM – ITS VISION AND APPROACH

To meet its strategic goals, the NIAP has initiated a program aimed at instigating both a new commercial security testing and evaluation industry with appropriate supports and a new security assessment paradigm. This new approach is unlike any of the other approaches summarized earlier. It capitalizes on the strengths, and overcomes the problems, of the earlier approaches.

The elements leveraged by NIAP for the new approach include

- newly emerging standards for specifying, testing, and evaluating security-enhanced products,
- many well-known and predefined security functional and assurance requirements profiles for specific areas of technology and specific vertical industries,
- commercial, competitive testing services that can be selected by customers based on their own, specific, business considerations,
- a scheme for accrediting testing laboratories, to foster testing quality and consistency, and for further increasing product trust by validating test and evaluation results from those laboratories,
- research and development to improve commercial security assessment quality,

- a framework for fostering international trade by recognizing validated test and evaluation results thereby fostering a "test once, buy/sell anywhere" marketplace,
- outreach to monitor the effectiveness of the approach, to tune the approach to evolving marketplace needs, and to promote development and enhancement of the quality of commercial, security-enhanced products.

The following subsections provide more details on each of these elements. Section 18.15 ("A New Common Criteria Scheme Ties Together the NIAP Elements") shows how these elements are related.

18.14.1 Relying on Standards

The NIAP approach relies on the use of international standards for specifying

- security requirements in products and systems,[6] and
- common security testing and evaluation methods.[7]

These standards are referred to as the CC (Common Criteria) and the CM (Common Methodology for Information Technology Security Evaluation), respectively. Use of the CC and CM standards is key to providing a common, internationally recognized understanding of IT security requirements and IT security assessment methods.

The CC provides a standard *language* for specifying security requirements. It also provides a flexible *method* for specifying security requirements of all sorts.

The standard language is contained in a pair of voluminous catalogs of elementary, re-usable, components of specific security functional and assurance requirements. Security functional requirements are organized into 11 major classes, such as auditing, cryptographic support, and security management. Similarly, security assurance requirements from several evaluation assurance classes form a set of seven defined levels of assurance. The elementary assurance requirements specify reasons to trust implemented security functionality to be effective and correct. These assurance levels articulate increasing rigor and formalism for ensuring increasing confidence in implementations of security functionality. The assurance levels range from a low assurance level, called Evaluation Assurance Level 1 (EAL1), to a high assurance level (EAL7).

The flexible method is based on the ability to use the CC to tailor-develop any of two different types of security requirements profiles.* A product-specific type of CC specification is called a security target (ST). It is typically developed by a vendor to describe the security-relevant portions of a single, specific product. The

---

* To those readers familiar with security testing and evaluation, the term *profile* is reserved for use only with the notion of a Protection Profile (defined later in the text above). In this chapter, the term is used in its traditional, less-constrained, colloquial sense of a selection of significant features from a larger set of features.

other type of CC specification profile is called a protection profile (PP). It is typically developed by

- a single user organization, or
- some broad user constituency with similar interests, or
- a consortium of vendors.

A PP is used to articulate the set of security requirements that define users' needs or that can define a class of desired products wherein any number of implementations may satisfy the stipulated requirements.

STs and PPs are constructed by selecting from the CC catalogs the set of elementary functional and assurance requirements that appropriately define the security aspects of a specific product or a generic class of products, respectively. The result is a tailored profile of standard security requirements. User needs and vendor product claims are profiled as specific subsets of standard security requirements from the CC catalogs. Some of the standard requirements may be refined from that which appears in the CC. Thus, solutions can be identified with exactly the degrees of security functionality and levels of assurance needed, no more and no less, for any particular situation. Being standard security requirements, they will generally be widely understood throughout the marketplace.

The CM defines assessment methodologies for CC-based testing and evaluations. It describes actions for conducting product tests and evaluations for a variety of assurance levels. Such common, well-recognized testing and evaluation approaches reduce the need for customer-unique and country-unique approaches.

Use of the CC and CM thus provides a common base for describing security-enhanced products and assessing whether they work as claimed. These standards form the foundation for international recognition of test results. They also form the basis for consumers of security-enhanced products to gain higher levels of confidence in the products they buy than has heretofore been generally available. The effect of these standards has been to raise the bar relative to trust in products.

The CC and CM are under various stages of public scrutiny and are thought to be technically fairly stable. Final standardization efforts for the CC are in progress in the joint International Organization for Standardization (ISO)/ International Electrotechnical Committee (IEC), Joint Technical Committee 1 (JTC 1) for Information Technology, Subcommittee 27 (SC27) for Security Techniques, Working Group 3 (WG3) for Security Criteria. Information about these standards and related activities is available on the World Wide Web.* The CM is under development in a multinational project and will likely be transitioned to the ISO/IEC community in the near future.**

---

* Public release versions of the CC and CM are available at http://csrc.nist.gov/cc/.
** The CC and CM are also available at http://ccse.cesg.gov.uk. This web site hosts the Common Criteria Support Environment (CCSE) which is expected to provide access to several CC-related and CM-related materials, such as Requests for Interpretations of the CC and CM Observation Reports, as well as access to newsgroups for discussing such materials.

18.14.2 GROWING THE SET OF SECURITY REQUIREMENTS PROFILES

As a way of jumpstarting the security-conscious community to begin using the new CC-based approach, NIAP, as well as its NIST and NSA parent organizations, have supported the development of a starter-set of PPs. Diverse user constituencies and vertical industry consortia are being encouraged to seed the marketplace with diverse, initial sets of PP requirements profiles. Vendors are also being encouraged to begin developing STs. Examples of the types of CC-based security requirements that existed, or were being completed, at the time of the writing of this chapter are indicated later.

Since early experience indicated that development of PPs and STs could be a daunting task for the uninitiated, NIAP has provided help in developing profiles and intends to continue providing help in a number of ways. The types of services that NIAP has provided to aid interested parties in specifying CC-based security requirements include

- profile development guidance,
- CC training,
- profile construction training,
- semi-automated profile analysis and construction tools,
- direct support to the initiation and construction of selected PPs,
- review (a.k.a. vetting) of selected draft PPs,
- validation services for formally evaluated PPs,
- a PP registry, and
- workshops, conferences, and forums to help produce, proliferate, and promote PPs.

For an understanding of the services that NIAP currently provides in this area, interested readers should visit the NIAP web site http://niap.nist.gov/.

18.14.3 SEEDING AND USING COMMERCIAL LABORATORIES

With the advent of the CC and CM and with growing proliferation of CC-based security requirements profiles, it became feasible to transition security assessment expertise and operations from current government facilities into approved, accredited, private sector laboratories that provide CC-based testing and evaluation.

In 1997, the NIAP began encouraging the initiation, growth, and development of a state-of-the-art, CC-based, commercial security testing and evaluation industry. Commercial laboratories operating under the auspices of NSA's Trust Technology Assessment Program (TTAP) provided initial CC-based testing and evaluation services.* The TTAP laboratories conduct CC-based testing using NSA's TTAP evaluation methodology. Commercial laboratories operating under the auspices of NIAP provide CC-based testing and evaluation using the CM.

The laboratories within this new industry have competitive flexibility to adjust their testing and evaluation services to accommodate different products and different security requirements. The laboratories operate by establishing private contracts

---

* Information on TTAP can be found at http://www.radium.ncsc.mil/tpep/ttap/index.htm.

with customers to provide such services as PP evaluations, ST development support, and assessments of the ST-specified features in security-enhanced network and IT products. As part of this initial effort, a number of market-dominating countries agreed[8] to recognize, multinationally, the results of this burgeoning U.S. testing and evaluation industry.

The types and degrees of testing and evaluation that need to be performed on products depend on the underlying security functional requirements and the degree of confidence desired in those products. Being based on the CC and CM, such tests and evaluation procedures are becoming well-known, repeatable, and credible.

18.14.4 ACCREDITING COMMERCIAL LABORATORIES

To increase trust in security assessments further, NIAP is instituting mechanisms for providing cost-conscious, government accreditation of commercial security testing laboratories. Such accreditation is in concert with international agreements regarding the multicountry mutual recognition[2] of security assessments. NIAP worked with NIST's internationally recognized National Voluntary Laboratory Accreditation Program (NVLAP) in 1998 to begin developing a laboratory accreditation process and procedures to accredit commercial testing laboratories. The process needed to be flexible so that laboratories could be accredited for exactly the types of security assessments they wanted to perform — no more or no less. The accreditation process and procedures are coming into place.

The accreditation mechanisms are being designed to assess a laboratory's ability to test products using test methods based on the CC and CM. More specifically, they are being used to ensure that commercial security assessment laboratories have the requisite capability to conduct quality security evaluations of network and IT products. They are ensuring consistency and quality among the different commercial testing laboratories both in terms of the quality of testing services they provide and the test results they produce.

According to the emerging accreditation mechanisms, laboratories are accredited, and periodically re-accredited, by NIST's National Voluntary Laboratory Accreditation Program (NVLAP). NVLAP ensures that laboratories meet specific international[9] and national[10] guidelines pertaining to laboratory competency. NVLAP ensures that laboratories meet additional, NIAP-specific requirements pertaining to security assessment procedures and requirements.[11] NVLAP also ensures that testing laboratories have all requisite, NIAP-specified proficiencies needed in order to facilitate subsequent government validation of test results.

Laboratories are accredited for a specific scope of security assessment activities and procedures. For example, a testing laboratory may limit its focus to products in only a specific range of claimed levels of assurance. Thus, a laboratory may choose to get accredited for a specific set of NIAP-approved test methods.

NIAP provides technical guidance, advice, support, and training standards to accredited testing laboratories. NIAP is working to ensure continuing quality within the private, security testing industry by monitoring the accredited laboratories. They are monitored for maintenance of competence and for their adherence to, application of, and interpretation of CC standards.

18.14.5 VALIDATING TEST AND EVALUATION RESULTS

In accordance with the multinational arrangement,[2] NIAP looked to establish independent validation of testing and evaluation results by an impartial third party. The purpose of such validation efforts is to

- increase trust even further in network and IT products that have undergone testing by an accredited testing laboratory,
- promote consistency and comparability among independently conducted assessments, and thereby
- facilitate the international trade for validated, security-assessed products.

NIAP is developing a scheme,[12] the CC Evaluation and Validation Scheme (CCEVS), that stipulates the details of the organization, operations, and management of such a validation concept within the U.S.. According to the NIAP CC scheme, a validation body reviews and provides independent confirmation that security assessments have been conducted according to procedures and guidelines stipulated by NIAP. The amount and depth of private industry oversight to be provided by the validation body is tailorable to the assurance requirements, i.e., the EAL level, claimed of the product under test, the complexity of the IT product, and the experience of the testing laboratory.

The NIAP Validation Body provides confirmation that

- the product was assessed by a testing and evaluation laboratory that is NVLAP accredited and NIAP-approved,
- the laboratory correctly and completely applied the evaluation methodology to verify conformance of the security functional and assurance aspects of the product to a PP or ST,
- the appropriate criteria, test methods, and procedures were used,
- the conclusions of the testing laboratory, as documented in the laboratory's evaluation report, are accurate and consistent with the facts presented in the security assessment.

The scheme stipulates that after the Validation Body has completed the requisite confirmations, the Validation Body facilitates the granting of a CC certificate and accompanying validation report.

The CC certificate is issued by NIAP as designated certificate issuing authorities, namely the NIST Information Technology Laboratory and the NSA Information Systems Security Organization.

The validation report provides information on how well the assessed product conforms to the security functionality and assurance level that it claimed. It indicates the configuration for which the product was assessed, the environment for which the product is intended to be used, the coverage and depth of security analyses, details of the testing approach used, the testing suites used, the testing environment used, the test tools used, and so on.

The NIAP scheme recognizes that other third parties, such as a professional society or a vertical industry association, may choose to implement other validation schemes that may or may not complement the government's scheme.

At the time of the writing of this chapter, NIAP was planning to complete a number of materials related to the scheme in early 1999, including, e.g., NIAP Validation Body policies and procedures, technical oversight and validation procedures, guidance to sponsors of security evaluations, and guidance to testing laboratories.

18.14.6 FOSTERING INTERNATIONAL TRADE

According to the multi-national arrangement,[2] the validation report and accompanying certificate issued by the government Validation Body are the only acceptable evidence that a product has undergone a security assessment that is recognized by the other country partners in the arrangement. Thus, a major benefit of the NIAP-advocated security testing, evaluation, and validation approach is that it opens global markets to vendors. All country partners recognize products that are tested, evaluated, and given certificates by any other country partner. This means that such products can be procured with a known degree of confidence and with no duplicative re-testing in foreign markets. The significant international competitiveness and market opportunities consequently afforded are powerful features that are working to increase the scope and availability of trusted products worldwide and to reduce their cost. The impact of the NIAP approach and the NIAP Validation Body is to help foster such improvements in international trade.

While validation is mandatory to obtaining an internationally-recognized certificate from the U.S. government, it is possible that obtaining such a certificate and its accompanying validation report may be an unnecessary final step for certain communities. For such communities, simply undergoing a security assessment by a government-accredited testing and evaluation laboratory may be sufficient.

18.14.7 PROMOTING R&D

During the first years of its existence, NIAP concentrated on fostering the establishment of the commercial security assessment industry, helping users articulate their security needs in Protection Profiles, and stimulating vendors to articulate their product's capabilities in Security Targets. NIAP is now focusing more attention on associated research and development (R&D).

NIAP is fostering public domain R&D. It intends to expand its support in key R&D areas. At a minimum, areas of interest include developing tools and techniques to help improve the efficiency, flexibility, quality, effectiveness, measurability of, and automation of commercial testing and evaluation methods and approaches. NIAP is especially interested in applied research that leads to quick, low-cost testing and evaluation solutions that can provide better assessment coverage and can be readily embraced within typical vendor product development cycles and product revision cycles.

In support of this, NIAP is investigating the feasibility of alternative assurance approaches, possibly to augment or to supplement its current focus on CC-based testing and evaluation. One such alternative assurance approach is the Systems Security Engineering Capability Maturity Model (SSE-CMM). Development of the SSE-CMM is progressing through active participation and corporate investment of the security engineering community, coupled with sponsorship from the National Security Agency, the Office of the Secretary of Defense, and the Canadian Communications Security Establishment.

The objective of the SSE-CMM efforts has been to advance security engineering as a defined, mature, and measurable discipline, with the effect of improving the quality, cost and availability of, and trust in, IT products, systems, and services. A project has been established* to provide a framework for measuring and improving performance in the application of security engineering principles. The model is in trial use on some government procurements. Its purpose is to enable

- selection of appropriately qualified providers of security engineering by being able to differentiate bidders by their capability levels and by the associated programmatic risks each presents,
- focused investments in security engineering tools, training, process definition, management practices, and improvements by engineering groups,
- capability-based assurance, i.e., development of system or product trustworthiness based on confidence in the measured competency and maturity of an engineering group's security practices and processes.

It is this latter focus that may be of interest to NIAP as a potential alternative approach for assessing the assurance that can be placed in products developed by measurably competent vendors. Follow-on efforts in this area will be focused on investigating the feasibility of extending the NIAP CC scheme to accommodate security assessed by such alternate means.

Another area of endeavor is to investigate how CC standards can be employed for large, distributed, evolving systems composed of many products. It is not clear how, or how well, the CC language can be used to describe the security features of such systems. How to apply the CM for testing and evaluating such systems is also in question. NIAP is teaming with the Federal Aviation Administration to investigate the issues associated with applying CC concepts and conventions for just such a system in the early stages of system planning, development, and acquisition.

18.14.8 Conducting Outreach

NIAP supports outreach as an important function. It is continually conducting outreach and associated education for a number of reasons, including:

- maintaining an up-to-date understanding of the marketplace and its needs and demands for security testing, evaluation, and validation services,
- raising general awareness of, confidence in, demand for, and use of the commercial security assessment industry,

---

* See http://www.sse_cmm.org or, duplicatively, http://constitution.ncsc.mil/wws/sse_cmm.

- stimulating user demand for and use of security-enhanced products,
- stimulating vendor investment in developing security-enhanced products,
- bolstering trust in such products so that manufacturers and consumers can build and buy with confidence, approaching non-governmental bodies, such as vertical industry trade groups and consortia, to encourage them to embrace the new security assessment approach by
  - encouraging the use of evaluated security-enhanced IT products, or
  - issuing their own certificates that may be based on either more lenient or more restrictive validation requirements than those supported by the NIAP certificate,
- promoting expansion in the base number of mutual recognition partner countries, and
- evangelizing for the need to enhance academic interest in
  - conducting R&D to support and to advance security testing and evaluation concepts, and
  - developing degree programs focused on matriculation of experts to help populate positions within the new commercial security testing and evaluation industry and applicable government oversight and validation bodies.

18.15 A NEW COMMON CRITERIA SCHEME TIES TOGETHER THE NIAP ELEMENTS

The elements of the NIAP initiative interact, in aggregate, to provide the internationally recognized, CC scheme[12] for conducting high quality security assessments within the U.S. The details of this scheme were being developed at the time of the writing of this chapter and thus there may be changes from what is indicated herein.

A summary of the scheme is portrayed in Figure 18.1.

According to the CC scheme, there are four types of activities that can be undertaken in conjunction with the various NIAP elements. These activities are

- developing and using basic CCEVS supports: standards, specifications, test and evaluation methods, and R & D (see lines numbered 1.1 through 1.4 in the diagram),
- developing a set of accredited testing and evaluation laboratories (see lines numbered 2.1 through 2.6 in the diagram),
- developing a set of validated products that have been granted certificates based on successfully undergoing testing, evaluation, and validation (see lines numbered 3.1 through 3.6 in the diagram), and
- mutual recognition interactions (see line numbered 4.1 in the diagram).

18.15.1 DEVELOPING AND USING THE BASIC SUPPORTS

The basis for all aspects of the scheme are the CC and CM standards. The CC provides the key input (line 1.1 in Figure 18.1) necessary for developing PPs. Validated PPs are entered into the PP registry. The PP registry identifies those PPs

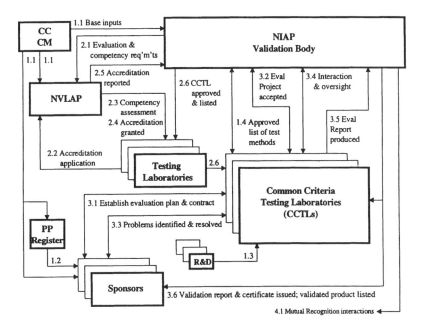

Figure 18.1 Summary of the CC Scheme.

that may serve as the basis for specifying products (line 1.2) that are submitted by product sponsors for testing, evaluation, and validation. Products that can be submitted may be PPs, or they may be hardware or software entities that implement STs. The CC and CM also provide the basic concepts (line 1.1) that drive the NIAP Validation Body and the laboratory accreditation efforts of the NVLAP. The CC and CM provides the basis for a list of approved test methods (line 1.4) that may be used during product testing and evaluation. NIAP-advocated R & D serves (line 1.3) to improve testing and evaluation concepts and methods approved by the Validation Body and used by accredited testing and evaluation laboratories.

18.15.2 ACCREDITING TESTING AND EVALUATION LABORATORIES

Accrediting commercial test and evaluation laboratories so that they can be approved as official CC Testing Laboratories (CCTLs) sanctioned by the NIAP Validation Body is a multistep process. The NIAP Validation Body provides security testing, evaluation, and competency requirements (line 2.1) to the NVLAP. These requirements are used by the NVLAP to assess (line 2.3) the technical, methodological and security testing and evaluation competency of laboratories that have applied (line 2.2) for accreditation. Upon successful laboratory assessment, the NVLAP grants accreditation (line 2.4) to testing and evaluation laboratories for a specific scope of approved testing and evaluation activities (such as the specific set of test methods that can be used by the CCTL, line 1.4). NVLAP reports (line 2.5) such accreditation to the Validation Body. The Validation Body then approves (line 2.6) the accredited laboratory to be recognized as an official CCTL. The Validation Body adds (line 2.6) the new CCTL to the list of approved laboratories maintained and publicized by NIAP. Through these processes the NIAP Validation Body expects to provide the marketplace with a set of competent and comparable private security testing and evaluation laboratories that can be used to assess the security-enhanced portions of any networking and IT product.

18.15.3 TESTING, EVALUATING, AND VALIDATING PRODUCTS

The actual testing, evaluation, and validation of specific products is a multistep process involving a continuous partnering of activities among the sponsor of a product seeking a NIAP certificate, a CC Testing Laboratory, and the NIAP Validation Body. A sponsor and a specific CCTL negotiate (line 3.1) a contract in which both parties agree to a testing and evaluation workplan and schedule for a specific product; the sponsor agrees to provide the product and other materials required for testing and evaluation efforts. The CCTL and Validation Body interact (line 3.2) and, if the work plan, sponsor documents, and other materials are in good order, the Validation Body approves (line 3.2) the initiation of the specific testing and evaluation project. As the testing and evaluation proceed, any problems encountered by the CCTL are shared with the sponsor and the Validation Body (line 3.3). The sponsor and CCTL work to resolve (line 3.3) such problems, and, as necessary, the Validation Body (line 3.4) engages in technical interactions and provides technical guidance and oversight to help handle the problems. If the sponsor desires that later releases and versions of the product should undergo testing, evaluation, and validation, the sponsor, CCTL, and Validation Body could collaborate in developing a certificate maintenance process to expedite subsequent security assessments of the later releases and versions of the product. Upon completing its testing and evaluation efforts, the CCTL writes a testing and evaluation report that is provided (line 3.5) to the Validation Body and the sponsor. The Validation Body drafts an associated validation report. After review (line 3.6) by the sponsor and CCTL, the Validation Body issues (line 3.6) a CC certificate to the sponsor for the specific product model and version that was assessed. The Validation Body also provides a final validation report to the sponsor and lists the specific product on the validated-products list that NIAP maintains and publicizes.

18.15.4 MUTUAL RECOGNITION MAINTENANCE

The NIAP Validation Body interacts with comparable organizations (line 4.1) in the other countries abiding by mutual recognition arrangements. The purposes of this interaction are to

* maintain and update the mutual recognition arrangements,
* synchronize any interpretations that may need to be made relative to, for example the CC, CM, approved test methods, or certificate issuance procedures, and
* exchange lists of validated products that are mutually recognized.

18.16 NIAP'S EARLY SUCCESSES

The NIAP initiative has had numerous, early successes. They attest to the expected longevity of the flexible, new approach NIAP advocates for assessing the trustworthiness and quality of security-enhanced network and IT products. They also attest to the robustness of the emerging marketplace associated with such products. Early successes, described more fully in subsequent sections of this chapter, include

- the rapid adoption of mutual recognition arrangements among many of the countries representing the bulk of the world's economy associated with building and buying trusted security-enhanced products,
- the rapid uptake of the international standards to proliferate the number of security requirements profiles,
- the emergence of tools to help automate the development of security requirements profiles,
- the unprecedented number of security testing and evaluation laboratories that rapidly emerged,
- the growing number of vendors that have engaged the new approach and the growing number of different products that have already undergone assessments according to the new approach, and
- the growing number of key, large user, and vendor consortia who are exploring the desirability of embracing the new approach.

These successes are mitigating the initially perceived risks that were thought to be barriers to achieving the NIAP vision. These earlier-perceived risks included overcoming the momentum and tradition ensconced in extant approaches, the timing of the introduction of a new approach relative to other large IT needs such as Y2K preparation, and the ability for the marketplace to achieve a critical mass for a new approach.

18.16.1 MUTUAL RECOGNITION ARRANGEMENTS GUIDE GLOBAL COOPERATION

One of the most significant early successes to which the NIAP contributed was the consummation of a CC mutual recognition arrangement among several countries. An initial, interim version of such a mutual recognition arrangement[8] was signed in early 1998 by government bodies within Canada, the U.K., and the U.S. Later in 1998, several countries (Canada, France, Germany, the U.K. and the U.S.) signed a more comprehensive mutual recognition arrangement,[2] with The Netherlands being able to sign somewhat later as soon as its national scheme was put into place. There is serious interest in other countries, such as Australia, Japan, New Zealand, and Sweden, to be added to these multicountry arrangements as soon as admittance procedures are finalized. Other countries appear to be in the wings. In total, the signing countries represent a very large share of the marketplace that produces and consumes security-enhanced network and IT products.

The purposes of the full mutual recognition arrangement[2] are many-fold. The signing countries acknowledge a strong mutual understanding and respect for each other's knowledge, abilities, and experiences with respect to CC-based, security assessments. They agree to maintain a mutual understanding and trust in each others' technical judgment, reliability, and competence pertaining to security testing, evaluation, and validation. They agree to put into place and to maintain comparable, national, CC-based security testing, evaluation, validation, and oversight processes that are to be carried out in a duly professional manner and that will ensure consistent results worldwide. Responsibilities of security testing and evaluation laboratories and national oversight and validation bodies are spelled out, along with requirements on country-specific processes and validation certificates. The signing countries also agree to formally recognize the results of security assessments in each other's countries. (The recognition does not apply to assessments of products claiming the highest levels of assurance.) The countries agree to harmonize any interpretations of the CC and the CM that they feel might be necessary in the future. They agree that each country will list products that receive validation certificates in other countries and that each country will recognize these products as if they were validated in their own country. In effect, each country accepts any other country's validation certificate as equal to its own. Also, agreements were made on procedures for adding or terminating membership in the arrangement.

These processes are established either under a law or an official administrative procedure valid in each of the signing countries.

18.16.2 PROFILES ARE PROLIFERATING

Another significant early success to which the NIAP contributed was the rapid uptake of the international CC standards by several networking-based and IT-based communities who have used these standards to develop and to proliferate the number of security requirements profiles. Examples of the types of security requirements profiles that were completed, or were being written, at the time of this chapter's writing, show a remarkable breadth of PP and ST development support, including the following:

Product-class-specific profiles

- Firewall (both router and packet filter) PPs
- Telecommunications switch PP
- Commercial DBMS PP
- Operating system PP equivalent to the TCSEC C2 class
- Operating system PP equivalent to the TCSEC B1 class
- Smart Card PP

General security profiles

- Role-Based Access Control (RBAC) PP
- CS2 PP for general computer security requirements at an EAL2 assurance level

Vertical-industry-specific or market-sector-specific profiles

- Federal Aviation Administration National Airspace System, Information Management System (FAA/NIMS) PP
- Telecommunications switch PP
- Department of Defense Warfighters' Intrusion Detection PP
- Health Open Systems and Trials (HOST) healthcare PPs (planned)
- Government Database Management System PP (G.DBMS PP, somewhat equivalent to a TCSEC C2 DBMS interpretation)
- Government Multi-Level Secure DBMS PP (G.MLS.DBMS PP, somewhat equivalent to a TCSEC B1 DBMS interpretation)

Vendor-specific security targets — A number of leading vendors in the database management system, firewall, and intrusion detection market sectors have written STs. (Readers should browse the validated products list available off the Product Testing page of the NIAP web site to get up-to-date information which indicates which vendors have written STs for which specific products.)

Additional information on many of the above security requirements profiles and profiling activities is available on the World Wide Web.*

18.16.3 R&D Makes Profile Development Faster, Better, Less Expensive

To maintain the momentum of the above early security requirements profiling efforts and to achieve maximum usage of the CC, the CC needed to be accessible to, and useable by, a wide swath of professionals throughout the various network and IT product communities. Since it is not practical to expect large numbers of potential PP and ST authors to acquire an in-depth knowledge and understanding of the CC, development of automated tools to help generate PPs and STs was needed.

Another early success to which NIAP contributed was the design, implementation, and delivery of a set of Java-based tools for profile development support that run under Windows95. The aggregate "CC Toolbox" is to assist both the PP author and the ST author in the basic tasks of generating the introductory documentation for PPs and STs. This is intended to simplify and to streamline the use of the CC for many profile writers. One of the main challenges for profile authors addressed by the CC Toolbox is to help find the appropriate components from within the CC requirements catalogs to apply to the profile being authored. Another major headache for profile authors that the CC Toolbox handles is the analysis of a draft profile for consistency and for resolution of dependencies that may exist among the various CC components tentatively selected to reflect the profiled security requirements.

The CC Toolbox consists of two basic tools. The PP tool is called PAA, Profile Authors Assistant, and the ST tool is CCDA, Common Criteria Developers Assistant. The PAA helps to define the security environment, security objectives, and security

---

* http://csrc.nist.gov/cc/pp/pplist.htm and
http://www.radium.ncsc.mil/tpep/library/protection_profiles/index.html.

requirements portion of a PP. The PAA will interview the PP developer regarding the security needs to be addressed. Based on the interview, the PAA is to produce a PP that specifies the appropriate functional and assurance requirements components from the CC catalogs.

The CCDA aids a product developer in creating an ST for an existing or new product. The CCDA interviews the developer about the security design features accompanying the product, or planned product, for which an ST is to be written. From the interview, the CCDA can produce an ST that specifies the applicable set of functional and assurance elements from the CC catalogs. These elements describe the security features, security environment, supporting documentation, and testing and evaluation activities of the product for which the ST is to be written. The CCDA also assists the ST author and evaluator (i.e., the product developer, the test and evaluation laboratory, or the evaluator) by performing some automated checking of functional security requirements.

Four release cycles are scheduled for the CC Toolbox, with the fourth scheduled for delivery in January 2000. Each release adds additional capabilities and is able to handle more complexities of the CC. Later releases are expected to incorporate artificial intelligence technologies so that a profile can be automatically generated in response to any specific threat scenario as input to the CC Toolbox.

18.16.4 TESTING AND EVALUATION LABORATORIES EMBRACE THE CC

Another early success indicator is that a significant number of private, security assessment laboratories have been rapidly entering the market under the interim auspices of the Trust Technology Assessment Program.* For them, there are clear, valid and profitable business cases for providing CC-based security testing and evaluation services in concert with NIAP's concepts. The TTAP was expanded in late 1997 to include CC-based evaluations. Recently, the TTAP has been operating in accordance with the Interim Mutual Recognition Agreement[8] using a TTAP Scheme** that could be thought of as an early, interim prototype of the NIAP scheme.[12] The procedures used by a commercial security assessment laboratory in order to become authorized to perform CC-based testing and evaluation within the TTAP are also spelled out in the TTAP Scheme. An up-to-date list of CC-based security testing and evaluation laboratories operating under the auspices of the TTAP can be found on the TTAP web site.

At the time of this chapter's writing, it was expected that in 1999 the NIAP CC scheme would be deployed and would operate in accordance with the final Mutual Recognition Arrangement,[2] and that TTAP would provide for a smooth transition to the CC-based testing program under NIAP. With the scheme fielded, the NIAP Validation Body would begin using NVLAP services to accredit laboratories. Laboratories would be approved as security assessment laboratories that could be used

---

* http://www.radium.ncsc.mil/tpep/ttap/
** The scheme can be viewed at http://www.radium.ncsc.mil/tpep/ttap/Scheme.html

by customers to obtain validated CC-based security testing and evaluation recognized worldwide under the final Common Criteria Mutual Recognition Arrangement.

It is expected that many of the above private laboratories that formerly operated under the auspices of TTAP and the Interim Mutual Recognition Arrangement would undertake NVLAP accreditation to become authorized under the NIAP CC scheme. These NVLAP-accredited laboratories will be electronically listed by NIAP on its web site as approved, accredited CCTLs (Common Criteria Testing Laboratories). Also listed on the web site will be the NIAP-approved test methods that can be used by the different CCTLs.

18.16.5 PRODUCTS EMERGING AMID MORE EFFICIENT TESTING AND EVALUATION PROCESSES

Another early success indicator is that CC-based, tested, and evaluated products are beginning to populate the marketplace. Industry-leading vendors in numerous networking and IT market sectors have been early adopters of the NIAP-advocated security testing and evaluation made available via the TTAP. Vendors have written STs to describe the security features of their products. They have had the security portions of such products assessed by TTAP-approved testing and evaluation laboratories using CC-based testing and TTAP evaluation methodologies. Some vendors have used comparable foreign, CC-based laboratories, such as UK CLEFs (Commercial Licensed Evaluation Facilities), whose results are recognized in the U.S. by the Interim Mutual Recognition Arrangement.

Products so evaluated according to CC concepts and conventions are appearing in the intrusion detection, firewall, guard, operating system, and database management system market sectors. An up-to-date list of products that have undergone such testing and evaluation within the U.S. is available on the World Wide Web.* Products evaluated in other countries appear on the evaluated products lists electronically maintained by other such countries.**

There are early indications that the new, NIAP-advocated, testing and evaluation approach is overcoming shortcomings inherent in earlier approaches. The new commercial-based testing and evaluation industry is providing ways to shorten testing cycles. For example, testing and evaluation laboratories are able to provide mechanisms such as double shifts for evaluators, and they are able to provide evaluation services on a vendor's premises. Providing less government oversight for evaluations targeted for lower level assurances is speeding up evaluations for those products claiming lower levels of assurance.

The list of prominent vendors in their fields showing early and strong support of the NIAP vision continues to grow. It is reasonable to expect that a rich choice of timely and cost-competitive, validated products will exist, and that customers worldwide will select such products over ones that are not tested and evaluated.

---

* The list of such CC-based products using the CCEVS scheme is posted on the Product Testing page of the NIAP web site (http://niap.nist.gov). The list of products using the TTAP scheme is posted in the Common Criteria Evaluated Products List at http://www.radium.ncsc.mil/tpep/epl/cc_st.html.

** For example, products that have been successfully tested and evaluated according to the national scheme of the United Kingdom are listed at http://www.itsec.gov.uk/

18.16.6   NIAP Reaches Out To Work Directly With Several Market Sectors

Another early success indicator is the number of prominent user and vendor consortia examining the utility of the new security testing, evaluation, and validation approach for their needs. Since 1998, NIAP has been reaching out to various networking, IT middleware, and IT infrastructure communities. NIAP personnel have begun working with numerous user, vendor, and vertical industry consortia to identify and assess areas of mutual interest regarding PP development and testing, evaluation, and validation of security-enhanced products of interest to their communities. Such outreach is useful to provide senior corporate management, CIOs, and other network and IT decision-makers with exposure to security and security testing and evaluation matters. Such individuals are not normally cognizant of directions being pursued within the traditional, more-confined, close-knit, security community.

At the time of the writing of this chapter, it seemed likely that several cooperative initiatives would soon be in the works or bearing fruit. The following are examples of initiatives that were being examined or initiated at that time. Efforts on engaging in any specific collaborative projects within any of these communities were not yet fully explored or defined. Interested readers should browse the NIAP web site to ascertain the status of any joint activities that may have been initiated as a result of these early discussions.

18.16.6.1   OMG (The Object Management Group)

At the time of the writing of this chapter, NIAP had begun discussions with senior, executive leadership of the OMG, its testing and security special interest groups, and several of its vertical industry domain task forces (medical, electronic commerce, financial, DOD, and telecommunications). Areas of potential mutual and complementary interest included

- development of a PP or PPs to accompany the OMG's base security and security interfacing specifications for the Common Object Request Broker Architecture (CORBA) so as to facilitate corresponding CM-based security testing and evaluation of CORBA interfaces to vendor-specific security mechanisms and services,
- OMG use of the accredited, commercial, CC test laboratories and governments' certificate issuing authorities to augment the OMG's CORBA testing and branding program provided by The Open Group,
- development of vertical-industry-specific, CORBA security interfacing PPs that stipulate mandatory security requirements to be addressed by CORBA implementations supporting specific vertical markets in which NIAP is initially focused (healthcare, financial services, electronic commerce, telecommunications and transportation), and
- cross fertilization of outreach, promotion, and education efforts.

18.16.6.2 IETF (Internet Engineering Task Force)

At the time of the writing of this chapter, NIAP had begun exploratory discussions with leadership from several of the IETF initiatives that have developed draft security standards specific to supporting network management (namely, the SNMPv3 standard) and to supporting general Internet working (namely, the IPsec standard). The purpose of these discussions was to explore IETF leadership interest in getting emerging IETF security standards tested and evaluated, especially before many implementations might begin appearing in the marketplace. As the CC-based approach advocated by NIAP had not yet been used to evaluate an extant, base standard, the notion of evaluating a standard was intriguing, and developing an appropriate approach was of significant interest to NIAP.

18.16.6.3 Financial Community

At the time of the writing of this chapter, NIAP had begun planning discussions with leadership from the financial community's networking and IT standardization groups in the ANSI Accredited Standards Committee (ASC) X9 standards arena. Areas of potential collaborations appeared to be centered on developing PPs as ANSI standards for the most stable of the financial community's needs, such as PPs for Certificate Authorities and smart cards. Such PPs would apparently not be impacted by evolution and changes in underlying protocols, such as SET (Secure Electronic Transfer) and SSL (Secure Sockets Layer), that are under various degrees of acceptance and trial use in numerous financial community pockets. To be of maximal utility and acceptability, development of such PPs would need the involvement of key, senior information security officers and senior auditors from major financial institutions and accounting firms. Workshops and forums were being planned for 1999 to educate such potential PP developers in the financial community about the NIAP-advocated security testing, evaluation, and validation approach and how to translate financial terminology (risk, exposure, prudent controls, etc.) into comparable CC terms (threats, vulnerabilities, etc.). Efforts were being planned to determine what synergy, if any, might be feasible between possible new CC-based activities in the ANSI ASC X9 world and the OMG's financial domain task force.

18.16.6.4 HOST (Healthcare Open Systems & Trials)

HOST is a nonprofit consortium created in 1994 to promote the development of networked IT to improve healthcare. At the time of the writing of this chapter, NIAP had begun planning discussions with HOST to establish forums beginning in mid 1999 to investigate the options of articulating certain healthcare security objectives and requirements in the form of PPs. It was envisioned that these objectives and requirements would be gathered in a format that would assist healthcare IT professionals in comparing and validating the security features of IT products and systems. In addition, there is interest in using the CC concepts and methodology for developing security requirements in an internationally standard format.

18.17 SEVERAL BENEFITS AND POSITIVE TRENDS CONTINUE TO BRIGHTEN THE FUTURE WITH NIAP

No barriers are slowing the transition from old, physical-supply-chain, business models to new electronic business models in virtually all industries. This seemingly inevitable migration will continue to fuel unprecedented growth in market-driven demand for security for supporting and managing business-critical and society-critical network and IT infrastructures.

There are several stakeholders in the security marketplace: security testing and evaluation laboratories, vendors of security-enhanced products, consumers of these products, and researchers developing commercially applicable security assessment advancements. NIAP is working to balance the interests of all these marketplace factions, to improve the quality of security testing and evaluation, and to foster expansion of the number of countries that recognize NIAP validated product evaluations. Positive trends are emerging throughout the marketplace.

18.17.1 SECURITY TESTING AND EVALUATION LABORATORIES

NIAP is helping bring commercial, accredited security testing and evaluation laboratories online for assessment of products claiming medium and lower levels of assurance. It is moving security testing, and evaluation and validation expertise and operations from the public to the private sector for products claiming these levels of assurance, while maintaining a notion of government oversight via the NIAP Validation Body. (The National Security Agency is still offering security assessment services for products claiming high levels of assurance, such as Orange Book B2-A1 levels or CC EAL 5-7 levels, and, importantly, these NSA security assessment services are becoming based on the CC and CEM.)

The private sector testing and evaluation laboratories are benefiting from the recognition they have received as government-accredited, CC-based, security assessment laboratories. General perceptions about testing and evaluation laboratories have improved by benefiting from NIAP efforts to maintain quality across the pool of private testing and evaluation laboratories.

Vendors' use of standard, CC-based, requirements specifications allowed testing and evaluation laboratories to use customized combinations of standard test and evaluation methods specifically tailored to each product being tested and evaluated. Familiarity with and use of such standard methods tend to shorten the lengths of testing and evaluation efforts.

18.17.2 VENDORS

NIAP is beginning to help vendors increase the value and competitiveness of their products through the use of formal, independent, validated, security assessment services recognized worldwide. NIAP is striving to ensure that such services are cost-effective. Vendor use of tailored, CC-based profiles of the security requirements addressed by their product is fostering rigorous and repeatable testing and evaluation.

Duplicate testing and evaluation for foreign markets is being minimized. It is also expected that duplicate, customer-specific testing is also being minimized.

By opening up a commercial security testing and evaluation industry and by qualifying security assessment laboratories, NIAP widened testing choices and alternatives. Vendors are now able to choose security assessment laboratories on business considerations as well as laboratory quality. Some vendors are looking to lower their overhead by reducing in-house security testing and evaluation resources. They also feel that the perception of increased trust arising from third party testing and evaluation will positively impact sales. NIAP's approach of tailoring the amount of validation oversight to the claimed assurance level of the product being tested is fostering sensitivity to time-to-market pressures for products undergoing security assessments.

The impacts of such NIAP efforts are facilitating the lowering of testing costs. Expectations are that vendors will continue to produce products that have undergone NIAP-advocated assessments as long as

- cost-avoidance continues due to minimal redundant testing,
- consumers demand security-assessed products, and
- security assessment processes stay reasonable (i.e., remain sensitive to cost and time-to-market pressures).

Expectations also continue to indicate that product retesting cost avoidance can facilitate more competitive product pricing and can lead to market share improvement.

From vendor marketing perspectives, it is expected that vendors who are among the early adopters of the NIAP-advocated security assessment approach will continue to be seen as leaders in their market sectors and that the validation certificates they receive will continue to provide a substantial sales tool.

It is also expected that acquisition authorities for large organizations of network and IT product users will begin using PPs and CC concepts in procurements of networking and IT products. Vendors will be able to quickly determine if their current products fulfill the needs of the procuring agency via PP to ST comparisons. By basing their responses on their products' STs, vendor responses will be more straightforward, less lengthy, and written in the common CC language and syntax that procurers will expect. Product procurement cycles might then be reduced in length, thereby potentially improving vendors' cash/profit flow. Even when a vendor's response is not based on COTS products, the vendor can still explain concepts using the language and syntax prescribed by the CC.

18.17.3 CONSUMERS

NIAP has collaborated with key consumer groups to help them develop PPs specific to their individual needs. These profiles are helping users — in both the public and private sectors — by providing a sound and reliable basis for adequate, appropriate, credible security testing and evaluation. Likewise, with the emergence of vendor-developed STs, buyers and manufacturers now get clearer mutual understandings of what security features were implemented, and what features are to be examined during accredited testing.

It is expected that consumer-specific PPs will be effective in ensuring that a consumer's security objectives are relevant to the policies and threats specific to the consumer's environment. It is also expected that PPs will be effective in demonstrating (potentially legally) that the consumer has taken steps to safeguard networking and information handling. PPs may prove to be effective in saving money since the consumer's focus can be narrowed to only those security requirements of essential need.

The availability of two independent and impartial third parties — one to conduct and another to validate security assessments — removed any sense of impropriety or partiality. It fostered comparability and consistency among testing efforts. It also seems to promote the expectation of additional product trust.

The evaluation reports from the security testing and evaluation laboratories are helping consumers by facilitating

- more meaningful understandings of what was evaluated, and
- comparison and selection of security-enhanced network and IT products.

Expectations continue to indicate that consumers will select COTS products that are assessed by the NIAP-advocated approach over products that are not assessed. Expectations also continue to indicate that consumers will buy with confidence, and with no duplicate testing, any product built in any one country and tested in another country.

Accordingly, it is expected that consumers will begin using PPs and CC concepts to help improve the cost, schedule, and performance of their processes to acquire network and IT products. It is expected that PPs will be used to state security requirements succinctly in a common language and syntax, and that acquisition authorities will expect vendor responses to use the same common language and syntax in the product STs they offer. Acquisitions will benefit from the lack of confusion that could have been caused by differences in terms and formats by consumers and the various responding vendors. Acquisition authorities will be able to verify quickly that bidders' STs match desired PPs, and they can expedite the selection of a winning vendor since all vendor responses can be easily compared because of their compliance with the CC.

18.17.4 RESEARCHERS

Academics and researchers are beginning to benefit from NIAP's interest in high priority R & D to advance the state of commercial security testing and evaluation practice. In time, NIAP-approved testing laboratories, vendors, and consumers may all benefit from research to make NIAP-advocated, commercial, security assessment quicker, less expensive, and better.

ACKNOWLEDGMENTS

The authors wish to acknowledge their appreciation to inputs provided by Ms. Daphne Willard, NSA and Ms. Mary Schanken, NSA, to reviews and assistance provided by NIAP staff and technical directors, and to assistance provided by Ms. Peggy Himes, NIST.

REFERENCES

1. William Clinton, Critical Infrastructure Protection, *Presidential Decision Directive*/NSC-63, PDD-63, The White House, Washington, D.C., May 22, 1998.
2. Arrangement on the Mutual Recognition of Common Criteria Certificates in the Field of Information Technology Security, The Sponsoring Organizations: Communications Security Establishment, Service Central de la Securite des Systems des Technologies de l'Information, Bundesamt fur Sicherheit in der Informationstechnik, Netherlands National Communications Security Agency, Communications-Electronics Security Group, National Institute of Standards and Technology, and National Security Agency, October 5, 1998.
3. Trusted Computer System Evaluation Criteria, DOD5200.28-STD, U.S. Department of Defense, December 1985.
4. Trusted Network Interpretation of the Trusted Computer System Evaluation Criteria, National Computer Security Center, National Security Agency, Ft. Meade, MD, July 31, 1987.
5. Trusted Database Management System Interpretation of the Trusted Computer System Evaluation Criteria, NCSC-TG-021, National Computer Security Center, National Security Agency, Ft. Meade, MD, April 1991.
6. Common Criteria for Information Technology Security Evaluation, Version 2, CCIB-98-026 through CCIB-98-028, May 1998, Common Criteria Implementation Board (CCIB), ISO/IEC Final Committee Draft (FCD) 15408-1 through FCD 15408-3, ISO/IEC, JTC1, SC27, WG3, N-1951 through N-1953, May 27, 1998.
7. Common Methodology for Information Technology Security Evaluation, Version 0.45, Common Evaluation Methodology Editorial Board, October 31, 1998.
8. Letter of Intent to Support Interim Mutual Recognition, NIST Computer Security Division, Gaithersburg, MD, October 1997.
9. General Requirements for the Competence of Calibration and Testing Laboratories, ISO/IEC Guide 25, 1990.
10. National Voluntary Laboratory Accreditation Program — Procedures and General Requirements, J. L. Cigler and V. R. White, Editors, NIST Handbook 150, U.S. Department of Commerce, Technology Administration, National Institute of Standards and Technology, U.S. Government Printing Office, Washington, D.C., March 1994.
11. Common Criteria Testing, Draft NIST Handbook 150-xx, U.S. Department of Commerce, National Institute of Standards and Technology, Computer Security Division, Gaithersburg, MD.
12. Common Criteria Evaluation and Validation Scheme for Information Technology Security — Organization, Management and Concept of Operations (Draft), Version 1, NIAP (a joint NIST/NSA initiative), Department of Commerce, National Institute of Standards and Technology, Computer Security Division, Gaithersburg, MD, August 1998.

Index

10baseT, defined 120

A

AAL. defined 120
AAL1, ATM StatMux trunks 157
AAL5 167
 multimedia networking 38
ABR
 defined 117, 120
 RSVP and ATM 19
AC-3 275
Access control, e-commerce 343
Active networks, adaptability for integrated
 services 233
Ad hoc network 53
Adaptation, dynamic 232
Additive white gaussian noise (AWGN) 55
Admission control
 IP and integrated services 219
 RSVP 4, 16, 17, 25, 33
ADSL
 and discrete multi-tone 108
 defined 120
AdSpec
 RSVP 25
 RSVP integrated services 18
 RSVP path message 11
Advanced TDMA 251
Advanced television *see* ATV, HDTV
AFC, satellite communications 246
Aggregate route-based IP switching *see* ARIS
Algorithm, defined 117
 see also Encoding, Compression
Aloha, contention resolution 114
AM/FM hum modulation 96
Amplifier, defined 117
ANM, defined 120
ANSI, defined 120
APS
 failure recovery 210
 W-BLSR 207
ARIS
 IP/ATM integration 183
Asymmetric video codec 62
Asynchronous transfer mode *see* ATM
ATM
 adaptation layer 51, 67

ATM-based cable modem 98, 106
 capacity and hybrid network 152
 cell, defined 117
 cell label 166
 compared to other networks 160
 defined 117, 120
 DOCSIS 97
 frame relay interoperation 169
 HDTV transmission 316
 integrated services 228
 IP integration, scalability 167, 169, 179
 IP routing (IP Navigator) 180
 IP via PVCS 169
 multimedia networking 31
 multipoint compared to RSVP multicast 19
 network capacity 158, 161
 RSVP 18
 StatMux trunking 148, 155
 VPI/VCI compared to MPLS label 186
 vs. MSN 192–194
ATMARP
 IP via 171
 NHRP-enhanced 176
ATP, IP integration 171
ATV and digital radio 325
Audio
 cards, preferred features 281
 data characteristics 272
 DVD 130
 gaps 284–285
 Internet characteristics 276
 low-cost communication 291
 Mbone 270–294
 quality via DVD 132
 reading on UNIX vs. Windows 281
Audio encoding
 aural masking 274
 modified discrete cosine transform 275
 MPEG 275
 perceptual model 275
Audio tool
 basic structure 282
 compression 284
 Internet 280
 real-time scheduling 288
 silence detection 284–285
Audio transmission
 acceptable packet loss 284
 described 38

383

O

P